395L

Springer Undergraduate Mathematics Series

G. Evans, J. Blackledge and P. Yardley

Analytic Methods for Partial Differential Equations

With 25 Figures

Springer

Gwynne A. Evans, MA, Dphil, DSc
Jonathan M. Blackledge, BSc, PhD, DIC
Peter D. Yardley, BSc, PhD

Faculty of Computing Sciences & Engineering, De Montfort University,
The Gateway, Leicester LE1 9BH, UK

Cover illustration elements reproduced by kind permission of:
Aptech systems, Inc., Publishers of the GAUSS Mathematical and Statistical System, 23804 S.E. Kent-Kangley Road, Maple Valley, WA 98038, USA. Tel.: (206) 432 - 7855 Fax (206) 432-7832 email: info@aptech.com URL.: www.aptech.com
American Statistical Association: Chance Vol 8 No 1,1995 articles by KS and KW Heiner 'Tree Rings of the Northern Shawangunks' page 32 fig 2
Springer-Verlag: Mathematica in Education and Research Vol 4 Issue 3 1995 article by Roman E Maeder, Beatrice Amrhein and Oliver Gloor 'Illustrated Mathematics: Visualization of Mathematical Objects' page 9 fig 11, originally published as a CD ROM 'Illustrated Mathematics' by TELOS: ISBN 0-387-14222-3, German edition by Birkhäuser: ISBN 3-7643-5100-4.
Mathematica in Education and Research Vol .4 Issue 3 1995 article by Richard J Gaylord and Kuzume Nishidate "Traffic Engineering with Cellular Automata' page 35 fig 2 Mathematica in Education and Research vol 5 issue 2 1996 article by Michael Trott 'The Implicitization of a Trefoil Knot' page 14.
Mathematica in Education and Research Vol 5 issue 2 1996 article by Lee de Cola 'Coins, Trees, Bars and Bells: Simulation of the Binomial PrOc-ess' page 19 fig 1), Mathematica in Education and Research Vol 5 Issue 2 1996 article by Richard Gaylord and Kazume Nishidate 'ContagiOus Spreading' page 33 fig 1. Mathematica in Education and Research Vol 5 Issue 2 1996 artide by Joe Buhler and Stan Wagon 'Secrets of the Madeung Constant' page 50 fig 1.

ISBN 3-540-76124-1 Springer-Verlag Berlin Heidelberg New York

British Library Cataloguing in Publication Data
Evans,G.
Analytic methods for partial differential equations..-(Springer undergraduate mathematics series)
1. Differential equations, Partial
I. Title II. Blackledge, J. M. (Jonathan M.) III. Yardley, P.
515'.3'53
ISBN 3540761241

Library of Congress Cataloging-in-Publication Data
Evans, G. (Gwynne), 1944-
Analytic methods for partial differential equations./ G. Evans.J. Blackledge and P.Yardley
p. cm. -- (Springer undergraduate mathematics series)
Includes bibliographical references and index.
ISBN 3-540-76124-1
1.Differential equations, Partial-Numerical solutions.
I. Blackledge,J.M.(Jonathan M.)II. Yardley, P. (Peter), 1948-
III. Title. IV. Series
QA377.E945 1999
515'.353-dc21 99-35689
CIP

2nd printing 2001

Evans et. al: Analytic Methods for Partial Differential Equations

First Indian Reprint 2010

ISBN 978-81-8489-667-1

This edition is published by Springer (India) Private Limited, (a part of Springer Science+Business Media), Registered Office: 3rd Floor, Gandharva Mahavidyalaya, 212, Deen Dayal Upadhyaya Marg, New Delhi – 110 002, India.

Printed in India by Rakmo Press (P) Ltd., New Delhi.

To our past and present students without whom this work would not have been developed

Preface

The subject of partial differential equations holds an exciting and special position in mathematics. Partial differential equations were not consciously created as a subject but emerged in the 18th century as ordinary differential equations failed to describe the physical principles being studied. The subject was originally developed by the major names of mathematics, in particular, Leonard Euler and Joseph-Louis Lagrange who studied waves on strings; Daniel Bernoulli and Euler who considered potential theory, with later developments by Adrien-Marie Legendre and Pierre-Simon Laplace; and Joseph Fourier's famous work on series expansions for the heat equation. Many of the greatest advances in modern science have been based on discovering the underlying partial differential equation for the process in question. James Clerk Maxwell, for example, put electricity and magnetism into a unified theory by establishing Maxwell's equations for electromagnetic theory, which gave solutions for problems in radio wave propagation, the diffraction of light and X-ray developments. Schrödinger's equation for quantum mechanical processes at the atomic level leads to experimentally verifiable results which have changed the face of atomic physics and chemistry in the 20th century. In fluid mechanics, the Navier–Stokes' equations form a basis for huge number-crunching activities associated with such widely disparate topics as weather forcasting and the design of supersonic aircraft.

Inevitably the study of partial differential equations is a large undertaking, and falls into several areas of mathematics. At one extreme the main interest is in the existence and uniqueness of solutions, and the functional analysis of the proofs of these properties. At the other extreme, lies the applied mathematical and engineering quest to find useful solutions, either analytically or numerically, to these important equations which can be used in design and construction. In both this text, and the companion volume (Evans, 1999), the emphasis is on the practical solution rather than the theoretical background, though this important work is recognised by pointers to further reading. This approach is based on courses given by the authors while at De Montfort University.

Hence in the first chapter, we start by covering some of the mathematical background including orthogonal polynomials, special functions such as Legendre Polynomials and Bessel functions and a brief coverage of complex variables. The use of characteristics to classify partial differential equations leads to specific techiques in the following chapters. This is supported by brief derivations of the wave equation, the heat equation and Laplace's equation. The chapter is concluded with some background to generalised functions for use in the final chapter on Green's functions.

Chapter 2 is a conventional coverage of separation of variables, applied to the heat equation and Laplace's equation in Cartesian, cylindrical polar and spherical polar coordinates. Chapter 3 is concerned with solutions involving characteristic curves, and seemed the natural place for first-order equations, including Charpit's method for nonlinear first-order equations. The chapter then moves on to second-order equations and D'Alembert's solution of the wave equation, including the method of characteristics in an analytic setting.

Integral transforms are covered in Chapter 4, with work on Fourier's integral theorem, Fourier sine and cosine transforms, Fourier complex transforms and Laplace transforms.

The final chapter is on Green's functions, and perforce covers the basic work in this field only. We have of course Green's birth place (Sneinton Windmill) and his grave very near to us here. In all these chapters, space limitations had to be considered and some cuts were made to this end. Topics here include Green's functions for the wave equation, the diffusion equation and Laplace's equation; Helmholtz and Schrödinger's equations with applications to scattering theory; Maxwell's equations; and Green's functions in optics with Kirchhoff diffraction theory. Approximation methods and Born series are also considered briefly.

Most sections have a set of exercises, and fairly complete solutions can be found in the appendix. Exceptions are small introductory sections and where a second section is required to make a topic viable for solution and further investigation by the reader. The exercises and solutions form an important part of the book and provide much insight to the ideas introduced in the text.

In the last stages of the preparation, the completed manuscript was read by Endre Süli (Oxford University), and we are very grateful for his general remarks and detailed comments.

Acknowledgements

We would like to express our thanks to Susan Hezlet who was our first point of contact with Springer-Verlag. She was instrumental in steering this book through to its conclusion, though the final stage is in the capable hands of David Ireland. We are also grateful for the continued support of De Montfort

University, Leicester, and particularly the technical staff who kept our computer systems running for the duration of the writing process.

Contents

1
Mathematical Preliminaries

1.1 Introduction

Partial differential equations emerged when shortcomings with the use of ordinary differential equations were found in the study of vibrations of strings, propagation of sound, waves in liquids and in gravitational attraction. Originally the calculus of partial derivatives was supplied by Euler in a series of papers concerned with hydrodynamics in 1734. This work was extended by D'Alembert in 1744 and 1745 in connection with the study of dynamics.

Partial differential equations are the basis of almost every branch of applied mathematics. Such equations arise from mathematical models of most real life situations. Hence quantum mechanics depends on Schrödinger's equations, fluid mechanics on various forms of Navier–Stokes' equations and electromagnetic theory on Maxwell's equations. Partial differential equations form a very large area of study in mathematics, and are therefore important for both analytical and numerical considerations. The analytical aspects are covered in this text and the numerical aspects in the companion volume, "Numerical methods for partial differential equations".

Inevitably there are many aspects of other branches of mathematics which are pertinent to this work, and the relevant material has been brought together in this chapter to save long digressions later, and to give an element of completeness. The first two sections should be covered at the first reading and form a general introduction to the book as a whole. The later sections deal with a range of related topics that will be needed later, and may be tackled as required.

When the differential equations involve only one independent variable such as $y(t)$ in the equation for simple harmonic motion given by

$$\frac{d^2y}{dt^2} + k^2 y = 0 \tag{1.1.1}$$

this is then called an ordinary differential equation. Standard methods are available for the analytic solution of particular classes of such equations such as those with constant coefficients, and these methods are familiar in references such as Nagle and Saff (1993), or the classic, Piaggio (1950). However, it is very easy to write an equation whose closed form solution is not expressible in simple terms such as

$$\frac{d^2y}{dx^2} = xy. \tag{1.1.2}$$

For such a problem the ordinary differential equation itself defines the solution function and is used to derive its analytic properties by such devices as series solutions. Numerical methods come into their own to obtain values of the solution function and again there is a vast literature on this topic which includes Lambert (1990) and Evans (1996).

Partial differential equations follow a similar line, but now the dependent variable is a function of more than one independent variable, and hence the derivatives are all partial derivatives. In view of ordinary differential equations, some types lend themselves to analytic solution, and there is a separate literature on numerical solutions. These aspects form the contents of this book and its companion volume.

The *order* of a partial differential equation is the order of the highest derivative. First-order equations can often be reduced to the solution of ordinary differential equations, which will be seen later in the considerations of characteristics. Second-order equations tend to demonstrate the numerical methods applicable to partial differential equations in general. For the most part, consideration here is limited to linear problems – the nonlinear ones constituting current research problems. Linear problems have the dependent variable and its partial derivatives occurring only to the first degree, hence there are no products of the dependent variable and its derivatives. Hence the equation

$$\frac{\partial^2 \phi}{\partial x^2} + \frac{\partial^2 \phi}{\partial y^2} = 0 \tag{1.1.3}$$

is linear. It is called Laplace's equation, and it will be a major topic in this book, whereas

$$\frac{\partial u}{\partial t} + 6u\frac{\partial u}{\partial x} + \frac{\partial^3 u}{\partial x^3} = 0 \tag{1.1.4}$$

is the nonlinear Korteweg–de Vries equation. For solutions of this equation, the method of inverse scattering is employed which is outside the scope of this book, and may be pursued in Ablowitz and Clarkson (1991). A linear equation is said to be *homogeneous* if each term contains either the dependent variable or one of its derivatives, otherwise it is said to be non-homogeneous or inhomogeneous.

The fundamental property of a homogeneous linear problem is that if f_1 and f_2 are solutions then so is $f_1 + f_2$. To begin the discussion, three specific physical applications which prove typical of the problems to be solved are introduced. A classification of second-order equations is then covered, and each of the three physical problems falls into a different class in this categorisation.

The first of these physical problems is the heat or diffusion equation which can be derived by considering an arbitrary volume V. The heat crossing the boundary will equate to the change of heat within the solid, which results in the equation

$$\int_V \rho c \frac{\partial \theta}{\partial t} dV = \int_S k \text{ grad } \theta \cdot d\mathbf{S} + \int_V H(\mathbf{r}, \theta, t)\, dV \tag{1.1.5}$$

where $d\mathbf{S} = \mathbf{n}dS$ with $\mathbf{n}$ the unit outward normal to the surface S of V and dS is a surface element, θ is the temperature, k is the thermal conductivity, ρ the density and c the specific heat. H represents any heat generated in the volume by such action as radioactive decay, electrical heating or chemical action. A short-hand notation, common in continuum mechanics is used here where grad is defined by

$$\text{grad } u = \left\{ \frac{\partial u}{\partial x}, \frac{\partial u}{\partial y}, \frac{\partial u}{\partial z} \right\} \tag{1.1.6}$$

which generates a vector from a scalar u. This is often written in the condensed form-grad $u = \nabla u$. A further short-hand notation that will be used where a condensed notation is acceptable is the use of subscripts to denote partial derivatives. Hence the above definition will become

$$\text{grad } u = \{u_x, u_y, u_z\}. \tag{1.1.7}$$

With this definition, the z dependence may be absent in the case of partial differential equations in two independent variables which will be the dominant case in this book. There are two other *vector operators* which are also used in this book, namely div and curl, defined by

$$\text{div}\, \mathbf{a} = \nabla \cdot \mathbf{a} = \frac{\partial a_1}{\partial x} + \frac{\partial a_2}{\partial y} + \frac{\partial a_3}{\partial z} \tag{1.1.8}$$

and

$$\text{curl}\, \mathbf{a} = \nabla \times \mathbf{a} = \begin{vmatrix} \mathbf{i} & \mathbf{j} & \mathbf{k} \\ \frac{\partial}{\partial x} & \frac{\partial}{\partial y} & \frac{\partial}{\partial z} \\ a_1 & a_2 & a_3 \end{vmatrix} \tag{1.1.9}$$

where $\mathbf{a} = \{a_1, a_2, a_3\}$. Hence the operator div operates on a vector to generate a scalar, and the operator curl operates on a vector and generates a vector. With these definitions in place the derivation of the main equations may now be continued.

The integral over the surface S can be converted to a volume integral by the divergence theorem (Apostol, 1974) to give

$$\int_V \rho c \frac{\partial \theta}{\partial t} dV = \int_V \text{div } (k \text{ grad } \theta) dV + \int_V H(\mathbf{r}, \theta, t)\, dV. \tag{1.1.10}$$

However, this balance is valid for an arbitrary volume and therefore the integrands themselves must satisfy

$$\rho c \frac{\partial \theta}{\partial t} = \text{div}\,(k \text{ grad } \theta) + H(\mathbf{r}, \theta, t). \tag{1.1.11}$$

In the special but common case in which k = constant, the diffusion equation reduces to

$$\frac{\partial \theta}{\partial t} = K\nabla^2\theta + Q(\mathbf{r}, \theta, t) \tag{1.1.12}$$

with

$$\left.\begin{aligned} K &= \frac{k}{\rho c} \\ Q &= \frac{H}{\rho c} \end{aligned}\right\}.$$

Typical boundary conditions for this equation would include the following.

(i) $\theta(\mathbf{r}, t)$ is a prescribed function of t for every point $\mathbf{r}$ on the boundary surface S.

(ii) The normal flux through the boundary $\frac{\partial \theta}{\partial n}$ is prescribed on S where $\mathbf{n}$ is a normal vector to the surface S.

(iii) The surface radiation is defined over S, for example, by

$$\frac{\partial \theta}{\partial n} = -a(\theta - \theta_0) \tag{1.1.13}$$

which is Newton's law of radiation.

The heat or diffusion equation applies to a very large number of other physical situations. The diffusion of matter such as smoke in the atmosphere, or a dye or pollutant in a liquid is governed by Fick's law

$$\mathbf{J} = -D \text{ grad } c \tag{1.1.14}$$

where D is the coefficient of diffusion and c is the concentration. The vector $\mathbf{J}$ is the diffusion current vector, and therefore c satisfies

$$\frac{\partial c}{\partial t} = \text{div}\,(D \text{ grad } c) \tag{1.1.15}$$

or

$$\frac{\partial c}{\partial t} = D\nabla^2 c \tag{1.1.16}$$

if D is a constant. Other physical situations, which are modelled by the diffusion equation include neutron slowing, vorticity diffusion and propagation of long electromagnetic waves in a good conductor such as an aerial.

The second of the fundamental physical equations is the wave equation. Consider a small length of a stretched string as shown in Figure 1.1.

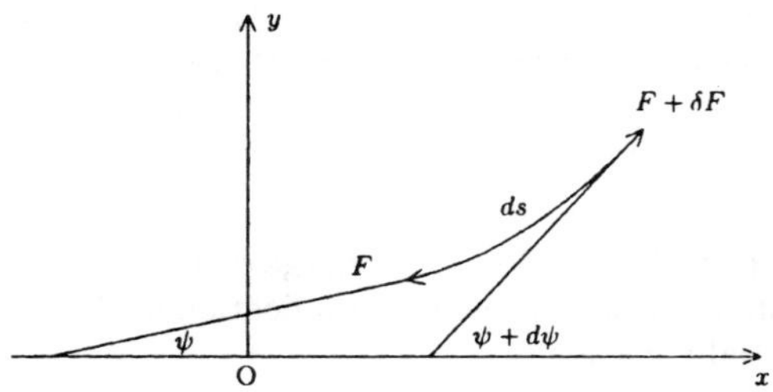

Fig. 1.1.

Then Newton's second law applied to the small length gives

$$F\sin(\psi + d\psi) - F\sin\psi = \rho\, ds\frac{\partial^2 y}{\partial t^2} = F\cos\psi d\psi \qquad (1.1.17)$$

to first order where F is the tension in the string, ρ is the string density, ψ is the tangential angle of the string to the x-axis, s is the distance coordinate along the string and y is the displacement from the neutral position. From elementary calculus,

$$\tan\psi = \frac{\partial y}{\partial x} \qquad (1.1.18)$$

and hence

$$\sec^2\psi d\psi = \frac{\partial^2 y}{\partial x^2}dx \qquad (1.1.19)$$

which yields

$$\begin{aligned}\rho\frac{\partial^2 y}{\partial t^2} &= F\cos^3\psi\frac{\partial^2 y}{\partial x^2}\frac{\partial x}{\partial s} \\ &= F\cos^4\psi\frac{\partial^2 y}{\partial x^2}\end{aligned} \qquad (1.1.20)$$

where $\frac{\partial x}{\partial s} = \cos\psi$ as ψ is the angle the tangent makes with the x-axis. However, for oscillations of small amplitude

$$\cos^2\psi = \left\{1 + \left(\frac{\partial y}{\partial x}\right)^2\right\}^{-1} \sim 1 \qquad (1.1.21)$$

to yield the wave equation in the form

$$\frac{\partial^2 y}{\partial x^2} = \frac{1}{c^2}\frac{\partial^2 y}{\partial t^2} \qquad (1.1.22)$$

with

$$c^2 = \frac{F}{\rho}. \qquad (1.1.23)$$

The third important example is Laplace's equation which is based on Gauss's law in electromagnetism (see Atkin, 1962) namely

$$\int_S \mathbf{E} \cdot d\mathbf{S} = \rho \tag{1.1.24}$$

where $\mathbf{E}$ is the electric field, and ρ is the charge density with S being the surface of an arbitary volume. If $\rho = 0$, then an application of the divergence theorem, gives the differential equation

$$\text{div } \mathbf{E} = 0. \tag{1.1.25}$$

However, Maxwell's equations with no time varying magnetic field yield

$$\text{curl } \mathbf{E} = 0 \tag{1.1.26}$$

which is the condition for the field to be irrotational. With this proviso, there exists a ϕ such that

$$\mathbf{E} = \text{ grad } \phi \tag{1.1.27}$$

and hence

$$\text{div grad } \phi = 0 \tag{1.1.28}$$

or

$$\nabla^2 \phi = 0 \tag{1.1.29}$$

which is Laplace's equation. The same equation holds for the flow of an ideal fluid. Such a fluid has no viscosity and being incompressible the equation of continuity is div $\mathbf{q} = 0$, where $\mathbf{q}$ is the flow velocity vector (see Acheson, 1990). For irrotational flows the equivalent of 1.1.26 holds to allow the use of the potential function $\mathbf{q} = \text{grad}\phi$ and again the defining equation is 1.1.29.

These three major physical problems (1.1.16, 1.1.21 and 1.1.28) are typical of the main types of second-order linear partial differential equations, and in the next section mathematical arguments will be used to establish a classification of such problems.

The following exercises cover the derivation of variations to the main equations above to allow further physical features to be considered. The mathematical and numerical solutions to these extended problems fall into the remit of the solutions for the basic equations.

EXERCISES

1.1 Establish that if a string is vibrating in the presence of air resistance which is proportional to the string velocity then the wave equation becomes

$$\frac{\partial^2 u}{\partial t^2} = c^2 \frac{\partial^2 u}{\partial x^2} - r\frac{\partial u}{\partial t} \qquad \text{with} \quad r > 0.$$

1.2 Show that if a vibrating string experiences a transverse elastic force (such as a vibrating coiled spring), then the relevant form of the wave equation is

$$\frac{\partial^2 t}{\partial t^2} = c^2 \frac{\partial^2 u}{\partial x^2} - ku \qquad \text{with} \quad k > 0.$$

1.3 If a vibrating string is subject to an external force which is defined by $f(x,t)$, then show that the wave equation takes the form

$$\frac{\partial^2 t}{\partial t^2} = c^2 \frac{\partial^2 u}{\partial x^2} + f(x,t).$$

1.4 If there is an external source or sink of heat given by $f(x,t)$ (such as that due to electrical heating of a rod, chemical heating or radioactive heating), then the diffusion equation becomes

$$\frac{\partial u}{\partial t} = \nabla(k\nabla u) + f(x,t).$$

1.5 If the end of a vibrating string is in a viscous liquid then energy is radiated out from the string to the liquid. Show that the endpoint boundary condition has the form

$$\frac{\partial u}{\partial n} + b\frac{\partial u}{\partial t} = 0$$

where n is the normal derivative and b is a constant which is positive if energy is radiated to the liquid.

1.6 When a current flows along a cable with leakage, G, the loss of voltage is caused by resistance and inductance. The resistance loss is Ri where R is the resistance and i is the current (Ohm's Law). The inductance loss is proportional to the rate of change of current (Gauss's Law), which gives the term Li_t where L is the inductance. Hence, the voltage equation is

$$v_x + Ri + Li_t = 0.$$

The current change is due to capacitance C, and leakage G. These terms yield

$$i_x + Cv_t + Gv = 0.$$

Deduce the telegraph equation in the form

$$LC\frac{\partial^2 v}{\partial t^2} + (GL + RC)\frac{\partial v}{\partial t} + RGv = \frac{\partial^2 v}{\partial x^2}.$$

Find the equation satisfied by the current i.

1.7 Show that the function

$$u(x,t) = \frac{1}{n}\sin nx\, e^{-n^2kt}$$

satisfies $u_t = ku_{xx}$. This is a typical separation of the variables solution and will be studied in great detail.

1.2 Characteristics and Classification

Characteristics were first used by Lagrange in two major papers in 1772 and 1779, in which first-order linear and nonlinear equations were considered. This work appears in Chapter 3. Gaspard Monge (1746–1818) introduced characteristic curves in his work "Feuilles d'analyse appliquie à la géométrie" completed in 1770, but not published until 1795, by which time he had considered second-order equations of the type to be discussed next.

Consider a general second-order quasilinear equation defined by the equation

$$Rr + Ss + Tt = W \tag{1.2.1}$$

where

$$R = R(x,y), \qquad S = S(x,y), \qquad T = T(x,y) \qquad \text{and} \qquad W = W(x,y,z,p,q). \tag{1.2.2}$$

with

$$p = \frac{\partial z}{\partial x}, \qquad q = \frac{\partial z}{\partial y}, \qquad r = \frac{\partial^2 z}{\partial x^2}, \qquad s = \frac{\partial^2 z}{\partial x \partial y} \qquad \text{and} \qquad t = \frac{\partial^2 z}{\partial y^2}. \tag{1.2.3}$$

Then the characteristic curves for this equation are defined as curves along which highest partial derivatives are not uniquely defined. In this case these derivatives are the second-order derivatives r, s and t. The set of linear algebraic equations which these derivatives satisfy can be written in terms of differentials, and the condition for this set of linear equations to have a non-unique solution will yield the equations of the characteristics, whose significance will then become more apparent. Hence the linear equations follow by the chain rule as $dz = pdx + qdy$ and also

$$\begin{aligned} dp &= rdx + sdy \\ \text{and} \quad dq &= sdx + tdy \end{aligned} \tag{1.2.4}$$

to give the linear equations

$$\left.\begin{aligned} Rr + Ss + Tt &= W \\ rdx + sdy &= dp \\ sdx + tdy &= dq \end{aligned}\right\} \tag{1.2.5}$$

and there will be no unique solution when

$$\begin{vmatrix} R & S & T \\ dx & dy & 0 \\ 0 & dx & dy \end{vmatrix} = 0 \tag{1.2.6}$$

which expands to give the differential equation

$$R\left(\frac{dy}{dx}\right)^2 - S\left(\frac{dy}{dx}\right) + T = 0. \tag{1.2.7}$$

However, when the determinant in 1.2.6 is zero, we require the linear system 1.2.5 to have a solution. Thus by Cramer's rule,

$$\begin{vmatrix} R & T & W \\ dx & 0 & dp \\ 0 & dy & dq \end{vmatrix} = 0 \tag{1.2.8}$$

also holds, and gives an equation which holds along a characteristic, namely

$$-R\frac{dy}{dx}\ dp - T\ dq + W\ dy = 0. \tag{1.2.9}$$

Returning now to 1.2.7, this equation is a quadratic in dy/dx and there are three possible cases which arise. If the roots are real, then the characteristics form two families of real curves. A partial differential equation 1.2.1 is then said to be of hyperbolic type. The condition is that

$$S^2 - 4RT > 0. \tag{1.2.10}$$

The second case is when

$$S^2 - 4RT = 0 \tag{1.2.11}$$

so the roots are real and equal; in this case we say that equation 1.2.1 is of the parabolic type. When the roots are complex the underlying equation is said to be of elliptic type corresponding to the condition

$$S^2 - 4RT < 0. \tag{1.2.12}$$

The importance of characteristics only becomes apparent at this stage. The first feature is the use of characteristics to classify equations. The methods that will be used subsequently to solve partial differential equations vary from type to type. In the case of hyperbolic equations, the characteristics are real and are used directly in the solution. Characteristics also play a role in reducing equations to a standard or canonical form. Consider the partial differential equation

$$R\frac{\partial^2 z}{\partial x^2} + S\frac{\partial^2 z}{\partial x \partial y} + T\frac{\partial^2 z}{\partial y^2} = G\left(x, y, z, \frac{\partial z}{\partial x}, \frac{\partial z}{\partial y}\right) \tag{1.2.13}$$

and put $\xi = \xi(x, y)$, $\eta = \eta(x, y)$ and $z = \zeta$ to see what a general change of variable yields. The result is the partial differential equation

$$A(\xi_x, \xi_y)\frac{\partial^2\zeta}{\partial\xi^2} + 2B(\xi_x, \xi_y, \eta_x, \eta_y)\frac{\partial^2\zeta}{\partial\xi\partial\eta} + A(\eta_x, \eta_y)\frac{\partial^2\zeta}{\partial\eta^2} = F(\xi, \eta, \zeta, \zeta_\zeta, \zeta_\eta) \tag{1.2.14}$$

where

$$A(u, \nu) = Ru^2 + Su\nu + T\nu^2 \tag{1.2.15}$$

and

$$B(u_1, \nu_1, u_2, \nu_2) = Ru_1u_2 + \frac{1}{2}S(u_1\nu_2 + u_2\nu_1) + T\nu_2\nu_2. \tag{1.2.16}$$

The question is now asked for what ξ and η do we get the simplest form? Certainly if ξ and η can be found to make the coefficients A equal to zero, then a simplified form will result. However the condition that A should be zero is a partial differential equation of first order which can be solved analytically (Sneddon (1957) or Vvedensky (1993)). This topic is treated in detail in Chapter 3, and at this stage the reader will need to accept the solution. Different cases arise in the three classifications. In the hyberbolic case when $S^2 - 4RT > 0$, let $R\alpha^2 + S\alpha + T = 0$ have roots λ_1 and λ_2 then $\xi = f_1(x, y)$ and $\eta = f_2(x, y)$ where $f_1(x, y)$ and $f_2(x, y)$ are the solutions of the two factors in the related ordinary differential equations

$$\left[\frac{dy}{dx} + \lambda_1(x, y)\right]\left[\frac{dy}{dx} + \lambda_2(x, y)\right] = 0. \tag{1.2.17}$$

Hence the required transformations are precisely the defining functions of the characteristic curves. With this change of variable, both $A(\xi_x, \xi_y)$ and $A(\eta_x, \eta_y)$ become zero, and the partial differential equation becomes

$$\frac{\partial^2\zeta}{\partial\eta\partial\xi} = \phi(\xi, \eta, \zeta, \zeta_\zeta, \zeta_\eta) \tag{1.2.18}$$

which is the canonical form for the hyperbolic case.

In the parabolic case, $S^2 - 4RT = 0$, there is now only one root, and any independent function is used for the other variable in the transformation. Hence $A(\xi_x, \xi_y) = 0$, but it is easy to show in general that

$$A(\xi_x, \xi_y)A(\eta_x, \eta_y) - B^2(\xi_x, \xi_y, \eta_x, \eta_y) = (4RT - S^2)(\xi_x\eta_y - \xi_y\eta_x)^2$$

and therefore as $S^2 = 4RT$, we must have $B(\xi_x, \xi_y, \eta_x, \eta_y) = 0$ and $A(\eta_x, \eta_y) \neq 0$ as η is an independent function of x and y. Hence when $S^2 = 4RT$, the transformation $\xi = f_1(x, y)$ and $\eta =$ any independent function yields

$$\frac{\partial^2\zeta}{\partial\eta^2} = \phi_1(\xi, \eta, \zeta, \zeta_\zeta, \zeta_n) \tag{1.2.19}$$

which is the canonical form for a parabolic equation. This reduction is shown in the example below.

In the elliptic case there are again two sets of characteristics but they are now complex. Writing $\xi = \alpha + i\beta$ and $\eta = \alpha - i\beta$ gives the real form

$$\frac{\partial^2 \zeta}{\partial \xi \partial \nu} = \frac{1}{4}\left(\frac{\partial^2 \zeta}{\partial \alpha^2} + \frac{\partial^2 \zeta}{\partial \beta^2}\right) \tag{1.2.20}$$

and hence the elliptic canonical form

$$\frac{\partial^2 \zeta}{\partial \alpha^2} + \frac{\partial^2 \zeta}{\partial \beta^2} = \psi(\alpha, \beta, \zeta, \zeta_\alpha, \zeta_\beta). \tag{1.2.21}$$

Note that Laplace's equation is in canonical form as is the heat equation, but the wave equation is not.

In cases where R, S and T are functions of x and y (quasilinear case), then the type of classification will depend on the values of x and y. Hence, there will be regions in the x, y plane in which the conditions for the equation to be parabolic, elliptic and hyperbolic hold. These are regions of parabolicity, ellipticity and hyperbolicity. (see for example Exercise 1.9 below).

As an example of reduction to canonical form consider the quasilinear second-order partial differential equation

$$\frac{\partial^2 u}{\partial x^2} + 2\frac{\partial^2 u}{\partial x \partial y} + \frac{\partial^2 u}{\partial y^2} + c^2\frac{\partial u}{\partial y} = 0. \tag{1.2.22}$$

Then the equation of the characteristic curves is

$$\left(\frac{dy}{dx}\right)^2 - 2\frac{dy}{dx} + 1 = 0 \tag{1.2.23}$$

or factorising

$$\left(\frac{dy}{dx} - 1\right)^2 = 0. \tag{1.2.24}$$

The single solution is $x - y =$ const, and therefore the transformation for the canonical form is:

$$\left.\begin{array}{rcl} \xi & = & x - y \\ \eta & = & x \end{array}\right\}. \tag{1.2.25}$$

The choice of η is a conveniently simple arbitrary function which is independent of ξ. The required partial derivatives after this change of variable are

$$\frac{\partial^2 u}{\partial x^2} = \frac{\partial^2 u}{\partial \xi^2} + 2\frac{\partial^2 u}{\partial \xi \partial \eta} + \frac{\partial^2 u}{\partial \eta^2} \tag{1.2.26}$$

and

$$\frac{\partial^2 u}{\partial x \partial y} = -\frac{\partial^2 u}{\partial \xi^2} - \frac{\partial^2}{\partial \xi \partial \eta} \tag{1.2.27}$$

which yields the reduced form

$$\frac{\partial^2 u}{\partial x^2} + 2\frac{\partial^2 u}{\partial x \partial y} + \frac{\partial^2 u}{\partial y^2} = \frac{\partial^2 u}{\partial \eta^2}. \qquad (1.2.28)$$

Since $\frac{\partial u}{\partial y} = \frac{\partial u}{\partial \xi}$, the transformed equation is

$$\frac{1}{c^2}\frac{\partial^2 u}{\partial \eta^2} = \frac{\partial u}{\partial \xi}. \qquad (1.2.29)$$

The reader may now wish to attempt the following exercises on this section.

EXERCISES

1.8 Find the equations of the characteristic curves for the following partial differential equations:

(i) $-\frac{\partial^2 u}{\partial x^2} + x\frac{\partial^2 u}{\partial x \partial y} + y^2\frac{\partial^2 u}{\partial y^2} = u + \frac{\partial u}{\partial x}$

(ii) $(1+x^2)\frac{\partial^2 u}{\partial x^2} + \frac{\partial^2 u}{\partial x \partial y} + (1+y^2)\frac{\partial^2 u}{\partial y^2} = \frac{\partial u}{\partial y}$

(iii) $x^2\frac{\partial^2 u}{\partial x^2} + 2x\frac{\partial^2 u}{\partial x \partial y} + \frac{\partial^2 u}{\partial y^2} = u$

and hence deduce which are parabolic, elliptic and hyperbolic.

1.9 Find the regions of parabolicity, ellipticity and hyperbolicity for the partial differential equation:

$$\frac{\partial^2 u}{\partial x^2} + 3xy\frac{\partial^2 u}{\partial x \partial y} + (x+y)\frac{\partial^2 u}{\partial y^2} = u$$

and sketch the resulting regions in the (x, y) plane.

1.10 Find the analytic form of the characteristic curves for the partial differential equation

$$\frac{\partial^2 u}{\partial x^2} + 2\left(x + \frac{1}{y}\right)\frac{\partial^2 u}{\partial x \partial y} + \frac{4x}{y}\frac{\partial^2 u}{\partial y^2} = xy$$

and hence categorise the equation.

1.11 Reduce the equation

$$\frac{\partial^2 z}{\partial x^2} + 2\frac{\partial^2 z}{\partial x \partial y} + 3\frac{\partial^2 z}{\partial y^2} = z$$

to canonical form. Make a further transformation to obtain a real canonical form.

1.3 Orthogonal Functions

There are a number of very important mathematical preliminaries which arise in the solution of partial differential equations, and indeed some of these concepts arise in other branches of mathematics. These preliminaries will be considered in the following sections. The first such topic is that of an orthogonal set of functions. A set of real-valued functions $g_1(x), g_2(x), \ldots$ is called an orthogonal set of functions on an interval (a, b), if they are defined there and if all the integrals $\int_a^b g_n(x)g_m(x)dx$ exist and are zero for all pairs of functions g_n and g_m with $m \neq n$.

The L_2 norm of g_m is

$$||g_m|| = \sqrt{\int_a^b [g_m(x)]^2 dx}. \tag{1.3.1}$$

As an example $\sin \frac{\pi x}{l}, \sin \frac{2\pi x}{l}, \ldots, \sin \frac{n\pi x}{l}, \ldots$ are orthogonal on $0 \leq x \leq l$, since

$$\int_0^l \sin \frac{n\pi x}{l} \sin \frac{m\pi x}{l} dx = 0, \qquad m \neq n \tag{1.3.2}$$

and

$$\left\| \sin \frac{m\pi x}{l} \right\|^2 = \frac{l}{2}. \tag{1.3.3}$$

An orthogonal set of functions $\{g_m\}$ is said to be *complete* if it is impossible to add to the set one other *non-zero* function which is orthogonal to each of the g_m. In other words, $\{g_m\}$ is a complete set if

$$\int_a^b g_m(x)f(x)dx = 0 \tag{1.3.4}$$

for all m implies $f(x) \equiv 0$ on (a, b).

Let $g_1(x), g_2(x), \ldots$ be any complete orthogonal set of functions on $a \leq x \leq b$ and let $f(x)$ be a given function such that

$$f(x) = \sum_{n=1}^{\infty} c_n g_n(x) \tag{1.3.5}$$

where the series converges in the L_2 norm, namely

$$\lim_{m \to \infty} \left\| f - \sum_{n=1}^{m} c_n g_n \right\| \to 0.$$

This is a generalised Fourier series, whose coefficients are found by multiplying both sides by $g_m(x)$ and integrating from a to b to give

$$\begin{aligned}\int_a^b f(x)g_m(x)dx &= \int_a^b \left(\sum_{n=1}^{\infty} c_n g_n(x)\right) g_m(x)dx \\ &= \sum_{n=1}^{\infty} c_n \int_a^b g_n(x)g_m(x)dx \\ &= c_m \int_a^b [g_m(x)]^2\, dx \\ &= c_m\|g_m\|^2. \end{aligned} \tag{1.3.6}$$

The interchange of the integral with the summation in the second line requires uniform convergence of the series. This happens for certain functions f, the conditions for which can be found in Apostol, (1974 Chapter 11). Hence, the Fourier coefficient c_n is given by

$$c_n = \frac{1}{\|g_n\|^2}\int_a^b f(x)g_n(x)dx. \tag{1.3.7}$$

For example, if $f(x) = x$, $0 < x < l$ and $g_m(x) = \sin\frac{m\pi x}{l}$, then if

$$f(x) = \sum_{n=1}^{\infty} c_n \sin\frac{n\pi x}{l} \tag{1.3.8}$$

then

$$c_n = \frac{2}{l}\int_0^l f(x)\sin\frac{n\pi x}{l}dx = -\frac{2l}{n\pi}(-1)^n. \tag{1.3.9}$$

Hence the series is

$$f(x) = \sum_{n=1}^{\infty} \frac{2l}{n\pi}(-1)^{n+1}\sin\frac{n\pi x}{l}, \qquad 0 < x < l. \tag{1.3.10}$$

A set of real-valued positive functions $g_1, g_2, \ldots$ is called an orthogonal set of functions with respect to the weight function $\phi(x)$ on the interval (a, b), if they are defined and if all the integrals $\int_a^b \phi(x)g_m(x)g_n(x)dx$ exist and are zero for all pairs of g_m and g_n with $m \neq n$.

The norm of g_m is then

$$\|g_m\| = \sqrt{\int_a^b \phi(x)[g_m(x)]^2 dx}. \tag{1.3.11}$$

Similary if $g_1, g_2, \ldots$ is a set of orthogonal functions with respect to a positive weight function ϕ on $a_1 < x < a_2$ and if $f(x) = \sum_{n=1}^{\infty} c_n g_n(x)$, then it can be shown that

$$c_n = \frac{1}{||g_n||^2} \int_a^b f(x)\phi(x)g_n(x)dx. \tag{1.3.12}$$

Consider the set $\{e^{-x} \sin nx : n = 1, 2, 3, \ldots\}$ which is an orthogonal set on $[0, \pi]$ with respect to the weight function e^{2x} since

$$\int_0^{\pi} e^{2x} \left(e^{-x} \sin nx\right) \left(e^{-x} \sin mx\right) dx = 0, \qquad m \neq n. \tag{1.3.13}$$

Moreover

$$||e^{-x} \sin nx||^2 = \int_0^{\pi} e^{2x} \left(e^{-x} \sin nx\right)^2 dx = \frac{\pi}{2}. \tag{1.3.14}$$

In general, most functions met in mathematical physics are sufficiently well-behaved to be expanded using Fourier series in this way, and this allows the wide use of the method of separation of variables which is covered in Chapter 2. A number of exercises on this work may now be attempted.

EXERCISES

1.12 Show that the following sets of functions form orthogonal sets on the stated intervals:

a) $1, \cos x, \cos 2x, \cos 3x, \ldots$ for $0 \leq x \leq \pi$

b) $\sin \pi x, \sin 2\pi x, \sin 3\pi x, \ldots$ for $-1 \leq x \leq 1$

c) $1, 1 - x, 1 - 2x + \frac{1}{2}x^2$ for $0 \leq x \leq \infty$ with respect to the weight function $w(x) = e^{-x}$.

1.13 Show that the functions $f(x) = 1$ and $g(x) = x$ are orthogonal on the interval $-1 < x < 1$. Determine the constants α and β such that the function $h(x) = 1 + \alpha x + \beta x^2$ is orthogonal to both $f(x)$ and $g(x)$.

1.14 Given that $\{1, \cos x, \ldots, \cos nx, \ldots\}$ form a complete set of orthogonal functions on $[0, \pi]$, find $||1||$, $||\cos nx||$, and express $x - 2$ in the form $a_0 + \sum_{n=1}^{\infty} a_n \cos nx$ in the interval $[0, \pi]$.

1.15 The Chebyshev polynomials are defined by

$$T_n(x) = \cos(n \arccos x).$$

Show that $T_3(x) = 4x^3 - 3x$ and prove that the orthogonality condition is

$$\int_{-1}^{1} \frac{T_r(x)T_s(x)}{\sqrt{1-x^2}}\,dx = \begin{cases} 0 & r \neq s \\ \pi/2 & r = s \neq 0 \\ \pi & r = s = 0. \end{cases}$$

1.4 Sturm–Liouville Boundary Value Problems

A second area of mathematics which will be treated here in some detail is the theory of Sturm–Liouville problems in ordinary differential equations. In solving

$$\frac{\partial^2 u}{\partial x^2} = \frac{1}{c^2}\frac{\partial^2 u}{\partial t^2} \tag{1.4.1}$$

the method of separation of variables is used as described in Chapter 2. This results in ordinary differential equations for two new functions $X(x)$ and $T(t)$. With the equation for X, there is a set of boundary conditions and in solving this system a set of orthogonal functions is found. The method of separation of variables, when applied to second-order linear partial differential equations frequently leads to an ordinary differential equation of the form

$$\frac{d}{dx}\left[p(x)\frac{d\phi}{dx}\right] + \{g(x) + \lambda r(x)\}\,\phi(x) = 0 \tag{1.4.2}$$

subject to the boundary conditions

$$\begin{aligned} \alpha_1\phi(a) + \beta_1\phi'(a) &= 0 \\ \alpha_2\phi(b) + \beta_2\phi'(b) &= 0 \end{aligned} \tag{1.4.3}$$

where $p(x), g(x)$ and $r(x)$ are given real continuous functions of x on $a \leq x \leq b$, p is continuously differentiable on (a, b), and such that $p(x) > 0$ (or $p(x) < 0$), $r(x) \geq 0$ (or $r(x) \leq 0$) for $a \leq x \leq b$, ($r(x)$ not identically zero for any part of the range of x considered) and $\alpha_1, \alpha_2, \beta_1, \beta_2$ are constants (at least one non-zero in each boundary condition), λ is a parameter. This system is called a Sturm–Liouville boundary value problem. It generally has non-trivial solutions only for certain λ, these solutions are called eigenfunctions and the corresponding λ are referred to as eigenvalues.

The solution of the linear equation 1.4.2 can be expressed in the form

$$\phi = c_1 u_1(x;\lambda) + c_2 u_2(x;\lambda) \tag{1.4.4}$$

where u_1 and u_2 are independent solutions and c_1 and c_2 are arbitrary constants.

Substituting into 1.4.3 gives

$$\begin{aligned} c_1[\alpha_1 u_1(a;\lambda) + \beta_1 u_1'(a;\lambda)] + c_2[\alpha_1 u_2(a;\lambda) + \beta_1 u_2'(a;\lambda)] &= 0 \\ c_1[\alpha_2 u_1(b;\lambda) + \beta_2 u_1'(b;\lambda)] + c_2[\alpha_2 u_2(b;\lambda) + \beta_2 u_2'(b;\lambda)] &= 0. \quad (1.4.5) \end{aligned}$$

To ensure that these equations are consistent without implying $c_1 = c_2 = 0$ we must have

$$\begin{vmatrix} \alpha_1 u_1(a;\lambda) + \beta_1 u_1'(a;\lambda) & \alpha_1 u_2(a;\lambda) + \beta_1 u_2'(a;\lambda) \\ \alpha_2 u_1(b;\lambda) + \beta_2 u_1'(b;\lambda) & \alpha_2 u_2(b;\lambda) + \beta_2 u_2'(b;\lambda) \end{vmatrix} = 0. \quad (1.4.6)$$

This equation in λ gives the eigenvalues and the corresponding eigenfunctions are obtained by solving 1.4.2 with this value of λ subject to 1.4.3.

The differential equation may be written $\{L + \lambda r(x)\}y = 0$ where $L = \frac{d}{dx}\left[p(x)\frac{d}{dx}\right] + g(x)$. The form L is called the *self-adjoint* form. A bounded operator is self-adjoint if $(Lu, v) = (u, Lv)$ for every u, v in the underlying function space (see Renardy and Rogers, 1993, p253, or Zauderer, 1983, p129). It may appear restrictive, but it is sufficiently general to include most second-order differential operators on $a \le x \le b$. For if

$$L_1 = a_2(x)\frac{d^2}{dx^2} + a_1(x)\frac{d}{dx} + a_0(x) \quad (1.4.7)$$

is such an operator then L_1 can be written as

$$\frac{d}{dx}\left[p(x)\frac{d}{dx}\right] + g(x) \quad (1.4.8)$$

by setting

$$p(x) = e^{\int \frac{a_1}{a_2} dx}, \qquad \text{and} \qquad g(x) = \frac{a_0}{a_2} e^{\int \frac{a_1}{a_2} dx}. \quad (1.4.9)$$

Hence for

$$L_1 = x^2 \frac{d^2}{dx^2} + \frac{d}{dx} + x^3 \quad (1.4.10)$$

the principle coefficients are

$$a_2 = x^2, \qquad a_1 = 1, \qquad a_0 = x^3$$

and

$$p(x) = e^{\int \frac{1}{x^2} dx} = e^{-\frac{1}{x}}, \qquad g(x) = \frac{x^3}{x^2} e^{-\frac{1}{x}} = x e^{-\frac{1}{x}}. \quad (1.4.11)$$

Therefore L_1 in self-adjoint form is

$$\frac{d}{dx}\left(e^{-\frac{1}{x}} \frac{d}{dx}\right) + x e^{-\frac{1}{x}}. \quad (1.4.12)$$

Two important theorems relating to Sturm–Liouville problems are as follows.

Theorem 1.1

If the Sturm–Liouville problem

$$\frac{d}{dx}\left[p(x)\frac{dy}{dx}\right] + \{g(x) + \lambda r(x)\}\, y = 0 \tag{1.4.13}$$

with

$$\alpha_1 y(a) + \beta_1 y'(a) = 0, \qquad \alpha_2 y(b) + \beta_2 y'(b) = 0 \tag{1.4.14}$$

has p, q and r real valued and continuous on $a \le x \le b$, p continuously differentiable on (a, b) and has $r(x)$ positive (or negative) everywhere in the whole interval then all the eigenvalues of the problem are real.

Theorem 1.2

The eigenfunctions that correspond to distinct eigenvalues of the Sturm–Liouville system are orthogonal with respect to the weight $r(x)$.

Proofs of these theorems can be found in Lomen and Mark (1988) or Burkill (1962).

1.5 Legendre Polynomials

The analytic solution of partial differential equations will result in the use of many transcendental functions which may be new to the reader. The main group of these form orthogonal sets on various intervals. Some of these functions will be considered here and the objective is to derive the usual relations which appear in compendia such as Abramowitz and Stegun (1964) such as recurrence relations, differential equations, generating functions, special values, formulae for derivatives, series and asymptotic forms.

Given a non-negative integer n, Legendre's equation for the Legendre polynomial of degree n is

$$(1 - x^2)\frac{d^2y}{dx^2} - 2x\frac{dy}{dx} + n(n+1)y = 0, \qquad x \in (-1, 1). \tag{1.5.1}$$

This equation can be written in the form

$$x^2\frac{d^2y}{dx^2} + xp(x)\frac{dy}{dx} + q(x)y = 0 \tag{1.5.2}$$

where

$$p(x) = \frac{-2x^2}{1 - x^2}, \qquad q(x) = \frac{n(n+1)x^2}{1 - x^2}. \tag{1.5.3}$$

These functions can be expanded as convergent series' in powers of x for $|x| < 1$, hence the solution y to 1.5.2 can itself be expressed as a convergent series in x, for x in the interval $-1 < x < 1$, and has the form:

$$P_n(x) = \sum_{m=0}^{[n/2]} \frac{(-1)^m (2n-2m)! x^{n-2m}}{2^n m!(n-m)!(n-2m)!} \tag{1.5.4}$$

where

$$[n/2] = \begin{cases} n/2 & n \text{ even} \\ (n-1)/2 & n \text{ odd.} \end{cases} \tag{1.5.5}$$

The polynomial $P_n(x)$ is called the Legendre polynomial of degree n.

The generating function for the Legendre polynomials has the form

$$(1-2tx+t^2)^{-1/2} = \sum_{t=0}^{\infty} t^l P_l(x) \tag{1.5.6}$$

with $|t| < 1$ and $|x| < 1$.

Proof

Expand $(1-2tx+t^2)^{-1/2}$ by the binomial theorem to yield

$$\begin{aligned} (1-2tx+t^2)^{-1/2} &= [1-t(2x-t)]^{-1/2} \\ &= \sum_{r=0}^{\infty} \frac{(2r)!}{2^{2r}(r!)^2} t^r (2x-t)^r. \end{aligned} \tag{1.5.7}$$

Now expanding $(2x-t)^r$ by the binomial theorem gives

$$(2x-t)^r = \sum_{p=0}^{r} \frac{r!(2x)^{r-p}(-t)^p}{p!(r-p)!} \tag{1.5.8}$$

which implies

$$(1-2tx+t^2)^{-1/2} = \sum_{r=0}^{\infty} \frac{(2r)!}{2^{2r}(r!)^2} \sum_{p=0}^{r} \frac{r!}{p!(r-p)!} (-1)^p t^{r+p} (2x)^{r-p}. \tag{1.5.9}$$

The coefficients of t^l are required, hence let $r+p=l$, and for fixed r we have $p=l-r$; now $0 \le p \le r$, so we must only consider r such that $0 \le l-r \le r$ or $l/2 \le r \le l$. Hence if l is even, r can take values between $l/2$ and l, while if l is odd r can take values between $(l+1)/2$ and l. For any r in these ranges the coefficient of t^l is obtained by taking $p=l-r$ to give

$$\frac{(2r)!r!(-1)^{l-r}(2x)^{2r-l}}{2^{2r}(r!)^2(l-r)!(2r-l)!} \tag{1.5.10}$$

and the total coefficient of t^l is

$$\sum_{r=\alpha}^{l} \frac{(2r)!r!(-1)^{l-r}(2x)^{2r-l}}{2^{2r}(r!)^2(l-r)!(2r-l)!} = \sum_{k=0}^{\beta}\left(\frac{(2l-2k)!(l-k)!(-1)^k(2x)^{l-2k}}{2^{2l-2k}[(l-k)!]^2k!(l-2k)!}\right) \tag{1.5.11}$$

where $k = l - r$ and

$$\alpha = \begin{cases} l/2 & l \text{ even} \\ (l+1)/2 & l \text{ odd} \end{cases} \qquad \beta = \begin{cases} l/2 & l \text{ even} \\ (l-1)/2 & l \text{ odd} \end{cases}$$

$$= \sum_{k=0}^{[l/2]} \frac{(-1)^k(2l-2k)!x^{l-2k}}{2^l(l-k)!(l-2k)!k!} = P_l(x). \tag{1.5.12}$$

Some explicit expressions for Legendre polynomials which follow from the series 1.5.4 above are:

$$P_0(x) = 1, \qquad P_1(x) = x, \qquad P_2(x) = \frac{1}{2}(3x^2 - 1),$$

$$P_3(x) = \frac{1}{2}(5x^3 - 3x), \qquad P_4(x) = \frac{1}{8}(35x^4 - 30x^2 + 3). \tag{1.5.13}$$

A commonly used relationship to form orthogonal functions is Rodrigues' formulae

$$P_l(x) = \frac{1}{2^l l!}\frac{d^l}{dx^l}(x^2-1)^l. \tag{1.5.14}$$

There are a number of special properties for Legendre Polynomials which arise from the definitions and these include

$$\text{(i) } P_l(1) = 1, \qquad \text{(ii) } P_l(-1) = (-1)^l, \qquad \text{(iii) } P_l'(1) = \frac{1}{2}l(l+1),$$

$$\text{(iv) } P_l'(-1) = (-1)^{l-1}\frac{l}{2}(l+1), \qquad \text{(v) } P_{2l}(0) = \frac{(-1)^l(2l)!}{2^{2l}(l!)^2},$$

$$\text{(vi) } P_{2l+1}(0) = 0 \tag{1.5.15}$$

and the derivatives of the Legendre functions satisfy recurrence relations of the form

$$\begin{array}{ll} \text{(i)} & (2l+1)xP_l(x) = (l+1)P_{l+1}(x) + lP_{l-1}(x) \\ \text{(ii)} & P_l(x) = P_{l+1}'(x) - 2xP_l'(x) + P_{l-1}'(x) \\ \text{(iii)} & P_{l+1}'(x) - P_{l-1}'(x) = (2l+1)P_l(x) \\ \text{(iv)} & xP_l'(x) - P_{l-1}'(x) = lP_l(x) \\ \text{(v)} & P_l'(x) - xP_{l-1}'(x) = lP_{l-1}(x) \\ \text{(vi)} & (x^2-1)P_l'(x) = lxP_l(x) - lP_{l-1}(x). \end{array} \tag{1.5.16}$$

Functions that are defined and are square integrable on the interval $(-1,1)$ can be expanded as series of Legendre Polynomials. Assume that $f(x)$ can be expressed in the interval $-1 \leq x \leq 1$ as an infinite series of Legendre polynomials of the form

$$f(x) = \sum_{n=0}^{\infty} A_n P_n(x). \tag{1.5.17}$$

The generalised Fourier coefficients A_n are required. If 1.5.17 is multiplied by $P_n(x)$ and integrated w.r.t. x from -1 to 1, the orthogonality of the set P_n on $[-1,1]$, namely

$$\int_{-1}^{1} P_n(x)P_m(x)\,dx = 0 \qquad (m \neq n), \tag{1.5.18}$$

may be utilised. This follows since $P_n(x)$ satisfies Legendre's equation

$$\frac{d}{dx}\left[(1-x^2)\frac{dP_n(x)}{dx}\right] + n(n+1)P_n(x) = 0 \tag{1.5.19}$$

and multiplying through by $P_m(x)$ and integrating w.r.t. x from -1 to 1 gives

$$\int_{-1}^{1} P_m(x)\frac{d}{dx}\left[(1-x^2)\frac{dP_n(x)}{dx}\right]dx + n(n+1)\int_{-1}^{1} P_m(x)P_n(x)\,dx = 0. \tag{1.5.20}$$

Integrate the first term by parts and since $(1-x^2) = 0$ when $x = \pm 1$, we obtain

$$-\int_{-1}^{1}(1-x^2)\frac{dP_n(x)}{dx}\frac{dP_m(x)}{dx}\,dx + n(n+1)\int_{-1}^{1} P_m(x)P_n(x)\,dx = 0. \tag{1.5.21}$$

On the other hand, starting with $P_m(x)$, and multipying through by $P_n(x)$ would give

$$-\int_{-1}^{1}(1-x^2)\frac{dP_n(x)}{dx}\frac{dP_m(x)}{dx}\,dx + m(m+1)\int_{-1}^{1} P_m(x)P_n(x)\,dx = 0. \tag{1.5.22}$$

Subtracting 1.5.21 from 1.5.22 gives

$$[n(n+1) - m(m+1)]\int_{-1}^{1} P_m(x)P_n(x)\,dx = 0 \tag{1.5.23}$$

and if $m \neq n$ we have

$$\int_{-1}^{1} P_m(x)P_n(x) = 0. \tag{1.5.24}$$

Hence the set $\{P_n\}$ is orthogonal on $[-1,1]$.

If $m = 0$ we have $P_0(x) = 1$ which gives

$$\int_{-1}^{1} P_n(x) = 0 \qquad \text{if } n \neq 0. \tag{1.5.25}$$

Since the highest power of x with a non-zero coefficient occurring in $P_n(x)$ is x^n, any polynomial of degree n in x, $Q_n(x)$ say, can be represented by

$$Q_n(x) = \sum_{r=0}^{n} B_r P_r(x) \tag{1.5.26}$$

where the B_r's are constants. Therefore

$$\int_{-1}^{1} P_m(x)Q_n(x)\,dx = \sum_{r=0}^{n} B_r \int_{-1}^{1} P_m(x)P_r(x)\,dx = 0 \qquad \text{if } m > n. \tag{1.5.27}$$

To evaluate

$$\int_{-1}^{1} [P_n(x)]^2\,dx \tag{1.5.28}$$

we make use of relation (iii) from 1.5.16, multiply through by $P_n(x)$ and integrate from -1 to 1 to obtain

$$(2n+1)\int_{-1}^{1} [P_n(x)]^2\,dx = \int_{-1}^{1} P'_{n+1}(x)P_n(x)\,dx - \int_{-1}^{1} P'_{n-1}(x)P_n(x)\,dx. \tag{1.5.29}$$

However, $P'_{n-1}(x)$ is a polynomial of degree $n-2$ in x, so the last integral is zero. Integrate first by parts to give

$$(2n+1)\int_{-1}^{1} [P_n(x)]^2 dx = P_{n+1}(x)P_n(x)|_{-1}^{1} - \int_{-1}^{1} P_{n+1}(x)P'_n(x)dx. \tag{1.5.30}$$

However, $P'_n(x)$ is a polynomial of degree $n-1$ in x and the last integral is zero, which leaves

$$(2n+1)\int_{-1}^{1} [P_n(x)]^2\,dx = 1-(-1)^{n+1}(-1)^n = 2. \tag{1.5.31}$$

Hence

$$\int_{-1}^{1} [P_n(x)]^2\,dx = \frac{2}{2n+1}. \tag{1.5.32}$$

Multiplying

$$f(x) = \sum_{n=0}^{\infty} A_n P_n(x) \tag{1.5.33}$$

by $P_m(x)$ and integrating from -1 to 1 gives

$$\begin{aligned} \int_{-1}^{1} f(x)P_m(x)\,dx &= \sum_{n=0}^{\infty} A_n \int_{-1}^{1} P_n(x)P_m(x)\,dx \\ &= A_m \int_{-1}^{1} [P_m(x)]^2\,dx = A_m \frac{2}{2m+1} \end{aligned} \tag{1.5.34}$$

and the required coefficient is

$$A_m = \frac{2m+1}{2} \int_{-1}^{1} f(x)P_m(x)\,dx. \tag{1.5.35}$$

The following exercises illustrate the above work and form a link with the section on orthogonal functions.

EXERCISES

1.16 Show that

$$\int_{-1}^{1} xP_l(x)P_{l-1}(x)\,dx = \frac{2l}{4l^2-1}.$$

1.17 Show that

$$\int_{-1}^{1} (1-x^2)P_l'(x)P_m'(x)\,dx = \frac{2l(l+1)}{2l+1}\delta_{ml}$$

where

$$\delta_{ml} = \begin{cases} 1 & m = l \\ 0 & m \neq l. \end{cases}$$

1.18 If

$$f(x) = \begin{cases} \frac{1}{2} & 0 < x < 1 \\ -\frac{1}{2} & -1 < x < 0 \end{cases}$$

expand $f(x)$ in the form

$$\sum_{r=0}^{\infty} c_r P_r(x).$$

1.19 If

$$u_n = \int_{-1}^{1} x^{-1} P_n(x) P_{n-1}(x)\, dx$$

show that

$$n u_n + (n-1) u_{n-1} = 2$$

and hence show that

$$u_n = \begin{cases} \frac{2}{n} & \text{if } n \text{ is even} \\ 0 & \text{if } n \text{ is odd.} \end{cases}$$

1.20 Show that

$$(1-x)\sum_{r=0}^{n}(2r+1)P_r(x) = (n+1)[P_n(x) - P_{n+1}(x)].$$

1.21 Show that

$$\sum_{r=0}^{n}(2r+1)P_r(x) = P'_{n+1}(x) + P'_n(x).$$

1.22 If n is a positive integer prove that

$$\int_{-1}^{1} P_n(x)(1-2xh+h^2)^{-1/2} dx = \frac{2h^n}{2n+1}.$$

1.23 Verify relations (iii)–(vi) of 1.5.16.

1.6 Bessel Functions

The second of our special functions to be considered is the Bessel function. The generating function for Bessel functions of integer order is

$$\exp\left[\frac{x}{2}\left(t - \frac{1}{t}\right)\right] = \sum_{n=-\infty}^{\infty} t^n J_n(x) \qquad (1.6.1)$$

where $J_n(x)$ denotes the Bessel function of the first kind of order n. We expand

$$\exp\left[\frac{x}{2}\left(t - \frac{1}{t}\right)\right] \qquad (1.6.2)$$

in powers of t to give:

$$\begin{aligned}\exp\left[\frac{x}{2}\left(t-\frac{1}{t}\right)\right] &= \exp\left(\frac{xt}{2}\right)\exp\left(-\frac{x}{2t}\right)\\ &= \sum_{r=0}^{\infty}\frac{\left(\frac{1}{2}xt\right)^r}{r!}\sum_{s=0}^{\infty}\frac{\left(-\frac{1}{2}\frac{x}{t}\right)^s}{s!}\\ &= \sum_{r=0}^{\infty}\sum_{s=0}^{\infty}\frac{(-1)^s}{r!s!}\left(\frac{x}{2}\right)^{r+s}t^{r-s}. \end{aligned} \tag{1.6.3}$$

Now pick out the coefficients of t^n, $n \geq 0$. For a fixed value of r we want $s = r - n$, and for this value we have

$$\frac{(-1)^{r-n}}{r!(r-n)!}\left(\frac{x}{2}\right)^{2r-n}. \tag{1.6.4}$$

Total coefficients of t^n are obtained by summing over all values of r. Since $s = r - n$ and we require $s \geq 0$ we must have $r \geq n$. Hence we have

$$\sum_{r=n}^{\infty}\frac{(-1)^{r-n}}{r!(r-n)!}\left(\frac{x}{2}\right)^{2r-n} = \sum_{p=0}^{\infty}\frac{(-1)^p\left(\frac{x}{2}\right)^{2p+n}}{(p+n)!p!} = J_n(x). \tag{1.6.5}$$

If $n < 0$, we still have the coefficient of t^n for fixed r given by

$$\frac{(-1)^{r-n}}{r!(r-n)!}\left(\frac{x}{2}\right)^{2r-n} \tag{1.6.6}$$

but now the requirement $s \geq 0$ with $s = r - n$ is satisfied by all r.

The coefficient of t^n is

$$\sum_{r=0}^{\infty}\frac{(-1)^{r-n}}{r!(r-n)!}\left(\frac{x}{2}\right)^{2r-n} = J_n(x) \tag{1.6.7}$$

for n negative.

Writing n positive in 1.6.7 gives

$$\begin{aligned}J_{-n}(x) &= \sum_{r=0}^{\infty}\frac{(-1)^{r+n}\left(\frac{x}{2}\right)^{2r+n}}{r!(r+n)!}\\ &= (-1)^n\sum_{r=0}^{\infty}\frac{(-1)^r\left(\frac{x}{2}\right)^{2r+n}}{r!(r+n)!}\\ &= (-1)^n J_n(x). \end{aligned} \tag{1.6.8}$$

Hence

$$J_n(x) = \sum_{p=0}^{\infty}\frac{(-1)^p\left(\frac{x}{2}\right)^{2p+n}}{p!(p+n)!} \tag{1.6.9}$$

and

$$J_{-n}(x) = (-1)^n J_n(x) \tag{1.6.10}$$

(n positive integer).

An integral representation for $J_n(x)$ is given by

$$J_n(x) = \frac{1}{\pi}\int_0^{\pi} \cos(n\phi - x\sin\phi)\,d\phi \tag{1.6.11}$$

for integer n, the proof of which can be found in Watson (1922).

The next set of formulae to be found are the three-term recurrence relations:

$$\text{(i)} \quad J_n(x) = \frac{x}{2n}[J_{n-1}(x) + J_{n+1}(x)] \tag{1.6.12}$$

$$\text{(ii)} \quad J_n'(x) = \frac{1}{2}[J_{n-1}(x) - J_{n+1}(x)]. \tag{1.6.13}$$

These formulae can be derived from the generating function 1.6.1.

Differentiate w.r.t. t to obtain

$$\exp\left[\frac{x}{2}\left(t - \frac{1}{t}\right)\right]\frac{x}{2}\left(1 + \frac{1}{t^2}\right) = \sum_{n=-\infty}^{\infty} nt^{n-1}J_n(x) \tag{1.6.14}$$

which gives

$$\frac{x}{2}\sum_{n=-\infty}^{\infty} t^n J_n(x) + \frac{x}{2}\sum_{n=-\infty}^{\infty} t^{n-2}J_n(x) = \sum_{n=-\infty}^{\infty} nt^{n-1}J_n(x). \tag{1.6.15}$$

The coefficient of t^{n-1} yields

$$J_n(x) = \frac{x}{2n}\left[J_{n-1}(x) + J_{n+1}(x)\right]. \tag{1.6.16}$$

Differentiating 1.6.1 w.r.t. x gives

$$\exp\left[\frac{x}{2}\left(t - \frac{1}{t}\right)\right]\frac{1}{2}\left(t - \frac{1}{t}\right) = \sum_{n=-\infty}^{\infty} nt^{n-1}J_n'(x) \tag{1.6.17}$$

which then results in

$$\frac{1}{2}\sum_{n=-\infty}^{\infty} t^{n+1}J_n(x) - \frac{1}{2}\sum_{n=-\infty}^{\infty} t^{n-1}J_n(x) = \sum_{n=-\infty}^{\infty} t^n J_n'(x). \tag{1.6.18}$$

The coefficient t^n gives

$$J_n'(x) = \frac{1}{2}\left[J_{n-1}(x) - J_{n+1}(x)\right] \tag{1.6.19}$$

Eliminating $J_{n+1}(x)$ from 1.6.16 and 1.6.19 gives

$$xJ_{n-1}(x) = nJ_n(x) + xJ_n'(x) \tag{1.6.20}$$

multiply by x^{n-1} to deduce

$$x^n J_{n-1}(x) = nx^{n-1}J_n(x) + x^n J_n'(x) \tag{1.6.21}$$

which gives

$$x^n J_{n-1}(x) = \frac{d}{dx}(x^n J_n(x)). \tag{1.6.22}$$

Similarly eliminating $J_{n-1}(x)$ gives

$$xJ_{n+1}(x) = nJ_n(x) - xJ_n'(x) \tag{1.6.23}$$

multiply by x^{-n-1} to get

$$x^{-n}J_{n+1}(x) = -\frac{d}{dx}(x^{-n}J_n(x)). \tag{1.6.24}$$

From 1.6.22 and 1.6.24,

$$x^{n+1}J_n(x) = \frac{d}{dx}\left[x^{n+1}J_{n+1}(x)\right] = -\frac{d}{dx}\left[x^{2n+1}\frac{d}{dx}[x^{-n}J_n(x)]\right] \tag{1.6.25}$$

which yields

$$x^2 J_n''(x) + xJ_n'(x) + (x^2 - n^2)J_n(x) = 0. \tag{1.6.26}$$

Hence $y = J_n(x)$ is a solution of the differential equation

$$x^2\frac{d^2y}{dx^2} + x\frac{dy}{dx} + (x^2 - n^2)y = 0 \tag{1.6.27}$$

which is called Bessel's equation for the Bessel function of order n.

Bessel's equation of order ν where ν is not necessarily an integer is

$$x^2\frac{d^2y}{dx^2} + x\frac{dy}{dx} + (x^2 - \nu^2)y = 0 \tag{1.6.28}$$

by the obvious extension of 1.7.27. This equation has a regular singular point at 0 and an irregular singular point at ∞, and a series solution gives

$$J_\nu(x) = \sum_{s=0}^{\infty}\frac{(-1)^s}{\Gamma(\nu+1+s)s!}\left(\frac{x}{2}\right)^{2s+\nu}. \tag{1.6.29}$$

This is Bessel's function of the first kind of order ν. If ν is an integer n we obtain the series as before. If ν is not an integer we can construct a series based on $c = -\nu$ giving a general solution

$$y = aJ_\nu(x) + bJ_{-\nu}(x) \tag{1.6.30}$$

where

$$J_{-\nu}(x) = \sum_{s=0}^{\infty} \frac{(-1)^s}{\Gamma(-\nu+1+s)s!} \left(\frac{x}{2}\right)^{2s-\nu}. \tag{1.6.31}$$

If ν is a positive integer or zero, $J_{-\nu}(x)$ is a multiple of $J_\nu(x)$. An independent second solution can be found that involves a log term.

Note that $J_0(0) = 1$, $J_n(0) = 0$ for $n \neq 0$ and integer. There are other types of Bessel functions such as spherical Bessel functions and modified Bessel functions. The relationships that exist between them are too numerous to mention and can be found in Abramowitz and Stegun (1964).

The reader is now in a position to consider the following exercises in which further properties of the Bessel functions are developed.

EXERCISES

Show that

1.24

$$J_{\frac{1}{2}}(x) = \left(\frac{2}{\pi x}\right)^{\frac{1}{2}} \sin x$$

1.25

$$J_{\frac{-1}{2}}(x) = \left(\frac{2}{\pi x}\right)^{\frac{1}{2}} \cos x$$

1.26

$$\frac{d}{dx}\left[xJ_n(x)J_{n+1}(x)\right] = x\left[J_n^2(x) - J_{n+3}^2(x)\right]$$

1.27

$$8J_n'''(x) = J_{n-3}(x) - 3J_{n-1}(x) + 3J_{n+1}(x) - J_{n+1}(x)$$

1.28

$$4J_0'''(x) + 3J_0'(x) + J_3(x) = 0$$

1.29

$$J_n(-x) = (-1)^n J_n(x).$$

1.30 Using the generating function prove that

$$1 = J_0^2(x) + 2J_1^2(x) + 2J_2^2(x) + \dots.$$

Note that

$$1 = \exp\left[\frac{x}{2}\left(t - \frac{1}{t}\right)\right] \exp\left[-\frac{x}{2}\left(t - \frac{1}{t}\right)\right].$$

1.31 By proving

$$\frac{d}{dx}\left[\frac{x^2}{2}(J_n^2(x) - J_{n-1}(x)J_{n+1}(x))\right] = xJ_n^2(x)$$

show that

$$\int_0^x tJ_n^2(t)dt = \frac{x^2}{2}(J_n^2(x) - J_{n-1}(x)J_{n+1}(x)).$$

1.32 Given that

$$\int_0^\infty J_n(x)dx = 1$$

by using

$$\frac{x}{2}[J_{n+1}(x) + J_{n-1}(x)] = nJ_n(x)$$

show that

$$\int_0^\infty \frac{J_n(x)}{x}dx = \frac{1}{n}.$$

1.33 Show that

$$\begin{aligned}\left(\frac{1}{x}\frac{d}{dx}\right)^r [x^n J_n(x)] &= x^{n-r}J_{n-r}(x)\\ \left(\frac{1}{x}\frac{d}{dx}\right)^r [x^{-n} J_n(x)] &= (-1)^r x^{-n-r}J_{n+r}(x).\end{aligned}$$

1.7 Results from Complex Analysis

Complex analysis is employed quite freely in certain sections of this book, in particular in Chapter 4 on transform methods. The intention here is to quote the most important theorems and explain some of the practical consequences which are used in this text. For a full account of this subject, the reader should refer to a standard text such as Copson (1935), Titchmarsh (1932), Needham (1997) or Stewart and Tall (1983).

A complex number z is an ordered pair, namely its real part and its imaginary part. This is conventionally written as

$$z = x + iy. \tag{1.7.1}$$

The modulus of a complex number z is written as $|z|$ and defined as $\sqrt{x^2+y^2}$. A complex number can be represented by a point in the (x, y) plane (called an Argand diagram), and hence can also be represented in polar form with

$$x = r\cos\theta \qquad \text{and} \qquad y = r\sin\theta. \tag{1.7.2}$$

The angle θ is called the argument of z. The complex conjugate of z is denoted by $\bar{z}$ and defined by

$$\bar{z} = x - iy. \tag{1.7.3}$$

The concepts of continuity and differentiability follow by analogy with the real case. A function which is one-valued and differentiable at every point of a domain of the Argand plane, except at a finite number of points, is said to be analytic in that domain. The exceptional points are called singularities. An analytic function with no singularities is said to be regular.

Some familiar complex functions then include:

$$\exp(z) = 1 + \sum_{n=1}^{\infty} \frac{z^n}{n!} \tag{1.7.4}$$

$$\sin(z) = \sum_{n=0}^{\infty} (-1)^n \frac{z^{2n+1}}{(2n+1)!} \tag{1.7.5}$$

$$\cos(z) = \sum_{n=0}^{\infty} (-1)^n \frac{z^{2n}}{(2n)!}, \tag{1.7.6}$$

from which it follows that

$$\sin(z) = (e^{iz} - e^{-iz})/2i \tag{1.7.7}$$

$$\cos(z) = (e^{iz} + e^{-iz})/2 \tag{1.7.8}$$

and the commonly used result that

$$e^{i\theta} = \cos(\theta) + i\sin(\theta). \tag{1.7.9}$$

The hyperbolic functions can be defined in the usual way as

$$\sinh z = \frac{1}{2}(e^z - e^{-z}) \tag{1.7.10}$$

$$\cosh z = \frac{1}{2}(e^z + e^{-z}) \tag{1.7.11}$$

to yield the relationships

$$\begin{aligned} \sin iz &= i\sinh z & \cos iz &= \cosh z \\ \sinh iz &= i\sin z & \cosh iz &= \cos z. \end{aligned} \tag{1.7.12}$$

As with real numbers, the logarithmic function can be defined as the solution $\log z$ of the equation $e^w = z$. There are an infinite number of solutions in the complex case. These have the form

$$\log z = \log |z| + i \arg z \tag{1.7.13}$$

and the multiplicity of solutions is realised by $\arg z$ having many values each differing from the next by 2π.

One of the most far-reaching theorems in complex analysis is Cauchy's theorem. If $z = x(t) + iy(t)$ then in the Argand plane z describes a curve as t varies. If the arc is replaced by n small straight lines segments, then the arc is said to be rectifiable if the sum of the lengths of these segments tends to a unique limit as $n \to \infty$. Hence the integration of a complex function along an arc in the Argand plane is reduced to the integration of two real functions. Cauchy's theorem states that if z_1 and z_2 are two points in a complex domain D joined by a rectifiable arc lying in D then the value of the integral from z_1 to z_2 is quite independent of the particular arc employed, provided the integrand is analytic in the region considered. The theorem is usually expressed in the form that if $f(z)$ is an analytic function, continuous within and on a simple closed rectifiable curve C, then if $f'(z)$ exists at each point within C, then

$$\int_C f(z)\,dz = 0. \tag{1.7.14}$$

A useful consequence of Cauchy's Theorem is Cauchy's integral theorem which states that if $f(z)$ is analytic, and is regular within a closed contour C (without self-intersections) and is continuous within and on C, then for any point a within C:

$$f(a) = \frac{1}{2\pi i}\int_C \frac{f(z)}{z-a}\,dz \tag{1.7.15}$$

where C is traversed in an anticlockwise sense. As $f(z)$ is regular at a then

$$f(z) = f(a) + (z-a)f'(a) + (z-a)\eta(z) \tag{1.7.16}$$

where $\eta \to 0$ as $z \to a$. Hence by Cauchy's theorem, if γ is a circle centred at a with radius $r < \delta$ where $|z-a| < \delta$ then

$$\begin{aligned}\int_C \frac{f(z)}{z-a}\,dz &= \int_\gamma \frac{f(z)}{z-a}\,dz \\ &= f(a)\int_\gamma \frac{dz}{z-a} + f'(a)\int_\gamma dz + \int_\gamma \eta\,dz \\ &= 2\pi i f(a) + \int_\gamma \eta\,dz. \end{aligned} \tag{1.7.17}$$

However $\left|\int_\gamma \eta\,dz\right| \le 2\pi r \max_{|z-a|<r} |\eta(z)|$, and hence the final integral tends to zero with r to yield the theorem. By a similar means it can be shown that

$$f'(a) = \frac{1}{2\pi i}\int_C \frac{f(z)}{(z-a)^2}\,dz \tag{1.7.18}$$

and more generally

$$f^{(n)}(a) = \frac{n!}{2\pi i} \int_C \frac{f(z)}{(z-a)^{n+1}}\, dz \tag{1.7.19}$$

where again C is traversed in the anticlockwise sense.

A consequence of these results is Taylor's theorem which states that if $f(z)$ is an analytic function regular in a neighbourhood of $z = a$, then $f(z)$ may be expanded in the form

$$f(z) = \sum_{n=0}^{\infty} a_n (z-a)^n. \tag{1.7.20}$$

and the series converges for all z s.t. $|z - a| < R$ for some $R > 0$. Consider a point $a + h$, then Cauchy's integral theorem gives

$$\begin{aligned} f(a+h) &= \frac{1}{2\pi i} \int_C \frac{f(z)}{z-a-h}\, dz \\ &= \frac{1}{2\pi i} \int_C f(z) \left\{ \frac{1}{z-a} + \frac{h}{(z-a)^2} + \cdots \right. \\ &\quad \left. + \frac{h^n}{(z-a)^{n+1}} + \frac{h^{n+1}}{(z-a)^{n+1}(z-a-h)} \right\} dz. \end{aligned} \tag{1.7.21}$$

Now use the derivative expressions in 1.7.20 to yield

$$f(a+h) = f(a) + \sum_{r=1}^{n} f^{(r)}(a) \frac{h^r}{r!} + \text{remainder}. \tag{1.7.22}$$

If a function is not regular in the domain $|z - a| < R_1$ but is regular in the annulus $R_2 < |z - a| < R_1$ then the expansion has the form

$$f(z) = \sum_{n=0}^{\infty} a_n (z-a)^n + \sum_{n=1}^{\infty} b_n (z-a)^{-n} \tag{1.7.23}$$

which is Laurent's theorem. If at $z = a$, the Laurent expansion is a terminating series in $1/(z-a)^n$ then the singularity is called a pole of order m if b_m is the last non-zero coefficient. The coefficient b_1 is called the residue of $f(z)$ at the pole a. This value has considerable importance in evaluating complex integrals. For a simple pole of order unity the residue is given by

$$b_1 = \lim_{z \to a} (z-a) f(z) \tag{1.7.24}$$

and the formula for a pole of order m is given in the exercises.

Being able to evaluate residues is important in the evaluation of contour integrals, and is called the calculus of residues, the important theorem being Cauchy's theorem of residues. This theorem states that if $f(z)$ is continuous within and on a closed contour C, and if $f(z)$ is regular except for a finite number of poles within C, then

$$\int_C f(z)\,dz = 2\pi i \text{ (sum of the residues of } f(z) \text{ at its poles within } C). \tag{1.7.25}$$

This result is easily seen by considering one specific pole of order m say at a. Then $f(z)$ has the form

$$f(z) = g(z) + \sum_{n=1}^{m} \frac{a_n}{(z-a)^n}, \tag{1.7.26}$$

where g is a regular function in the neighbourhood of a. The contour C may be deformed by Cauchy's theorem to a circle defined by $z = a + \epsilon e^{i\theta}$ centred on the pole. Then

$$\int_C f(z)\,dz = \sum_{n=1}^{m} a_n \epsilon^{1-n} \int_0^{2\pi} e^{(1-n)i\theta} i\,d\theta = 2\pi i a_1 \tag{1.7.27}$$

as required. For a set of poles, the contour C may be distorted round each pole to be summed to give the theorem.

As an example consider the integral

$$\int_0^\infty \frac{dx}{1+x^2} = \frac{\pi}{2}. \tag{1.7.28}$$

To evaluate this integral using the calculus of residues consider

$$\int \frac{dz}{1+z^2} \tag{1.7.29}$$

around a contour which is defined by the real axis from $-R$ to R closed by the semicircle in the upper half plane. The integrand has one pole inside this contour at $z = i$, and its residue is $1/2i$, making the contour integral have value π. However, the contour integral is equal to twice the required integral plus the integral round the semicircle as $R \to \infty$. Putting $z = Re^{i\theta}$ gives for the integral round the semicircle

$$\int_0^\pi \frac{Rie^{i\theta}}{1+R^2e^{2i\theta}}\,d\theta \tag{1.7.30}$$

which tends to zero as $R \to \infty$. Some exercises are now presented to ensure familiarity with these ideas.

EXERCISES

1.34 Prove the the residue b_1 for pole of order m at $z = a$ is given by

$$b_1 = \lim_{z\to a} \frac{d^{m-1}}{dz^{m-1}} (z-a)^m f(z).$$

1.35 Show by contour integration that

$$\int_0^\infty \frac{\sin x}{x}\,dx = \frac{\pi}{2}$$

by considering the contour integral

$$\int_C \frac{e^{iz}}{z}\,dz$$

over a contour consisting of the real axis from r to R, a semicircle centre the origin radius R in the upper half plane, the real axis from $-R$ to $-r$ and a small semicircle radius r back to $z = r$. Let $R \to \infty$ and $r \to 0$.

1.36 Use contour integration to evaluate the integral

$$\int_0^\infty \frac{\cos x}{a^2 + x^2}\,dx.$$

1.8 Generalised Functions and the Delta Function

This section discusses the background to generalised functions which is used extensively in Chapter 5 on Green's functions. The reader is advised to consider this work as a prelude to Chapter 5, rather than consider this material at the first reading.

Since the mid 1930s, engineers and physicists have found it convenient to introduce fictitious functions having idealised properties that no physically significant functions can possibly possess. The main reason for this was their use in solving engineering problems often compounded in terms of a partial differential equation. For example, if a partial differential equation is taken to represent some dynamical engineering system, it is often useful to know how this system behaves when it is disturbed by an impulse. The delta function is a commonly used example of a generalised function, and can be thought of as an idealised mathematical representation of an impulse. The solution to the partial differential equation using a δ-impulse can therefore provide a model for the behaviour of the system.

The Dirac delta function is defined by some authors as the function having the properties

$$\delta(x) = \begin{cases} 0, & x \neq 0; \\ \infty, & x = 0; \end{cases}$$

$$\int_{-\infty}^{\infty} f(x)\delta(x)\,dx = f(0). \tag{1.8.1}$$

This function was first introduced by PAM Dirac in the 1930s in order to facilitate the analysis of partial differential equations in quantum mechanics; although the idea had been around for a century or more in mathematical circles. Clearly, such a function does not exist in the sense of classical analysis. Dirac called it an "improper function" and its use in analysis is recommended only when it is obvious that no inconsistency will follow from it (self-consistency being one of the principal foundations of any mathematical and scientific discipline). It is interesting to note that Dirac was also the first to postulate the existence of antimatter; like the delta function, another very abstract concept considered to be rather absurd at the time.

When attempting to provide a rigorous interpretation of the above equations, it is necessary to generalise the concept of a function. It was the work of L Schwartz and MJ Lighthill in the 1950s which put the theory of $\delta(x)$, and another fictitious functions, on a firm foundation. The mathematical apparatus developed by Schwartz and Lighthill is known as the "Theory of distributions". Today, there exist other approaches which put the Dirac delta function on a firm basis including "non-standard analysis".

1.8.1 Definition and Properties of a Generalised Function

To obtain an idea of what a generalised function is, it is convenient to use as an analogy, the notion of an irrational number ξ being a sequence $\{\rho(n)\}$ of rational numbers $\rho(n)$ such that

$$\xi = \lim_{n\to\infty} \rho(n),$$

where the limit indicates that the points $\rho(n)$ on the real line converge to the point representing ξ. All arithmetic operations performed on the irrational number ξ are actually performed on the sequence $\{\rho(n)\}$ defining ξ. A generalised function can be thought of as being a sequence of functions, which when multiplied by a test function and integrated over the whole sequence yields a finite limit. This approach is analogous to the one developed by K. Weierstrass in the 18th century who expressed a differential in terms of the limit of a variable approaching zero (but not actually reaching zero), and gives

$$\frac{df}{dx} = \lim_{\delta x\to 0} \frac{f(x+\delta x) - f(x)}{\delta x}.$$

This limiting argument avoided the issue associated with the fact that when $\delta x = 0$ we have 0/0! – one of the principal criticisms of calculus until the concept of a limit was introduced. A list of definitions and results from the theory of distributions is now given.

(i) Test Functions

The definition of test functions is as follows, where **R** and **C** are taken to denote a set of real and complex numbers respectively. The term "iff" denotes the statement "if and only if".

A function $\phi: \mathbf{R} \to \mathbf{C}$ is said to be a test function iff:

(i) $\phi \in C^\infty(\mathbf{R}, \mathbf{C})$ (i.e. ϕ is infinitely differentiable);

(ii) $|x^i \phi^{(j)}(x)| \leq M_{ij}$, for all integers $i, j \geq 0$ and all x in **R**.

Here, $C^\infty(\mathbf{R}, \mathbf{C})$ denotes the linear space of all **C**-valued and continuous functions defined on **R** such that their derivatives, $\phi^{(j)}$, of all orders are continuous.

The set of test functions is denoted by $S(\mathbf{R}, \mathbf{C})$. However, for certain generalised functions, this class can be extended, e.g. for the delta function it is sufficient to assume that the test functions are continuous.

An example of a function belonging to $S(\mathbf{R}, \mathbf{C})$ is the Gaussian function

$$\text{gauss}(x) = \frac{1}{\sqrt{\pi}} \exp(-x^2). \tag{1.8.2}$$

(ii) Regular Sequences

A sequence $\{\phi(x; n)\} \subset S(\mathbf{R}, \mathbf{C})$ is said to be regular iff the following limit exists for any $f \in S(\mathbf{R}, \mathbf{C})$:

$$\lim_{n \to \infty} \langle \phi(x; n), f(x) \rangle \equiv \lim_{n \to \infty} \int_{-\infty}^{\infty} \phi(x; n) f(x)\, dx. \tag{1.8.3}$$

For this limit to exist, it is not necessary that the sequence converges pointwise. For example, the sequence $\{n\, \text{gauss}(nx)\}$ approaches infinity as $n \to \infty$ at the point $x = 0$. However, the above limit exists. Also, even when the pointwise limit of a regular sequence does exist everywhere, it does not need to have any connection with the above limit. For example, the sequence $\{\exp(-1/n^2x^2)/n^2x^2\}$ approaches zero everywhere (except at $x = 0$ where the entries of the sequence are undefined), as $n \to \infty$, whereas the above limit approaches to $f(0)$.

(iii) Equivalent Regular Sequences

Two regular sequences $\{\phi(x; n)\}, \{\psi(x; n)\}$ are said to be equivalent iff for all $f \in S(\mathbf{R}, \mathbf{C})$

$$\lim_{n \to \infty} \langle \phi(x; n), f(x) \rangle = \lim_{n \to \infty} \langle \psi(x; n), f(x) \rangle. \tag{1.8.4}$$

For example, the sequences $\{n\,\mathrm{gauss}(nx)\}$ and $\{2n\,\mathrm{gauss}(2nx)\}$ are equivalent, leading to the limit $f(0)$. Generalised functions are now defined in terms of equivalent regular sequences.

(iv) A Generalised Function

ψ is a generalised function iff ψ is defined as the total, or complete class of equivalent regular sequences.

The term total in the above definition means that there exists no other equivalent regular sequences not belonging to this class. Any member of the class, for example $\{\psi(x;n)\}$, is sufficient to represent both ψ and the total class of equivalent regular sequences defining ψ. This is symbolically denoted as $\psi \sim \{\psi(x;n)\}$.

(v) The Functional $\langle\psi,\phi\rangle$

The functional $\langle\psi,\phi\rangle$ is defined as

$$\langle\psi,\phi\rangle \equiv \lim_{n\to\infty}\int_{-\infty}^{\infty}\psi(x;n)\phi(x)\,dx \equiv \int_{-\infty}^{\infty}\psi(x)\phi(x)\,dx. \tag{1.8.5}$$

where ψ is a generalised function and $\phi \in S(\mathbf{R},\mathbf{C})$.

The last integrand on the right-hand side of the above definition is used symbolically and does not imply actual integration. For some purposes it is also convenient to introduce the set whose elements are called functions of slow growth.

(vi) Functions of Slow Growth

$f:\mathbf{R}\to\mathbf{C}$ is said to be of slow growth iff

(i) $f \in C^\infty(\mathbf{R},\mathbf{C})$;

(ii) for each $j, j = 0,1,2,\ldots$, there exists a $B_j > 0$ such that $|f^{(j)}(x)| = O(|x|^{B_j})$, as $|x|\to\infty$.

The set of functions of slow growth will be denoted as $N(\mathbf{R},\mathbf{C})$. From this definition it is clear that any polynomial is an element of $N(\mathbf{R},\mathbf{C})$. Moreover, if $a \in N(\mathbf{R},\mathbf{C})$ and $\phi \in S(\mathbf{R},\mathbf{C})$, then $a\phi \in S(\mathbf{R},\mathbf{C})$. The elements of $S(\mathbf{R},\mathbf{C})$ are known as good functions and those of $N(\mathbf{R},\mathbf{C})$ as fairly good functions.

The algebraic operations for generalised functions are now defined. In this definition, ϕ and ψ are generalised functions represented by the sequences

$\phi(x) \sim \{\phi(x;n)\}$ and $\psi(x) \sim \{\psi(x;n)\}$ respectively.

(vii) Algebra of Generalised Functions

1 Addition: $\phi(x) + \psi(x) \sim \{\phi(x;n) + \psi(x;n)\}$.

2 Multiplication by a scalar: $\alpha\phi(x) \sim \{\alpha\phi(x;n)\}$, $\alpha \in \mathbf{C}$.

3 Derivative: $\phi'(x) \sim \{\phi'(x;n)\}$.

4 Shifting similarity: $\phi(\alpha x + \beta) \sim \{\phi(\alpha x + \beta; n)\}$, $\alpha, \beta \in \mathbf{C}$.

5 Multiplication by elements of $N(\mathbf{R}, \mathbf{C})$: $a(x)\phi(x) \sim \{a(x)\phi(x;n)\}$.

Note that the operation of multiplication between two generalised functions is not defined in general. From the above definition the following properties can be derived which are presented without proof.

(viii) Properties of Generalised Functions

If $\phi \in S(\mathbf{R}, \mathbf{C})$, ψ is a generalised function and $a(x) \in N(\mathbf{R}, \mathbf{C})$, then

1 $\langle \psi^{(l)}, f \rangle = (-1)^l \langle \psi, f^{(l)} \rangle$, l positive integer.

2 $\langle \psi(\alpha x + \beta), f(x) \rangle = |\alpha|^{-1} \langle \psi(x), f(x - \beta/\alpha) \rangle$, $\alpha, \beta \in \mathbf{R}$, $\alpha \neq 0$,

3 $\langle a(x)\psi(x), f(x) \rangle = \langle \psi(x), a(x)f(x) \rangle$.

(ix) Ordinary Functions as Generalised Functions

We now consider an important theorem which is presented without proof and enables us to represent any ordinary function by an equivalent generalised function.

If

1 $f: \mathbf{R} \to \mathbf{C}$,

2 $(1 + x^2)^{-M} |f(x)| \in L(\mathbf{R}, \mathbf{C})$, for some $M \geq 0$

where $L(\mathbf{R}, \mathbf{C})$ is the set of Lesbegue integrable functions, then there is a generalised function $\psi(x) \sim \{f(x;n)\}$ such that

$$\langle \psi, \phi \rangle = \langle f, \phi \rangle, \qquad \phi \in S(\mathbf{R}, \mathbf{C}).$$

In other words, an ordinary function satisfying Condition (2) is equivalent, in the sense of generalised functions, to a generalised function. If, in addition, f is continuous in an interval, then

$$\lim_{n \to \infty} f(x;n) = f(x)$$

is pointwise in that interval. This theorem increases the range of generalised functions available by using not only ordinary functions satisfying Condition (2), but also the new generalised functions which can be obtained by differentiation. A good example of this is the Heaviside step function defined by the expression

$$H(x) = \begin{cases} 1, & \text{if } x > 0; \\ 0, & \text{if } x < 0; \end{cases} \tag{1.8.6}$$

which is a generalised function. In fact, H satisfies the conditions of the theorem with $M = 1$. Moreover, if $\{H(x;n)\}$ is a sequence defining the generalised function H, then for any $\phi \in S(\mathbf{R}, \mathbf{C})$

$$\langle H, \phi\rangle = \int_0^\infty \phi(x)\, dx. \tag{1.8.7}$$

Thus the generalised function H is the process of assigning to a function ϕ a number equal to the area of ϕ from zero to infinity.

In the above example, the condition $\phi \in S(\mathbf{R}, \mathbf{C})$ is a sufficient, but not a necessary, condition to ensure that the integral on the right-hand side exists. Another condition could be

$$\int_0^\infty \phi(x)\, dx < \infty.$$

This is one case in which the set of test functions can be extended to consider test functions which do not belong to $S(\mathbf{R}, \mathbf{C})$.

Now there is sufficient background to consider in detail the most commonly used generalised function, namely the Delta function. The Dirac delta function is defined as

$$\delta(x) \sim \{H'(x;n)\} \tag{1.8.8}$$

where $\{H(x;n)\}$ is the sequence defining the Heaviside step function. δ should be called a delta generalised function instead, but the name delta function is now part of a long tradition. It should be stressed that δ is merely the symbolic representation for all classes of equivalent regular sequences represented by $\{H'(x;n)\}$. Thus

$$\int_{-\infty}^{\infty} \delta(x)\phi(x)\, dx, \qquad \phi \in S(\mathbf{R}, \mathbf{C})$$

actually means

$$\int_{-\infty}^{\infty} \delta(x)\phi(x)\, dx = \langle \delta, \phi\rangle = \lim_{n\to\infty} \int_{-\infty}^{\infty} H'(x;n)\phi(x)\, dx.$$

For the delta function, the set of test functions can be extended to consider bounded and piecewise continuous functions. Since the Heaviside step function defines a generalised function, then for $\phi \in S(\mathbf{R}, \mathbf{C})$ it follows that

$$\begin{aligned} \langle \delta, \phi \rangle &= \langle H', \phi \rangle \\ &= -\langle H, \phi' \rangle \\ &= -\int_0^\infty \phi'(x)\, dx \\ &= \phi(0). \end{aligned} \tag{1.8.9}$$

This property of the delta function is called the sampling property and is arguably its most important property which is why some authors define the δ function via its sampling property alone. In other words, we define the delta function in terms of the role it plays in a mathematical operation rather than in terms of what it actually is. Thus, we should always bear in mind that, strictly speaking, the delta function is not really a function even though it is referred to as one. It is actually just one of infinitely many different distributions but its sampling property is unique and is the main reason why it has such a wide range of applications. A more general expression for this property is

$$\begin{aligned} \langle \delta(x-\alpha), \phi(x) \rangle &= \int_{-\infty}^{\infty} \delta(x-\alpha)\phi(x)\, dx \\ &= \phi(\alpha). \end{aligned} \tag{1.8.10}$$

Let $m = 1, 2, 3, \ldots$, then the mth derivative of δ, denoted as $\delta^{(m)}$, is defined symbolically as $\delta^{(m)}(x) \sim \{H^{(m+1)}(x; n)\}$. Thus, there also exists a derivative-sampling property which can be stated as follows:

$$\begin{aligned} \langle \delta^{(m)}(x-\alpha), \phi(x) \rangle &= \int_{-\infty}^{\infty} \delta^{(m)}(x-\alpha)\phi(x)\, dx \\ &= (-1)^m \phi^{(m)}(\alpha). \end{aligned}$$

The main properties of δ are collected together in the following expressions, where $\phi \in S(\mathbf{R}, \mathbf{C})$, $a \in N(\mathbf{R}, \mathbf{C})$, and $\alpha, \beta \in \mathbf{R}$ with $\alpha \neq 0$. These results follow directly from the properties of generalised functions.

$$\begin{aligned} \langle \delta(-x), \phi(x) \rangle &= \langle \delta(x), \phi(x) \rangle; \\ \langle \delta'(-x), \phi(x) \rangle &= \langle -\delta'(x), \phi(x) \rangle; \\ \langle x\delta(x), \phi(x) \rangle &= \langle 0, \phi(x) \rangle; \\ \langle x\delta'(x), \phi(x) \rangle &= \langle -\delta(x), \phi(x) \rangle; \\ \langle \delta(\alpha x), \phi(x) \rangle &= \langle |\alpha|^{-1}\delta(x), \phi(x) \rangle; \\ \langle \delta(\alpha x + \beta), \phi(x) \rangle &= \langle |\alpha|^{-1}\delta(x + \beta/\alpha), \phi(x) \rangle; \\ \langle a(x)\delta(x), \phi(x) \rangle &= \langle a(0)\delta(x), \phi(x) \rangle. \end{aligned} \tag{1.8.11}$$

The respective symbolic notation for the above expressions are:

$$\begin{aligned}
\delta(-x) &= \delta(x); \\
\delta'(-x) &= -\delta'(x); \\
x\delta(x) &= 0; \\
x\delta'(x) &= -\delta(x); \\
\delta(\alpha x) &= |\alpha|^{-1}\delta(x); \\
\delta(\alpha x+\beta) &= |\alpha|^{-1}\delta(x+\beta/\alpha); \\
a(x)\delta(x) &= a(0)\delta(x).
\end{aligned} \tag{1.8.12}$$

Note that

$$xf(x) = xg(x) \Longrightarrow f(x) = g(x) + \alpha\delta(x),$$

$$\begin{aligned}
\phi(\alpha) &= \int_{-\infty}^{\infty} \delta(x-\alpha)\phi(x)\,dx \\
&= \int_{-\infty}^{\infty} \delta(\alpha-x)\phi(x)\,dx,
\end{aligned}$$

and

$$\begin{aligned}
\delta(x)\otimes\phi(x) &\equiv \int_{-\infty}^{\infty} \delta(x-\alpha)\phi(\alpha)\,d\alpha \\
&= \phi(x).
\end{aligned}$$

Finally, the following interesting property of the delta function is quoted. Let $f \in C^1(\mathbf{R}, \mathbf{C})$ be a function which does not vanish identically and suppose that the roots x_n, $n = 1, 2, \ldots$, of the equation $f(x) = 0$ are such that $f'(x_n) \neq 0$; then

$$\delta[f(x)] = \sum_n \frac{1}{|f'(x_n)|}\delta(x - x_n).$$

Note that, for descriptive purposes only, it is often useful to "visualise" the delta function in terms of the limit of a sequence of ordinary functions or regular sequences $\{\phi(x; n)\}$ which are known as delta sequences, namely

$$\delta(x) = \lim_{n\to\infty} \phi(x; n).$$

1.8.2 Differentiation Across Discontinuities

Let $f \in PC(\mathbf{R}, \mathbf{C})$ (a linear space of all $\mathbf{C}$-valued and piecewise continuous functions defined on $\mathbf{R}$) be a function satisfying the conditions of ordinary functions being generalised functions. Let $x_1, x_2, \ldots,$ be points at which f is discontinuous and let

$$\sigma_m = f(x_m^+) - f(x_m^-), \qquad m = 1, 2, \ldots,$$

be the jumps of the discontinuities at those points, respectively. The function

$$f_1(x) = f(x) - \sum_m \sigma_m H(x - x_m),$$

where H is the Heaviside step function, is obviously a continuous function everywhere. Moreover, f_1 also satisfies the same conditions as f. Hence f_1 defines a generalised function. Taking derivatives in the sense of generalised functions of both sides of the equation above yields

$$f_1{}'(x) = f'(x) - \sum_m \sigma_m \delta(x - x_m),$$

and therefore it follows that

$$f'(x) = f_1{}'(x) + \sum_m \sigma_m \delta(x - x_m). \tag{1.8.13}$$

Thus, the derivative, in a generalised sense, of a piecewise continuous function across a discontinuity is the derivative of the function in the classical sense plus the summation of the jumps of each discontinuity multiplied by a delta function centred at those discontinuities. This is a generalisation of the concept of a derivative in the classical sense, because if the function were continuous everywhere then σ_m would be zero for all m and the derivative in a generalised sense would coincide with the derivative in the classical sense.

1.8.3 The Fourier Transform of Generalised Functions

The definition of the Fourier transform of a generalised function is based in the following theorem.

If $\phi \in S(\mathbf{R}, \mathbf{C})$ and $\phi(x) \leftrightarrow \Phi(u)$ then $\Phi \in S(\mathbf{R}, \mathbf{C})$ where $\leftrightarrow$ denotes that $\Phi(u)$ is the Fourier transform of $\phi(x)$ which is the inverse Fourier transform of $\Phi(u)$ (so that ϕ and Φ are Fourier transform pairs).

From this result, if $\{\psi(x;n)\} \subset S(\mathbf{R}, \mathbf{C})$ then its Fourier transform $\{\Psi(x;n)\}$ also belongs to $S(\mathbf{R}, \mathbf{C})$. Similarly, if $\phi \in S(\mathbf{R}, \mathbf{C})$, then its Fourier transform $\Phi \in S(\mathbf{R}, \mathbf{C})$ and by Parseval's theorem (see Apostol, 1974)

$$\int_{-\infty}^{\infty} \Psi(u;n)\Phi(u)\,du = \int_{-\infty}^{\infty} \psi(x;n)\phi(x)\,dx$$

so that if the limit

$$\lim_{n\to\infty}\int_{-\infty}^{\infty}\psi(x;n)\phi(x)\,dx$$

exists for an arbitrary member of $S(\mathbf{R},\mathbf{C})$, then

$$\lim_{n\to\infty}\int_{-\infty}^{\infty}\Psi(u;n)\Phi(u)\,du$$

also exists for any $\Phi\in S(\mathbf{R},\mathbf{C})$. In other words, if $\{\psi(x;n)\}$ is a regular sequence, then so is the sequence $\{\Psi(u;n)\}$.

Let ψ be a generalised function represented by the regular sequence $\psi\sim\{\psi(x;n)\}$, then from the above discussion it follows that the Fourier transform of ψ, denoted by Ψ, is a generalised function represented by the regular sequence $\Psi\sim\{\Psi(u;n)\}$ of the Fourier transform of $\{\psi(x;n)\}$. The inverse Fourier transform can be defined in exactly the same way since $\{\Psi(u;n)\}$ is also a regular sequence.

A result that can also be derived from Parseval's Theorem and from the symmetry of the Fourier transform as follows.

If ψ is a generalised function and ψ and Ψ are Fourier transform pairs and if $\phi\in S(\mathbf{R},\mathbf{C})$ and ϕ and Φ are Fourier transform pairs, then

$$\langle\Psi,\phi\rangle=\langle\psi,\Phi\rangle.$$

From this theorem, using the algebra of generalised functions, it can be shown that the usual properties of the classical Fourier transform are preserved. Also, the following Fourier transform pairs can be derived where H denotes the Heaviside step function; $\alpha,\beta\in\mathbf{R}$ and $m=0,1,2,\ldots$.

$$\begin{aligned}
H(x) &\leftrightarrow \pi\delta(u)+(iu)^{-1};\\
\delta(x) &\leftrightarrow 1;\\
1 &\leftrightarrow 2\pi\delta(u);\\
\delta(x-\alpha) &\leftrightarrow \exp(-iu\alpha);\\
\exp(i\alpha x) &\leftrightarrow 2\pi\delta(u-\alpha);\\
\delta(\alpha x+\beta) &\leftrightarrow |\alpha|^{-1}\exp(-iu\beta/\alpha);\\
\delta^{(m)}(x) &\leftrightarrow (iu)^m;\\
x^m &\leftrightarrow 2\pi i^m\delta^{(m)}(u).
\end{aligned}\tag{1.8.14}$$

The following symbolic integral representation of $\delta(x-\alpha)$ follows directly from the second of the Fourier transform pairs given above;

$$\delta(x-\alpha)=\frac{1}{2\pi}\int_{-\infty}^{\infty}\exp(iux-iu\alpha)\,du.\tag{1.8.15}$$

Finally, the following Fourier transform pair is useful in sampling theory where $0 \neq b \in \mathbf{R}$:

$$\sum_{n=-\infty}^{\infty} \delta(x - nb) \leftrightarrow \frac{2\pi}{b} \sum_{n=-\infty}^{\infty} \delta(u - 2\pi n/b). \tag{1.8.16}$$

The function $\sum\limits_{n=-\infty}^{\infty} \delta(x - nb)$ is called the "comb function" by some authors.

1.8.4 Convolution of Generalised Functions

For generalised functions, the definition of convolution depends on the concept of a direct product; therefore, this section starts by giving the definition of this product.

Let ψ_1 and ψ_2 be generalised functions. The expression

$$\langle \psi_1 \times \psi_2, \phi \rangle \equiv \langle \psi_1(x), \langle \psi_2(y), \phi(x, y) \rangle \rangle, \qquad \phi \in S(\mathbf{R}^2, \mathbf{C}),$$

is called the direct product of ψ_1 and ψ_2.

The direct product is of a particularly simple form when $\phi(x, y)$ is separable. If ψ_1 and ψ_2 are generalised functions and $\phi(x, y) = \phi_1(x)\phi_2(y)$, $\phi_1, \phi_2 \in S(\mathbf{R}, \mathbf{C})$, then

$$\langle \psi_1(x), \langle \psi_2(y), \phi(x, y) \rangle \rangle = \langle \psi_1(x), \phi_1(x) \rangle \langle \psi_2(x), \phi_2(x) \rangle.$$

Thus, for example, the direct product of δ with itself yields the delta function over $\mathbf{R}^2$,

$$\delta(x) \times \delta(y) = \delta(x, y).$$

Over $\mathbf{R}^3$ the result is

$$\delta(x) \times \delta(y) \times \delta(z) = \delta(x, y, z).$$

In order to define the convolution of two generalised functions, let $f, g: \mathbf{R} \to \mathbf{C}$ be two functions for which their convolution exists. If ϕ is a function belonging to $S(\mathbf{R}, \mathbf{C})$ then with $\otimes$ denoting the convolution integral,

$$\begin{aligned}
\langle f \otimes g, \phi \rangle &= \int_{-\infty}^{\infty} (f \otimes g)(x)\phi(x)\, dx \\
&= \int_{-\infty}^{\infty} \int_{-\infty}^{\infty} f(\alpha)g(x - \alpha)\, d\alpha\, \phi(x)\, dx \\
&= \int_{-\infty}^{\infty} \int_{-\infty}^{\infty} f(\alpha)g(y)\phi(\alpha + y)\, d\alpha\, dy
\end{aligned}$$

$$= \int_{-\infty}^{\infty} f(\alpha) \int_{-\infty}^{\infty} g(y)\phi(\alpha + y)\, dy\, d\alpha$$
$$= \langle f(x), \langle g(y), \phi(x + y)\rangle\rangle.$$

In other words, the desired result is equivalent to applying the direct product of f and g to the function $\phi(x + y)$. This result suggests that we define the convolution of two generalised functions ψ_1 and ψ_2 as follows:

$$\begin{aligned}\langle \psi_1 \otimes \psi_2, \phi\rangle &= \langle \psi_1(x), \langle \psi_2(y), \phi(x + y)\rangle\rangle \\ &= \langle \psi_1(x) \times \psi_2(y), \phi(x + y)\rangle, \quad \phi \in S(\mathbf{R}, \mathbf{C}).\end{aligned}$$

Some examples of such convolutions are:

$$\begin{aligned}\delta(x - \alpha) \otimes \psi(x) &= \psi(x - \alpha); \\ \delta^{(m)}(x - \alpha) \otimes \psi(x) &= \psi^{(m)}(x - \alpha); \\ \delta'(x) \otimes H(x) &= \delta(x); \\ \delta(x - \alpha) \otimes \delta(x - \beta) &= \delta(x - \alpha - \beta).\end{aligned}$$

An important consequence of the second of the above results is that every linear differential equation with constant coefficients can be represented as a convolution. Thus if $a_i \in \mathbf{R}$, $i = 0, 1, \ldots, n$, it follows that

$$\sum_{i=0}^{n} a_i f^{(i)}(x) = \left[\sum_{i=0}^{n} a_i \delta^{(i)}(x)\right] \otimes f(x).$$

Note that this statement cannot be made if the convolution operation is restricted to ordinary functions. If ψ_1 and ψ_2 are generalised functions and either ψ_1 or ψ_2 has bounded support then $\psi_1 \otimes \psi_2$ exists.

The convolution of generalised functions preserves the basic properties of the classical convolution except that it is not generally associative. Even if $\psi_1 \otimes (\psi_2 \otimes \psi_3)$ exists as a generalised function it need not to be the same as $(\psi_1 \otimes \psi_2) \otimes \psi_3$; it does not even follow that this will co-exist with $\psi_1 \otimes (\psi_2 \otimes \psi_3)$. An example of this situation is as follows:

$$1 \otimes [\delta'(x) \otimes H(x)] = 1 \otimes \delta(x) = 1,$$

whereas

$$[1 \otimes \delta'(x)] \otimes H(x) = 0 \otimes H(x) = 0.$$

For some classes of generalised functions, e.g. the delta function, it is possible to formulate a convolution theorem. In order to state this theorem the following definitions are needed.

Convergence in $S(\mathbf{R}, \mathbf{C})$

$\{\phi(x;n)\} \in S(\mathbf{R}, \mathbf{C})$ is said to converge in $S(\mathbf{R}, \mathbf{C})$ iff $\{|x|^l \phi^{(m)}(x;n)\}$ converges uniformly over $\mathbf{R}$ for $l > 0$, $m \geq 0$. If the limit function of the sequence $\{\phi(x;n)\}$ is $\phi(x)$ then it may be proved that $\phi \in S(\mathbf{R}, \mathbf{C})$, or the linear space $S(\mathbf{R}, \mathbf{C})$ is closed under convergence.

Multipliers

$\phi: \mathbf{R} \to \mathbf{C}$ is called a multiplier in $S(\mathbf{R}, \mathbf{C})$ iff

1 $\psi \in S(\mathbf{R}, \mathbf{C}) \Longrightarrow \phi \otimes \psi \in S(\mathbf{R}, \mathbf{C})$,

2 $\{\psi(x;n)\} \subset S(\mathbf{R}, \mathbf{C})$, and $\lim_{n\to\infty} \psi(x;n) = 0$
$\Longrightarrow \phi(x)\psi(x;n) \to 0$ as $n \to \infty$ in $S(\mathbf{R}, \mathbf{C})$.

Not all generalised functions are multipliers and so these functions cannot in general be multiplied; the convolution of two arbitrary generalised functions may therefore not necessarily be formed.

The convolution theorem for generalised functions therefore takes the following form:

If

1 ψ_1 and ψ_2 generalised functions,

2 $\psi_1 \leftrightarrow \Psi_1$, $\psi_2 \leftrightarrow \Psi_2$,

3 either ψ_1 or ψ_2 has bounded support,

then either Ψ_1 or Ψ_2 is a multiplier, respectively, and $\psi_1 \otimes \psi_2 \leftrightarrow \Psi_1 \Psi_2$.

The convolution of two arbitrary generalised functions does not always exist. However, in practical problems, such as in electrodynamics, multiplications of arbitrary generalised functions can be used. The formal Fourier transform of these products sometimes give rise to the appearance of convolution integrals which diverge.

1.8.5 The Discrete Representation of the Delta Function

The discrete representation of the delta function $\delta(x - x_0)$ is given by the so-called *Kronecker's delta* δ_{jr} defined by the expression

$$\delta_{jr} = \begin{cases} 1 & \text{if } j = r; \\ 0 & \text{if } j \neq r. \end{cases} \tag{1.8.17}$$

It is a sequence which contains only one non-zero valued element, the value of that element being unity.

The above representation, namely Kronecker's delta, has the discrete-version properties of the delta function. For example, it follows that an arbitrary sequence f_j can be written as a weighted sum of Kronecker delta's:

$$f_j = \sum_{r=-\infty}^{\infty} f_r \delta_{jr}. \tag{1.8.18}$$

This equation is the discrete analogue of the result

$$\phi(\alpha) = \int_{-\infty}^{\infty} \delta(\alpha - x)\phi(x)dx. \tag{1.8.19}$$

Furthermore, δ_{jr} defines the following discrete Fourier transform pairs:

$$\begin{aligned} \delta_{j0} &\leftrightarrow 1; \\ 1 &\leftrightarrow N\delta_{0l}; \\ \delta_{jr} &\leftrightarrow \exp(-2\pi irl/N); \\ \exp(-2\pi ijm/N) &\leftrightarrow N\delta_{jl}. \end{aligned}$$

Finally, the following result is the discrete analogue of the integral representation of the delta function:

$$\frac{1}{N}\sum_{m=0}^{N-1} \exp(2\pi ijm/N)\exp(-2\pi irm/N) = \delta_{jr}.$$

EXERCISES

1.37 By using the sequence $\{\frac{n}{\pi}\text{sinc}(nx)\}$ where sinc$x$ is the Cardinal sine function $(\frac{\sin x}{x})$ show that in the limit as $n \to \infty$

$$\delta(x) = \frac{1}{2\pi}\int_{-\infty}^{\infty} \exp(ikx)dk.$$

Hence evaluate the Fourier transforms of $\cos x$ and $\sin x$.

1.38 Find the limits (in the distributional sense) of the following

$$\lim_{n\to\infty}\left[\frac{\sin^2(nx)}{nx^2}\right], \qquad \lim_{n\to\infty}[\sin(n!x)].$$

1.39 Show that

$f(x)\delta(x-\alpha) = f(\alpha)\delta(x-\alpha)$; $x\delta(x) = 0$; $\delta(\alpha-x) = \delta(x-\alpha)$; $\delta(\alpha x) = \frac{1}{|\alpha|}\delta(x), \alpha \neq 0$; $\int\limits_{-\infty}^{\infty} f(x)\delta'(x)dx = -f'(0)$; $\int\limits_{-\infty}^{\infty} f(x)\delta^{(n)}(x)dx = (-1)^n f^{(n)}(0)$;

$\delta(\alpha^2 - x^2) = \frac{1}{2\alpha}[\delta(x+\alpha) + \delta(x-\alpha)]$; $\delta(\sin x) = \sum\limits_{n=-\infty}^{\infty} \delta(x - n\pi)$.

1.40 Show that

$$\int\limits_{-\infty}^{\infty} e^{-|x|}\delta(x)dx = e^0 = 1$$

by using the regular sequence $\{n\,\text{gauss}(nx)\}$ and considering

$$\lim_{n\to\infty} \langle n\,\text{gauss}(nx), e^{-|x|}\rangle.$$

1.41 Show that

$\delta(x) \leftrightarrow 1$; $1 \leftrightarrow 2\pi\delta(u)$; $\delta(x-\alpha) \leftrightarrow \exp(-iu\alpha)$; $\exp(i\alpha x) \leftrightarrow 2\pi\delta(u-\alpha)$;

$\delta(\alpha x + \beta) \leftrightarrow | \alpha |^{-1} \exp(-iu\beta/\alpha)$; $\delta^{(m)}(x) \leftrightarrow (iu)^m$; $x^m \to 2\pi i^m \delta^{(m)}(x)$.

1.42 The sign function $\text{sgn}(x)$ is defined by

$$\text{sgn}(x) = \begin{cases} 1, & x > 0; \\ -1, & x < 0. \end{cases}$$

By computing the Fourier transform of $\exp(-\epsilon \mid x \mid)\text{sgn}(x)$ over the interval $[-a, a]$ and then letting $a \to \infty$ and $\epsilon \to 0$, show that

$$\text{sgn}(x) \leftrightarrow \frac{2}{iu}$$

given that the Fourier transform of a function $f(x)$ is defined by

$$F(u) = \int\limits_{-\infty}^{\infty} f(x)\exp(-iux)dx.$$

Hence show that

$$H(x) \leftrightarrow \pi\delta(u) - \frac{i}{u}$$

where $H(x)$ is the step function.

2
Separation of the Variables

2.1 Introduction

In the previous chapter, some of the mathematical preliminaries were discussed. The principal physical equations which we will consider are those for heat flow, the wave equation and the potential equation of Laplace and are derived in the companion volume, Numerical methods for partial differential equations. These equations are very typical of second-order linear equations. It was shown in the previous chapter, §1.2 that there are just three canonical forms for the general second-order quasilinear equation of which the three physical problems are linear examples. Hence, in order to introduce methods of solution, these three equations will be considered in their own light and various methods will be studied.

There are in fact only a limited number of methods available to solve partial differential equations analytically without introducing approximate or numerical techniques. For a few equations, such as the wave equation, a general solution can be written which involves arbitrary functions. These arbitrary functions are the partial differential equation equivalent of the arbitrary constants which arise in the solution of ordinary differential equations. To obtain the solution of an ordinary differential equation the boundary conditions are used to fix the arbitrary constants. In the partial differential case, the boundary conditions lead to a functional equation for the arbitrary function. The solution of such functional equations can be as difficult as the original partial differential equation, although some progress can be made along these lines and this is pursued in Chapter 3.

A powerful method is the method of separation of the variables. Here the solution of an equation, say $u(x,t)$, is expressed in the form

$$u(x,t) = X(x)T(t) \tag{2.1.1}$$

in which the solution is separated into a product of a function purely of x and another purely of y. This form is substituted into the differential equation and results in ordinary differential equations for the single variable functions $X(x)$ and $T(t)$. A set of such solutions is obtained and these can be summed because of linearity to give a "general solution". The boundary conditions are applied to this solution and these restrict the summed functions to a subset, and yield the coefficients of the series. The latter process is effectively one of expressing the solution as a series of orthogonal functions which have been covered in the previous chapter. The process is the expansion of the solution as a generalised Fourier series with close links to Fourier expansions in signal analysis. There are considerable theoretical grounds for expecting such a method to be successful. Primarily, the theory of Fourier series and orthogonal functions indicates that quite general classes of functions can be approximated by convergent series of orthogonal functions (Apostol, 1974), and hence one might expect the series of separated functions to be effective. Clearly if no such split as in 2.1.1 exists for a given problem the approach will break down.

A third powerful approach to be considered in the fourth chapter is the use of integral transforms, in which once again, partial derivatives are reduced to algebraic terms and ordinary derivatives. The transformed problem is then soluble and the original function space can be recovered by applying the inverse transform which is an integration process. Solutions often turn out in integral form.

To describe the separation of variables method, each of the standard problems will be taken in turn, and a fairly general problem will be solved by the method in each case. A series of exercises based on the previous work should then be attempted by the reader. The wave equation is the first to be considered.

2.2 The Wave Equation

The study of the wave equation was the motivation for the first work on partial differential equations. Jean Le Rond D'Alembert (1717–1783) published "Fraité de dynamiquie" in 1743 in which the equation of a stretched string is considered. Considerable arguments arose between D'Alembert and Leonard Euler (1707–1783) who published important papers in 1734 and 1748. In these papers, periodic solutions had been established and initial conditions involving "discontinuous" functions. In fact these functions had discontinuous first derivatives. The argument was later joined by Jean Bernoulli (1700–1782), one of the two famous mathematical brothers (the other was Jacques), who was studying waves in bars and considerably extended the set of allowable initial conditions. A little later Joseph-Louis Lagrange considered the propagation of sound and came to the threshold of discovering Fourier series in 1759. By 1762 and 1763, both Euler and D'Alembert had moved to solving waves in strings of varying thickness, and in 1759, Euler had considered waves in a membrane.

The problem of an elastic string stretched to a length l and then fixed at its endpoints constitutes a simple problem on which to consider the separation of variables method. The string is deformed and then released with a known velocity. The wave equation is solved to find the deflection of the string at time t. From the physical derivation in the previous chapter, the vibrations of an elastic string are governed by the wave equation in the form

$$\frac{\partial^2 u}{\partial t^2} = c^2 \frac{\partial^2 u}{\partial x^2}, \tag{2.2.1}$$

where $u(x,t)$ is the deflection of the string at a distance x along the string at a time t. The string is fixed at its endpoints $x = 0$ and $x = l$, yielding the boundary conditions $u(0,t) = u(l,t) = 0$ for $t \geq 0$. For the initial conditions, let the deflection at $t = 0$ be described by the function $f(x)$, so that $u(x,0) = f(x)$, and let the velocity at $t = 0$ be similarly described by the function $g(x)$. The latter condition yields $\frac{\partial u}{\partial t}|_{t=0} = g(x)$, and both of these conditions are valid for $0 \leq x \leq l$. The separation of variables method now assumes the solution may be expressed in the form $u(x,t) = X(x)T(t)$, where $X(x)$ and $T(t)$ are functions of only x and t respectively. Hence

$$\frac{\partial^2 u}{\partial x^2} = T\frac{d^2X}{dx^2} \quad \text{and} \quad \frac{\partial^2 u}{\partial t^2} = X\frac{d^2T}{dt^2} \tag{2.2.2}$$

and the partial differential equation becomes

$$\frac{1}{c^2T}\frac{d^2T}{dt^2} = \frac{1}{X}\frac{d^2X}{dx^2}. \tag{2.2.3}$$

It is clear that the left-hand side of this equation is a function of t and the right-hand side is independent of t. Hence the equation can only have a solution of the form 2.1.1 if both sides have the same constant value. Let this constant value be k to give

$$\frac{1}{X}\frac{d^2X}{dx^2} = \frac{1}{c^2T}\frac{d^2T}{dt^2} = k \tag{2.2.4}$$

where k is arbitrary. Then the solution of 2.2.1 has reduced to solving the ordinary differential equations

$$\frac{d^2X}{dx^2} = kX \quad \text{and} \quad \frac{d^2T}{dt^2} = kc^2T \tag{2.2.5}$$

for $X(x)$ and $T(t)$, which will yield a set of solutions for varying k. The solution $u(x,t)$ must now be forced to satisfy the boundary conditions, $u(0,t) =$

$X(0)T(t) = 0$ and $u(l,t) = X(l)T(t) = 0$. Therefore either $T(t) \equiv 0$ for all t giving $u(x,t) \equiv 0$ for all t which is a trivial solution, or $X(0) = X(l) = 0$ for all t.

Hence for non-trivial solutions, consider

$$\frac{d^2X}{dx^2} = kx \qquad \text{with } X(0) = X(l) = 0,$$

which yields three cases, namely $k > 0$, $k = 0$ and $k < 0$ which each gives a different form of solution. As this is the first example of this type, the three cases will be pursued in detail, but with experience the valid choice for the given boundary conditions is usually written without explicitly considering each avenue in detail. Hence the cases are as follows.

(i) $k = 0$ gives $\frac{d^2X}{dx^2} = 0$ with solution $X(x) = Ax+B$. The boundary condition $X(0) = 0$ then gives $B = 0$ and the condition $X(l) = 0$ gives $0 = Al + B$, so forcing $A = 0$. Hence this case has generated the trivial solution $X = 0$. Although the pedantry employed here would normally be omitted, it is instructive to observe that these alternative solutions are not just being ignored, but they genuinely do not satisfy the boundary conditions without yielding trivial solutions. In some examples such solutions are necessary and care must be taken not to ignore alternatives.

(ii) For the case $k > 0$, let $k = h^2$ to give $\frac{d^2X}{dx^2} = h^2X$. The trial solution $X(x) = e^{mx}$ gives $m^2 = h^2$ or $m = \pm h$ with the corresponding solution

$$X(x) = Ae^{hx} + Be^{-hx}. \tag{2.2.6}$$

The boundary conditions $X(0) = 0$ and $X(l) = 0$ then give

$$0 = A + B \qquad \text{and} \qquad 0 = Ae^{hl} + Be^{-hl}$$

which again yield the trivial solution $A = B = 0$. With a finite string, it is unlikely that an exponential solution in x is relevant as we would physically expect some form of oscillatory solution which will yield differing frequencies associated with, for example, stringed musical instruments. Indeed this is borne out by case (iii) below.

(iii) The third case with $k < 0$, can be written as $k = -p^2$ giving $\frac{d^2X}{dx^2} = -p^2x$, with the trial solution $X = e^{mx}$ yielding $m^2 = -p^2$, or $m = \pm ip$. Hence this solution is

$$X(x) = A\cos px + B\sin px \tag{2.2.7}$$

and the above boundary conditions then give

$$A = 0 \qquad \text{and} \qquad 0 = B\sin pl.$$

Hence either $B = 0$ or $\sin pl = 0$. $B = 0$ gives the trivial solution once again, and $\sin pl = 0$ gives $pl = n\pi, n = 1,2,3,\ldots$ or $p = \frac{n\pi}{l}, n = 1,2,3,\ldots$.

The k's are called eigenvalues and the corresponding solution $X(x)$ is an eigenfunction. In the case of a vibrating string, these different eigenvalues correspond to the different frequencies with which the string may vibrate, giving the effect violinists use to sound overtones by stopping.

Hence non-trivial solutions only exist for k restricted to the values $-\left(\frac{n\pi}{l}\right)^2$, $n = 1, 2, 3, \ldots$. The T equation has the general solution

$$T(t) = E_n \cos\left(\frac{cn\pi t}{l}\right) + D_n \sin\left(\frac{cn\pi t}{l}\right), \tag{2.2.8}$$

giving the full separated solution

$$u_n(x,t) = \left[F_n \cos\left(\frac{cn\pi t}{l}\right) + G_n \sin\left(\frac{cn\pi t}{l}\right)\right] \sin\left(\frac{n\pi x}{l}\right), \tag{2.2.9}$$

where $n = 1, 2, 3, \ldots, F_n = B_n E_n$ and $G_n = D_n B_n$. There is an infinite set of discrete values of k and consequently to each value of k will correspond a solution having the above form. However, the wave equation is a linear equation, and so any linear combination of such solutions is also a solution, and hence

$$u(x,t) = \sum_{n=1}^{\infty} \left(F_n \cos\left(\frac{cn\pi t}{l}\right) + G_n \sin\left(\frac{cn\pi t}{l}\right)\right) \sin\left(\frac{n\pi x}{l}\right) \tag{2.2.10}$$

is the separated solution of 2.2.1 with $u(0,t) = u(l,t) = 0$.

The final part of the solution is the fixing of the arbitrary constants F_n and G_n so that the initial conditions $u(x,0) = f(x)$ and $\frac{\partial u}{\partial t}(x,0) = g(x)$ are satisfied. The first condition gives

$$f(x) = \sum_{n=1}^{\infty} F_n \sin\left(\frac{n\pi x}{l}\right) \tag{2.2.11}$$

and the second using

$$\frac{\partial u}{\partial t} = \sum_{n=1}^{\infty} \left[-F_n \frac{cn\pi}{l} \sin\left(\frac{cn\pi t}{l}\right) + \frac{cn\pi}{l} G_n \cos\left(\frac{cn\pi t}{l}\right)\right] \sin\left(\frac{n\pi x}{l}\right) \tag{2.2.12}$$

gives

$$g(x) = \frac{c\pi}{l} \sum_{n=1}^{\infty} n G_n \sin\left(\frac{n\pi x}{l}\right). \tag{2.2.13}$$

The coefficients F_n and G_n may now be determined by an orthogonal (Fourier) series technique (of the previous chapter) and are given by the Fourier coefficients

$$F_n = \frac{2}{l} \int_0^l f(x) \sin\frac{n\pi x}{l} dx \tag{2.2.14}$$

and

$$G_n = \frac{2}{n\pi c}\int_0^l g(x)\sin\frac{n\pi x}{l}dx, \qquad n = 1,2,3,\ldots. \tag{2.2.15}$$

Substituting these constants into 2.2.10 gives

$$\begin{aligned} u(x,t) &= \sum_{n=1}^{\infty}\left[\left[\frac{2}{l}\int_0^l f(x')\sin\frac{n\pi x'}{l}dx'\right]\cos\frac{cn\pi t}{l}\sin\frac{n\pi x}{l}\right. \\ &\quad \left.+\left[\frac{2}{n\pi c}\int_0^l g(x')\sin\frac{n\pi x'}{l}dx'\right]\sin\frac{cn\pi t}{l}\sin\frac{n\pi x}{l}\right] \end{aligned} \tag{2.2.16}$$

which is the solution of the wave equation subject to the given conditions.

Before leaving this solution to consider some examples it is useful at this stage to link this work to the more general solution which will be obtained in Chapter 3. To this end, the solution may be rewritten as

$$u(x,t) = \sum_{n=1}^{\infty}\left(F_n\cos\frac{cn\pi t}{l} + G_n\sin\frac{cn\pi t}{l}\right)\sin\frac{n\pi x}{l} \tag{2.2.17}$$

which yields

$$\begin{aligned} u(x,t) &= \sum_{n=1}^{\infty}\left[\frac{F_n}{2}\left(\sin\frac{n\pi}{l}(x+ct) + \sin\frac{n\pi}{l}(x-ct)\right)\right. \\ &\quad \left.+\frac{G_n}{2}\left(\cos\frac{n\pi}{l}(x-ct) - \cos\frac{n\pi}{l}(x+ct)\right)\right] \\ &= \sum_{n=1}^{\infty}\left[\left(\frac{F_n}{2}\sin\frac{n\pi}{l}(x+ct) - \frac{G_n}{2}\cos\frac{n\pi}{l}(x+ct)\right)\right. \\ &\quad \left.+\left(\frac{G_n}{2}\cos\frac{n\pi}{l}(x-ct) + \frac{F_n}{2}\sin\frac{n\pi}{l}(x-ct)\right)\right] \\ &= f(x+ct) + g(x-ct) \end{aligned}$$

which we will see is the form of the general solution of 2.2.1 which is precisely the form used in Chapter 3.

At this point the reader should consider working through the following examples whose solutions are given in the Solution appendix, and then attention can turn to the second of the standard equations, the heat equation.

EXERCISES

2.1 Use the method of separation of the variables to obtain a solution of the wave equation

$$\frac{\partial^2 u}{\partial x^2} = \frac{1}{c^2}\frac{\partial^2 u}{\partial t^2}$$

on the interval $x \in [0, L]$ subject to the conditions that

$$\begin{aligned} u(0,t) &= 0, \quad u(L,t) = 0 \\ u(x,0) &= \sin(\pi x/L) + \sin(2\pi x/L) \\ u_t(x,0) &= 0. \end{aligned}$$

2.2 Find a separable solution of the equation

$$\frac{\partial^2 u}{\partial x^2} = a^2\frac{\partial^2 u}{\partial t^2}$$

on the interval $x \in [0, 1]$ which satisfies the conditions

$$\begin{aligned} u(0,t) &= 0, \quad u(1,t) = 0 \\ u_t(x,0) &= 0 \end{aligned}$$

and the initial displacement $u(x,0)$ consists of the string being pulled a small distance δ from the equilibrium position at a point one third of the way along, the two sections of the string being straight lines.

2.3 Solve the wave equation in the form

$$\frac{\partial^2 u}{\partial x^2} = \frac{1}{c^2}\frac{\partial^2 u}{\partial t^2}$$

using separation of the variables on the interval $x \in [0, L]$ in which the end points are fixed at zero, and the initial displacement of the string is also zero. The motion is set in place by the string having an initial velocity given by

$$u_t(x,0) = \sin(n\pi x/L)$$

for an integer n.

2.4 Use the method of separation of variables to solve

$$\frac{\partial^2 u}{\partial t^2} = a^2\frac{\partial^2 u}{\partial x^2}$$

on the interval $x \in [0, 2]$ satisfying the boundary conditions

$$\frac{\partial u}{\partial x}(0,t) = \frac{\partial u}{\partial x}(2,t) = 0$$

for all t and the initial conditions

$$\frac{\partial u}{\partial t} = 0, \qquad t = 0, \qquad 0 < x < 2$$

and

$$u(x,0) = \begin{cases} kx & 0 \le x \le 1 \\ k(2-x) & 1 \le x \le 2. \end{cases}$$

2.5 The problems so far have been based purely on the standard wave equation $u_{tt} = c^2 u_{xx}$ with effective changes in the functions $f(x)$ and $g(x)$ which give the initial displacement and velocity. Some very simple extensions can be made which allow problems with extra complications to be solved. Hence use separation of the variables to solve the damped wave equation:

$$c^2 \frac{\partial^2 u}{\partial x^2} = \frac{\partial^2 u}{\partial t^2} + \mu \frac{\partial u}{\partial t}$$

on the interval $x \in [0, L]$ subject to conditions:

$$\begin{aligned} u(0,t) &= 0, \qquad u(L,t) = 0 \\ u(x,0) &= \sin(\pi x/L) \\ u_t(x,0) &= 0. \end{aligned}$$

Further extensions, including what happens when the boundary conditions are inhomogeneous and lend themselves less well to the separation method, are considered in section 2.5. The use of non-Cartesian co-ordinates also gives rise to important separation methods, and these are considered in section 2.6. In the meantime, expertise in the use of the separation method can now be extended by consideration of the second of the canonical forms, namely the heat equation.

2.3 The Heat Equation

The first work in the study of the heat equation was "Theorie analytique de la chaleur" by Joseph Fourier (1768–1830). This paper was rejected in 1807 due to lack of rigour! It was later resubmitted in 1811 and led to the development of Fourier series, although at this stage Fourier was convinced that any function could be expanded as a convergent Fourier series. Simeon-Denis Poisson (1781–1840) took up the work and developed it further, but still believed such series included the "most general solution".

The heat equation is the second of the canonical forms derived in the previous chapter. A typical problem would be to consider the case of a bar of finite length l as shown in Figure 2.1.

Fig. 2.1.

Suppose the ends $x = 0$ and $x = l$ are perfectly insulated. The flux of heat through the endpoints is proportional to the values of $\frac{\partial u}{\partial x}$ at each endpoint. Let the initial temperature distribution be

$$u(x, 0) = f(x) = \beta x, \qquad 0 < x < l \tag{2.3.1}$$

making the full problem to solve

$$\frac{\partial u}{\partial t} = c^2 \frac{\partial^2 u}{\partial x^2} \tag{2.3.2}$$

subject to

$$\frac{\partial u}{\partial x}(0, t) = \frac{\partial u}{\partial x}(l, t) = 0 \tag{2.3.3}$$

and

$$u(x, 0) = f(x) = \beta x, \qquad 0 < x < l. \tag{2.3.4}$$

In the spirit of the separation of variables method, put $u(x, t) = X(x)T(t)$ where X and T are functions of x and t respectively, then the heat equation gives

$$XT' = c^2 TX'' \qquad \text{or} \qquad \frac{T'}{c^2 T} = \frac{X''}{X} = k \tag{2.3.5}$$

as each side is a function of a different independent variable and must therefore be a constant. Hence

$$T' = c^2 kT \qquad \text{and} \qquad X'' = kX. \tag{2.3.6}$$

Since $u(x, t) = X(x)T(t)$, the boundary conditions give $X'(0)T(t) = X'(l)T(t) = 0$ for all t which implies $X'(0) = X'(l) = 0$ for all t, otherwise $T(t) \equiv 0$ giving the trivial solution.

Hence $X'' = kX$ is solved subject to $X'(0) = X'(l) = 0$ and again three cases arise, the non-trivial solution being when $k < 0$ or $k = 0$. Writing $k = -p^2$, the solution is

$$X(x) = A \cos px + B \sin px, \qquad X'(x) = -pA \sin px + pB \cos px \tag{2.3.7}$$

and the boundary condition $X'(0) = 0$ gives $B = 0$ and the condition $X'(l) = 0$ gives $0 = -pA \sin pl$. For a non-trivial solution, $p > 0$ and $A \neq 0$, which implies $\sin pl = 0$ or $pl = n\pi$, $n = 1, 2, 3, \ldots$ or $p = \frac{n\pi}{l}$, $n = 1, 2, 3, \ldots$. Hence $k < 0$ gives non-trivial solutions if $k = -\left(\frac{n\pi}{l}\right)^2$, and the solutions are

$$X_n(x) = A_n \cos \frac{n\pi x}{l}. \tag{2.3.8}$$

If $k = 0$ the $X = Ax + B$ and the boundary conditions require $A = 0$.

The non-trivial solutions are given by $k = 0$ and $k = -\left(\frac{n\pi}{l}\right)^2$, $n = 1, 2, 3, \ldots$ and are $X_0 = B$ and $X_n = A_n \cos \frac{n\pi x}{l}$.

If $k = 0$, then $T' = 0$ implies that $T_0 = C_0$, and if $k = -\left(\frac{n\pi}{l}\right)^2$, then

$$T' = -\left(\frac{nc\pi}{l}\right)^2 T \tag{2.3.9}$$

or

$$T' = -\lambda_n T, \qquad \text{with} \qquad \lambda_n = \left(\frac{cn\pi}{l}\right)^2 \tag{2.3.10}$$

which solves to give

$$T_n(t) = C_n e^{-\lambda_n t}, \qquad n = 1, 2, 3, \ldots . \tag{2.3.11}$$

The various solutions are then

$$u_0(x,t) = X_0(x)T_0(t) = BC_0 = D_0 \tag{2.3.12}$$

and

$$u_n(x,t) = T_n(t)X_n(x) = D_n \cos \frac{n\pi x}{l} e^{-\lambda_n t} \tag{2.3.13}$$

where $D_n = A_n C_n$, $n = 1, 2, 3, \ldots$.

As in the wave equation solution, the initial conditions need to be satisfied by the full separated solution which is:

$$u(x,t) = \sum_{n=0}^{\infty} u_n(x,t) = D_0 + \sum_{n=1}^{\infty} D_n \cos \frac{n\pi x}{l} e^{-\lambda_n t}. \tag{2.3.14}$$

Hence setting $t = 0$, gives

$$u(x,0) = D_0 + \sum_{n=1}^{\infty} D_n \cos \frac{n\pi x}{l} = f(x). \tag{2.3.15}$$

The Fourier methods of Chapter 1 give

$$D_0 = \frac{1}{l} \int_0^l f(x)\, dx \tag{2.3.16}$$

and

$$D_n = \frac{2}{l} \int_0^l f(x) \cos \frac{n\pi x}{l}\, dx \tag{2.3.17}$$

which for the particular case $f(x) = \beta x$ gives

$$D_0 = \frac{1}{l}\int_0^l \beta x\,dx = \frac{\beta}{l}\frac{x^2}{2}\Big|_0^l = \frac{\beta l}{2} \tag{2.3.18}$$

and

$$D_n = \frac{2}{l}\int_0^l \beta x \cos\frac{n\pi x}{l}\,dx = \frac{2\beta l}{n^2\pi^2}\left[(-1)^n - 1\right] \tag{2.3.19}$$

with

$$D_n = \begin{cases} -\dfrac{4l\beta}{n^2\pi^2} & n = 1, 3, 5, \ldots \\ 0 & n = 2, 4, 6, \ldots \end{cases} \tag{2.3.20}$$

which can be written as

$$D_{2m-1} = -\frac{4l\beta}{(2m-1)^2\pi^2}, \qquad m = 1, 2, 3, \ldots. \tag{2.3.21}$$

Hence

$$u(x,t) = \frac{\beta l}{2} - \sum_{m=1}^{\infty} \frac{4l\beta}{(2m-1)^2\pi^2}\cos\frac{(2m-1)\pi x}{l} e^{\frac{-(2m-1)^2\pi^2c^2t}{l^2}}. \tag{2.3.22}$$

Note that as $t \to \infty$, $u(x,t) \to \frac{\beta l}{2}$ which is the mean value of $f(x) = \beta x$ on $[0, 1]$.

The following exercises may now be attempted on the above separation method for heat and diffusion problems.

EXERCISES

2.6 Find separable solutions to the diffusion equation

$$\frac{\partial^2 u}{\partial x^2} = a^2\frac{\partial u}{\partial t}$$

on the interval $0 < x < \pi$ subject to the boundary conditions:

(i) $u = 0$ when $x = 0$ and $x = \pi$ and $u \to 0$ as $t \to \infty$;

(ii) $u_x = 0$ when $x = 0$ and $x = \pi$ and $u \to 0$ as $t \to \infty$.

2.7 Use separation of the variables to solve the heat equation

$$\frac{\partial^2 u}{\partial x^2} = a^2\frac{\partial u}{\partial t}$$

in the region $0 < x < \pi$ and $t > 0$ satisfying the boundary conditions

$$\begin{aligned} u &= 0 \quad \text{when} \quad x = 0 \quad \text{and} \quad x = \pi \\ & u \to 0 \quad \text{as} \quad t \to \infty \\ & u = x + 1 \quad \text{when} \quad t = 0 \quad \text{for} \quad 0 < x < \pi. \end{aligned}$$

2.8 The function $\theta(x,t)$ satisfies

$$\frac{\partial^2 \theta}{\partial x^2} = \frac{1}{c^2}\frac{\partial \theta}{\partial t}$$

on the interval $0 < x < a$ subject to

$$\frac{\partial \theta}{\partial x}(0,t) = \theta(a,t) = 0, \qquad t > 0$$

and

$$\theta(x,0) = \theta_0, \qquad 0 < x < a,$$

(θ_0 const). Use separation of variables to obtain θ.

2.9 Solve the dimensionless heat equation

$$\frac{\partial u}{\partial t} = \frac{\partial^2 u}{\partial x^2}$$

with boundary conditions $u = 0$ at $x = 0$ and $x = 1$ for all $t > 0$, and with initial conditions that

$$\begin{aligned} u &= 2x \qquad \text{for} \quad 0 \le x \le \frac{1}{2} \\ u &= 2(1-x) \qquad \text{for} \quad \frac{1}{2} \le x \le 1 \end{aligned}$$

in the form

$$u(x,t) = \frac{8}{\pi^2}\sum_{n=1}^{\infty}\frac{1}{n^2}\left(\sin\frac{1}{2}n\pi\right)(\sin n\pi x)e^{-n^2\pi^2 t}.$$

2.10 Show that the solution of

$$\frac{\partial u}{\partial t} = \frac{\partial^2 u}{\partial x^2}$$

with initial condition $u = 0$ for $0 \le x \le \frac{1}{2}$ at $t = 0$ and with

$$\frac{\partial u}{\partial x} = 0 \qquad \text{at } x = 0 \quad \forall t > 0$$

and

$$\frac{\partial u}{\partial x} = 1 \qquad \text{at } x = \frac{1}{2} \quad \forall t > 0$$

has the solution

$$u(x,t) = \frac{-1}{12} + x^2 - \sum_{n=1}^{\infty} \frac{(-1)^n}{n^2\pi^2} e^{-(2\pi n)^2 t} \cos 2n\pi x.$$

Solutions of the final of the three canonical forms are sought in the next section, where some further exercises will be suggested.

2.4 Laplace's Equation

The concept of a potential function seems to have been first used by Daniel Bernoulli (1700–1782), son of the more famous Jean Bernoulli, in "Hydrodynamica" in 1738, and Euler wrote Laplace's equation in 1752, from the continuity equation for incompressible fluids. The real progress was made by two of the three L's, Adrien-Marie Legendre (1752–1833) and Pierre-Simon Laplace (1749–1827). (The other L was Lagrange.) Legendre looked at the gravitational attraction of spheroids in 1785 and developed the Legendre polynomials as part of this work. Laplace used expansions in spherical functions to solve the equation since named after him, and both mathematicians continued their work into the 1790s.

The physical derivation of Laplace's equation in Chapter 1 resulted in the equation:

$$\nabla^2 u = \frac{\partial^2 u}{\partial x^2} + \frac{\partial^2 u}{\partial y^2} = 0. \tag{2.4.1}$$

For this equation, the boundary conditions are given on a closed curve. This is analogous to the boundary value problems of ordinary differential equations, which means that the solution has to be found simultaneously over the whole region, as opposed to the step-by-step methods which are feasible in the time direction in the heat equation and in both variables in the wave equation.

An important principle for certain elliptic second-order equations, and for Laplace's equation in particular, is the maximum principle. The principle states that the solution of Laplace's equation cannot have a maximum (or a minimum) in the interior of the defining domain. This principle leads to easy proofs of uniqueness and existence which will not be pursued here in accordance with our aims of seeking practical methods of solution, but can be seen in Renardy and Rogers (1993, chapter 4), for example.

In the numerical aspects in the companion volume, the fundamental differences in the three canonical forms reflect closely in the methods which can be successfully implemented. These differences are masked in the type of analytic solution being attempted at this stage. Suppose for the sake of this example that $u(x,y) = 0$ on $x = 0$, $y = 0$ and $y = b$. Also that $u(d,y) = f(y)$ for $0 < y < b$. Then the requirement is to solve for $u(x,y)$ in the closed region $0 \le x \le d$, $0 \le y \le b$.

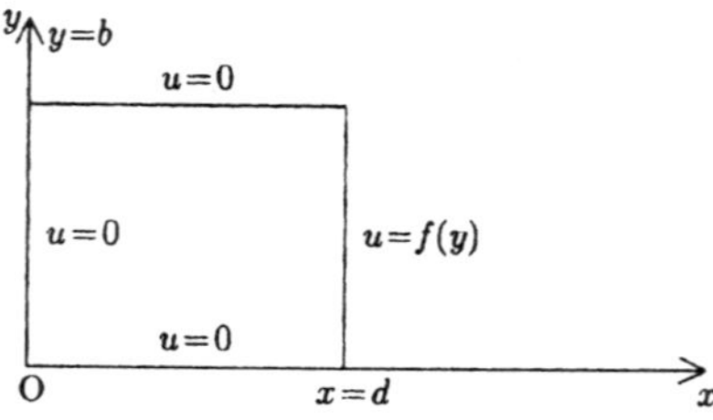

Fig. 2.2.

To solve the equation, the separation of variables approach suggests the trial solution $u(x,y) = X(x)Y(y)$ giving $X''Y + XY'' = 0$ which results in the differential equations:

$$\frac{X''}{X} + \frac{Y''}{Y} = 0 \qquad \text{or} \qquad \frac{X''}{X} = -\frac{Y''}{Y} = k \tag{2.4.2}$$

where k is arbitrary, which gives the problems

$$X'' = kX \qquad \text{and} \qquad Y'' = -kY \tag{2.4.3}$$

with zero boundary conditions

$$\left.\begin{array}{l} u(0,y) = 0 \Rightarrow X(0)Y(y) = 0 \Rightarrow X(0) = 0 \\ u(x,0) = 0 \Rightarrow X(x)Y(0) = 0 \Rightarrow Y(0) = 0 \\ u(x,b) = 0 \Rightarrow X(x)Y(b) = 0 \Rightarrow Y(b) = 0 \end{array}\right\} \tag{2.4.4}$$

for a non-trivial solution. Consider first the two boundary conditions associated with the function $Y(y)$, then once more there are three cases which may arise. The non-trivial solution arises when $k > 0$ or $k = p^2$, and gives the equation $Y'' = -p^2Y$ with solution

$$Y(y) = A\cos py + B\sin py. \tag{2.4.5}$$

Now the boundary condition $Y(0) = 0$ gives $0 = A$, and the condition $Y(b) = 0$ gives $0 = B\sin pb$. For non-trivial solutions $\sin pb = 0$ which implies that $pb = n\pi$, $n = 1,2,3,\ldots$ or $p = \frac{n\pi}{b}$, $n = 1,2,3,\ldots$. Hence if

$$k = \left(\frac{n\pi}{b}\right)^2, \qquad n = 1,2,3,\ldots \tag{2.4.6}$$

then

$$Y_n(y) = B_n \sin\frac{n\pi y}{b}. \tag{2.4.7}$$

Now return to the $X(x)$ equation with $k = \left(\frac{n\pi}{b}\right)^2$ then $X'' = \left(\frac{n\pi}{b}\right)^2 X$ with solution

$$X_n(x) = C_n e^{\frac{n\pi x}{b}} + D_n e^{\frac{-n\pi x}{b}}. \tag{2.4.8}$$

The boundary condition here is $X(0) = 0$ or $0 = C_n + D_n$, to give

$$X_n = C_n \left(e^{\frac{n\pi x}{b}} - e^{\frac{-n\pi x}{b}}\right) = 2C_n \sinh \frac{n\pi x}{b}. \qquad (2.4.9)$$

The full solution is then

$$u_n(x, y) = X_n(x)Y_n(y) = E_n \sinh \frac{n\pi x}{b} \sin \frac{n\pi y}{b} \qquad (2.4.10)$$

where

$$E_n = 2C_n B_n.$$

To satisfy $u(d, y) = f(y)$ the solutions are summed over n to give

$$u(x, y) = \sum_{n=1}^{\infty} E_n \sinh \frac{n\pi x}{b} \sin \frac{n\pi y}{b} \qquad (2.4.11)$$

to require for the remaining boundary condition that

$$u(d, y) = \sum_{n=1}^{\infty} E_n \sinh \frac{n\pi d}{b} \sin \frac{n\pi y}{b} = f(y). \qquad (2.4.12)$$

Hence by the usual Fourier approach

$$E_n \sinh \frac{n\pi d}{b} = \frac{2}{b} \int_0^b f(y) \sin \frac{n\pi y}{b}\, dy \qquad (2.4.13)$$

which yields the complete solution

$$u(x, y) = \sum_{n=1}^{\infty} \frac{2}{b \sinh \frac{n\pi d}{b}} \left\{ \int_0^b f(y') \sin \frac{n\pi y'}{b}\, dy' \right\} \sinh \frac{n\pi x}{b} \sin \frac{n\pi y}{b}. \qquad (2.4.14)$$

For a specific function $f(y)$, the integral in 2.4.14 can be evaluated to yield the solution as an infinite series.

A second example will now be considered with boundary conditions involving partial derivatives of u, and an extra term in the equation. Consider the problem of solving

$$\frac{\partial^2 u}{\partial x^2} + \frac{\partial^2 u}{\partial y^2} + u = 0 \qquad (2.4.15)$$

in $0 < x < 1$, $0 < y < 1$ subject to the boundary conditions $u(0, y) = 0$, $\frac{\partial u}{\partial x} = 0$ when $x = 1$, $u(x, 0) = 0$ and $u(x, 1) = x$. Again in the spirit of separation of the variables, let $u(x, y) = X(x)Y(y)$ where $X(x)$ and $Y(y)$ are functions of x and y respectively, to give

$$\frac{X''}{X} + \frac{Y''}{Y} + 1 = 0. \qquad (2.4.16)$$

Now $\frac{Y''}{Y}$ is a function of y alone and the rest of the expression is a function of x only, hence

$$\frac{Y''}{Y} = k, \qquad \text{and} \qquad \frac{X''}{X} + 1 = -k \tag{2.4.17}$$

to yield the two equations

$$X'' + (k+1)X = 0 \qquad \text{and} \qquad Y'' = kY. \tag{2.4.18}$$

The zero boundary conditions give

$$\left.\begin{aligned} u(0,y) &= 0 \Rightarrow X(0)Y(y) = 0 \Rightarrow X(0) = 0 \\ \left.\frac{\partial u}{\partial x}\right|_{x=1} &= 0 \Rightarrow X'(1)Y(0) = 0 \Rightarrow X'(1) = 0 \\ u(x,0) &= 0 \Rightarrow X(x)Y(0) = 0 \Rightarrow Y(0) = 0 \end{aligned}\right\} \tag{2.4.19}$$

for the non-trivial solution, leaving

$$X'' + (k+1)X = 0, \qquad X(0) = X'(1) = 0 \tag{2.4.20}$$

whose solution depends on the sign of $k+1$. The non-trivial solution arises when $k+1 > 0$, hence setting $k+1 = p^2$ gives $X'' + p^2X = 0$ with solution

$$X(x) = A\cos px + B\sin px. \tag{2.4.21}$$

The boundary condition $X(0) = 0$ then gives $0 = A$, and $X'(1) = 0$ gives $0 = pB\cos p$. But $p > 0$ and hence $\cos p = 0$ to give $p = (2n+1)\frac{\pi}{2}$, for $n = 0, 1, 2, 3, \ldots$. Hence the non-trivial solution arises when

$$k = (2n+1)^2\frac{\pi^2}{4} - 1 \tag{2.4.22}$$

and has the form

$$X_n(x) = B_n \sin(2n+1)\frac{\pi}{2}x. \tag{2.4.23}$$

The final step is to solve the equation for $Y(y)$ which assumes the form

$$Y'' = \left((2n+1)^2\frac{\pi^2}{4} - 1\right)Y = \frac{1}{4}\left[(2n+1)^2\pi^2 - 4\right]Y = \alpha_n^2 Y \tag{2.4.24}$$

which solves to give

$$Y_n = C_n e^{\alpha_n y} + D_n e^{-\alpha_n y} \tag{2.4.25}$$

where

$$\alpha_n = \frac{1}{2}\left[(2n+1)^2\pi^2 - 4\right]^{1/2}. \tag{2.4.26}$$

The boundary conditions then give for $Y(0) = 0$, $0 = C_n + D_n$ and hence

$$Y_n = C_n\left(e^{\alpha_n y} - e^{-\alpha_n y}\right) = 2C_n \sinh \alpha_n y. \tag{2.4.27}$$

The general separated solution is then

$$u(x,y) = \sum_{n=0}^{\infty} E_n \sin(2n+1)\frac{\pi}{2}x \sinh \alpha_n y \tag{2.4.28}$$

and hence the final boundary conditions require

$$u(x,1) = \sum_{n=0}^{\infty} E_n \sin(2n+1)\frac{\pi}{2}x \sinh \alpha_n. \tag{2.4.29}$$

Hence the usual Fourier expansion has coefficients given by

$$\begin{aligned} E_n \sinh \alpha_n &= 2\int_0^1 x \sin(2n+1)\frac{\pi}{2}x\,dx \\ &= \frac{8}{(2n+1)^2\pi^2}\sin(2n+1)\frac{\pi}{2} \end{aligned} \tag{2.4.30}$$

which leaves

$$E_n = \frac{1}{\sinh \alpha_n}\frac{8}{(2n+1)^2\pi^2}(-1)^n. \tag{2.4.31}$$

Hence the complete solution is

$$\begin{aligned} u(x,y) &= \sum_{n=0}^{\infty} \frac{8}{(2n+1)^2\pi^2}(-1)^n \frac{1}{\sinh \frac{1}{2}\sqrt{(2n+1)^2\pi^2-4}} \\ &\quad \sin(2n+1)\frac{\pi}{2}x \sinh \frac{1}{2}\sqrt{(2n+1)^2\pi^2-4}y. \end{aligned} \tag{2.4.32}$$

Again, some exercises are presented now to complete the straightforward use of the separation method for this third type of second-order equation.

EXERCISES

2.11 A square plate $0 \le x \le a$, $0 \le y \le a$ is heated along the edge $x = 0$ with the other sides held at 0 °C, when the plate has reached its steady state, the side at $x = 0$ has the temperature distribution

$$u = f(y) = \begin{cases} y, & 0 < y < a/2 \\ a - y, & a/2 < y < a. \end{cases}$$

The temperature in the steady state condition satisfies Laplace's equation

$$\frac{\partial^2 u}{\partial x^2} + \frac{\partial^2 u}{\partial y^2} = 0.$$

Find the temperature distribution over the plate in the steady state condition.

2.12 Solve the equation

$$\frac{\partial^2 u}{\partial x^2} + c^2 \frac{\partial^2 u}{\partial y^2} = 0$$

given $u \to 0$ as $x \to \infty$, $u = 0$ at $y = 0$, $\frac{\partial u}{\partial y} = 0$ at $y = l$ (for all x),$u = 2y$ for $0 \leq y \leq l$ at $t = 0$.

2.13 Find the solution of the equation

$$\frac{\partial^2 u}{\partial x^2} + \frac{\partial^2 u}{\partial y^2} = u$$

valid in the region $0 < x < \pi$ and $0 < y < a$ which satisfies the boundary conditions

$u = 0$ when $x = 0$, $x = \pi$ and $y = 0$

and

$u = 1$ when $y = a$ for $0 < x < \pi$.

2.14 Extend the method of separation of the variables to solve the three-dimensional Laplace equation

$$\frac{\partial^2 \phi}{\partial x^2} + \frac{\partial^2 \phi}{\partial y^2} + \frac{\partial^2 \phi}{\partial z^2} = 0$$

on the region $0 \leq x \leq a$, $0 \leq y \leq b$ and $0 \leq z \leq c$ with $\phi = 0$ on $x = 0$, $x = a$, $y = 0$, $y = b$ and $z = 0$, and $\phi = f(x)$ on $z = c$ for $0 \leq x \leq a$ and $0 \leq y \leq b$.

The method of separation of the variables has been applied to each of the standard canonical forms. Some modifications are necessary when the boundary conditions are not homogeneous, these issues are dealt with in the next section.

2.5 Homogeneous and Non-homogeneous Boundary Conditions

In the previous sections, the boundary conditions have taken the form

$$\alpha_1 u(0,t) + \beta_1 \frac{\partial u}{\partial x}(0,t) = 0 \tag{2.5.1}$$

and

$$\alpha_2 u(l,t) + \beta_2 \frac{\partial u}{\partial x}(l,t) = 0 \tag{2.5.2}$$

which are said to be homogeneous. These conditions led, on separating the variables to Sturm–Liouville problems of the type considered in Chapter 1 which gave sets of orthogonal functions as solutions to the resulting ordinary differential equations. When non-homogeneous boundary conditions of the form

$$\alpha_1 u(0,t) + \beta_1 \frac{\partial u}{\partial x}(0,t) = u_1(t) \tag{2.5.3}$$

and

$$\alpha_2 u(l,t) + \beta_2 \frac{\partial u}{\partial x}(l,t) = u_2(t) \tag{2.5.4}$$

arise some modifications are required. Consider first the case when $u_1(t)$ and $u_2(t)$ are constants. In this case, let $u(x,t) = v(x) + w(x,t)$ where $v(x)$ is a time independent function representing the "steady state" solution and $w(x,t)$ the function representing the deviation from the "steady state". For the combined solution, it is required that $v(x)$ and $w(x,t)$ must satisfy the given differential equation separately and $v(x)$ must satisfy the given boundary conditions. This makes $w(x,t)$ the solution of a homogeneous differential equation with homogeneous boundary conditions, which is the problem already solved in the previous sections. In effect, the solution now involves both the separated solution of products of functions of x and t, and a solution just in x, which is just another valid solution.

As a worked example consider solving

$$\frac{\partial^2 u}{\partial x^2} = \frac{1}{c^2}\frac{\partial u}{\partial t} \tag{2.5.5}$$

subject to the boundary conditions

$$u(0,t) = u_0 \qquad \text{and} \qquad \frac{\partial u}{\partial x}(l,t) = u_1, \qquad t \geq 0 \tag{2.5.6}$$

and the initial conditions

$$u(x,0) = f(x), \qquad 0 \leq x \leq l, \tag{2.5.7}$$

where u_0, u_1, l and c are given constants, and $f(x)$ is a given function. Following the above discussion, write $u(x,t) = v(x) + w(x,t)$. Then the functions $v(x)$ and $w(x,t)$ are required to satisfy 2.5.5 separately to give

$$\frac{d^2 v}{dx^2} = 0, \qquad \text{and} \qquad \frac{1}{c^2}\frac{\partial w}{\partial t} = \frac{\partial^2 w}{\partial x^2}. \tag{2.5.8}$$

In addition $v(x)$ has to satisfy the given boundary conditions

$$v(0) = u_0, \qquad \text{and} \qquad \frac{\partial v}{\partial x}(l) = u_1 \tag{2.5.9}$$

which require

$$u(0,t) = v(0) + w(0,t) = u_0 + w(0,t) = u_0 \tag{2.5.10}$$

as $u(0,t) = u_0$ giving $w(0,t) = 0$. Also

$$\frac{\partial u}{\partial x}(l,t) = \frac{dv}{dx}(l) + \frac{\partial w}{\partial x}(l,t) = u_1 + \frac{\partial w}{\partial x}(l,t) = u_1 \tag{2.5.11}$$

as $\frac{\partial u}{\partial x}(l,t) = u_1$ giving $\frac{\partial w}{\partial x}(l,t) = 0$. The initial condition gives

$$u(x,0) = f(x) = v(x) + w(x,0), \qquad \text{and} \qquad w(x,0) = f(x) - v(x). \tag{2.5.12}$$

The solution for $v(x)$ is simple and yields $v(x) = u_1 x + u_0$. For $w(x,t)$, let $w = X(x)T(t)$ where X and T are functions of x and t only respectively, to give the ordinary differential equations:

$$\frac{T'}{c^2 T} = \frac{X''}{X} = k \tag{2.5.13}$$

which solve in the conventional separation of variables way to give

$$w(x,t) = \sum_{n=0}^{\infty} E_n \sin\left(n + \frac{1}{2}\right)\frac{\pi x}{l} e^{-\alpha_n t}. \tag{2.5.14}$$

However, again $\sin\left(n+\frac{1}{2}\right)\frac{\pi x}{l}$ forms a complete orthogonal set on $0 < x < l$ with norm $\sqrt{\frac{l}{2}}$, giving Fourier coefficients,

$$E_n = \frac{2}{l}\int_0^l [f(x) - u_1 x - u_0] \sin\left(n + \frac{1}{2}\right)\frac{\pi x}{l}\, dx \tag{2.5.15}$$

from which E_n can be found. Then finally

$$\begin{aligned} u(x,t) &= v(x) + w(x,t) \\ &= u_1 x + u_0 + \sum_{n=0}^{\infty} E_n \sin\left(n + \frac{1}{2}\right)\frac{\pi x}{l} e^{-\left(n+\frac{1}{2}\right)^2 \frac{\pi^2 c^2}{l^2} x}. \end{aligned} \tag{2.5.16}$$

The next question to answer is what happens if $u_1(t)$ and $u_2(t)$ (in the boundary conditions) are not constants as in the previous case. In this situation try $u(x,t) = v(x,t) + w(x,t)$ where $v(x,t)$ is chosen so that it satisfies the boundary conditions. By substituting this expression into the differential equation the following occurs, as the boundary conditions for $w(x,t)$ are then homogeneous.

Consider the problem of solving

$$\frac{\partial^2 u}{\partial x^2} = \frac{1}{c^2}\frac{\partial u}{\partial t} \tag{2.5.17}$$

subject to the boundary conditions

$$u(0,t) = u_0(t), \qquad t \geq 0 \quad \text{and} \qquad u(l,t) = u_1(t),\ t \geq 0$$

and the initial condition

$$u(x,0) = f(x), \quad 0 < x < l,$$

so that in this case the boundary conditions are functions of time.

Write $u(x,t) = v(x,t) + w(x,t)$ where $v(x,t)$ is chosen to satisfy the boundary conditions, namely

$$v(0,t) = u_0(t), \quad \text{and} \quad v(l,t) = u_1(t).$$

One obvious choice of $v(x,t)$ would be

$$v(x,t) = u_0(t) + \frac{x}{l}[u_1(t) - u_0(t)]. \tag{2.5.18}$$

Substituting into 2.5.17 gives:

$$\frac{\partial^2 v}{\partial x^2} + \frac{\partial^2 w}{\partial x^2} = \frac{1}{c^2}\left[\frac{\partial v}{\partial t} + \frac{\partial w}{\partial t}\right]. \tag{2.5.19}$$

Now

$$\frac{\partial^2 v}{\partial x^2} = 0 \qquad \text{and} \qquad \frac{\partial v}{\partial t} = u_0'(t) + \frac{x}{l}[u_1'(t) - u_0'(t)]. \tag{2.5.20}$$

Hence

$$\frac{\partial^2 w}{\partial x^2} - \frac{1}{c^2}\frac{\partial w}{\partial t} = \frac{1}{c^2}\left[u_0'(t) + \frac{x}{l}(u_1'(t) - u_0'(t)\right]. \tag{2.5.21}$$

This equation is inhomogeneous and is a special case of the generalised heat equation

$$\frac{\partial^2 w}{\partial x^2} - \frac{1}{c^2}\frac{\partial w}{\partial t} = F(x,t) \tag{2.5.22}$$

where $F(x,t)$ is a known function.

The function w is subject to the conditions

$$w(0,t) = 0, \qquad w(l,t) = 0$$

and

$$w(x,0) = f(x) - v(x,0) = f(x) - \left[u_0(0) + \frac{x}{l}(u_1(0) - u_0(0))\right]. \tag{2.5.23}$$

The result is that the non-homogeneous boundary conditions, which are functions of t, imposed on the homogeneous equation lead to the need to solve an inhomogeneous equation. This did not happen in the previous example where the non-homogeneous boundary values were constants.

To solve inhomogeneous equations with homogeneous and inhomogeneous boundary conditions, the first step is to make the boundary conditions homogeneous, by:

(i) letting $u(x,t) = v(x) + w(x,t)$, if the boundary values are constants and making $v(x)$ satisfy the homogeneous equation and the boundary conditions, or

(ii) letting $u(x,t) = v(x,t) + w(x,t)$, if the boundary values are functions of t, and choosing $v(x,t)$ to satisfy the boundary conditions, then substituting into the differential equation to give an equation for $w(x,t)$ subject to homogeneous boundary conditions.

If the resulting equation is non-homogeneous, the following method is used.

Consider the solution of the equation

$$\frac{\partial w}{\partial t} = \frac{1}{c^2}\frac{\partial^2 w}{\partial x^2} + f(x,t) \tag{2.5.24}$$

where c is a constant and $f(x,t)$ a given function, subject to the boundary conditions

$$w(0,t) = w(l,t) = 0, \qquad t \geq 0 \tag{2.5.25}$$

and the initial condition

$$w(x,0) = \psi(x), \qquad 0 \leq x \leq l. \tag{2.5.26}$$

Consider the homogeneous equation

$$\frac{\partial \phi}{\partial t} = \frac{1}{c^2}\frac{\partial^2 \phi}{\partial x^2} \tag{2.5.27}$$

subject to $\phi(0,t) = \phi(l,t) = 0$. Separating the variables by trying $\phi(x,t) = X(x)T(t)$ gives that

$$X_n = B_n \sin\frac{n\pi x}{l}, \qquad n = 1,2,3,\ldots. \tag{2.5.28}$$

Hence write

$$w(x,t) = \sum_{n=1}^{\infty} w_n(t)\sin\frac{n\pi x}{l} \tag{2.5.29}$$

which ensures that the boundary conditions on w are satisfied.

Suppose that $f(x,t)$ is expanded in the same way

$$f(x,t) = \sum_{n=1}^{\infty} f_n(t)\sin\frac{n\pi x}{l} \tag{2.5.30}$$

then as the set of functions $\sin\frac{n\pi x}{l}$, $n = 1,2,3,\ldots$ form a complete set of orthogonal functions on $0 \leq x \leq l$,

$$f_n(t) = \frac{2}{l}\int_0^l f(x',t)\sin\frac{n\pi x'}{l}\,dx'. \tag{2.5.31}$$

Substituting the series expansions for $w(x,t)$ and $f(x,t)$ into the given partial differential equations leads to

$$\frac{\partial}{\partial t}\left[\sum_{n=1}^{\infty} w_n(t)\sin\frac{n\pi x}{l}\right] = \frac{1}{c^2}\frac{\partial^2}{\partial x^2}\left[\sum_{n=1}^{\infty} w_n(t)\sin\frac{n\pi x}{l}\right] + \sum_{n=1}^{\infty} f_n(t)\sin\frac{n\pi x}{l}$$

$$\text{or} \sum_{n=1}^{\infty}\left\{\left[\frac{dw_n(t)}{dt} + \frac{1}{c^2}\left(\frac{n\pi}{l}\right)^2 w_n(t) - f_n(t)\right]\sin\frac{n\pi x}{l}\right\} = 0 \qquad (2.5.32)$$

which gives the first order ordinary differential equation

$$\frac{d}{dt}w_n(t) + \left(\frac{n\pi}{cl}\right)^2 w_n(t) = f_n(t) \qquad (2.5.33)$$

for $w_n(t)$.

The initial condition to be imposed on this equation is determined by putting $t = 0$ in

$$w(x,t) = \sum_{n=1}^{\infty} w_n(t)\sin\frac{n\pi x}{l} \qquad (2.5.34)$$

and equating to $u(x,0) = \psi(x)$ expanded as a Fourier series. Hence

$$\psi(x) = \sum_{n=1}^{\infty} B_n \sin\frac{n\pi x}{l} \qquad (2.5.35)$$

where

$$B_n = \frac{2}{l}\int_0^l \psi(x')\sin\frac{n\pi x'}{l}\,dx' \qquad (2.5.36)$$

giving

$$w(x,0) = \sum_{n=1}^{\infty} w_n(0)\sin\frac{n\pi x}{l} = \psi(x) = \sum_{n=1}^{\infty} B_n \sin\frac{n\pi x}{l}. \qquad (2.5.37)$$

Comparing the two forms gives $w_n(0) = B_n$. Hence $w_n(t)$ is the solution of

$$\frac{dw_n}{dt} + \left(\frac{n\pi}{cl}\right)^2 w_n = f_n \qquad (2.5.38)$$

subject to $w_n(0) = B_n$.

This value of w_n inserted into

$$w(x,t) = \sum_{n=1}^{\infty} w_n(t)\sin\frac{n\pi x}{l}$$

gives the required solution. Clearly the validity of this method depends on the assumption that the necessary functions are all expandable as Fourier series.

As a worked example consider solving

$$\frac{\partial w}{\partial t} = \frac{1}{c^2}\frac{\partial^2 w}{\partial x^2} + xt \tag{2.5.39}$$

subject to

$$w(0,t) = w(\pi,t) = 0, \qquad t \geq 0 \tag{2.5.40}$$

and

$$w(x,0) = \psi(x) = 1, \qquad 0 < x < \pi. \tag{2.5.41}$$

Consider

$$\frac{\partial \phi}{\partial t} = \frac{1}{c^2}\frac{\partial^2 \phi}{\partial x^2} \tag{2.5.42}$$

subject to

$$\phi(0,t) = \phi(\pi,t) = 0, \qquad t \geq 0. \tag{2.5.43}$$

Separating the variables gives

$$X_n(x) = B_n \sin nx, \qquad n = 1, 2, 3, \ldots \tag{2.5.44}$$

where

$$\phi(x,t) = X(x)T(t).$$

The details here are left as an exercise for the reader.

Let

$$w(x,t) = \sum_{n=1}^{\infty} w_n(t) \sin nx \tag{2.5.45}$$

and let

$$xt = \sum_{n=1}^{\infty} f_n(t) \sin nx \tag{2.5.46}$$

then

$$f_n(t) = \frac{2}{\pi}\int_0^{\pi} xt \sin nx \, dx = (-1)^{n+1}\frac{2t}{n}. \tag{2.5.47}$$

Putting (2.5.45) and (2.5.47) into equation 2.5.39 gives

$$\frac{d}{dt}w_n(t) + \frac{n^2}{c^2}w_n(t) = (-1)^{n+1}\frac{2t}{n}. \tag{2.5.48}$$

Now let

$$\psi(x) = \sum_{n=1}^{\infty} B_n \sin nx \tag{2.5.49}$$

where

$$B_n = \frac{2}{\pi}\int_0^\pi \sin nx\, dx \qquad (2.5.50)$$

as $\psi(x) = 1$. Hence

$$B_n = -\frac{2}{\pi}\frac{\cos nx}{n}\Big|_0^\pi = \frac{2}{n\pi}[1-(-1)^n]. \qquad (2.5.51)$$

The initial condition gives

$$\begin{aligned} w(x,0) &= \sum_{n=1}^{\infty} w_n(0)\sin nx = \psi(x) \\ &= \sum_{n=1}^{\infty} \frac{2}{n\pi}[1-(-1)^n]\sin nx \end{aligned} \qquad (2.5.52)$$

and hence

$$w_n(0) = \frac{2}{n\pi}[1-(-1)^n]. \qquad (2.5.53)$$

Thus $w_n(t)$ is required such that

$$\frac{dw_n}{dt} + \frac{n^2}{c^2}w_n = (-1)^{n+1}\frac{2t}{n} \qquad (2.5.54)$$

with initial condition

$$w_n(0) = \frac{2}{n\pi}[1-(-1)^n]. \qquad (2.5.55)$$

This is a standard ordinary differential equation with integrating factor

$$e^{\frac{n^2 t}{c^2}} \qquad (2.5.56)$$

to give

$$w_n = (-1)^{n+1}\frac{2}{n^3}c^2\left[t-\frac{c^2}{n^2}\right]C_n' e^{\frac{-n^2 t}{c^2}}. \qquad (2.5.57)$$

However

$$w_n(0) = \frac{2}{n\pi}[1-(-1)^n] = (-1)^{n+1}\frac{2c^2}{n^3}\left(\frac{-c^2}{n^2}\right) + C_n' \qquad (2.5.58)$$

to fix C_n' as

$$C_n' = \frac{2}{n\pi}[1-(-1)^n] + (-1)^{n+1}2\frac{c^4}{n^5}. \qquad (2.5.59)$$

Hence finally

$$\begin{aligned} w_n(t) &= (-1)^{n+1}\frac{2}{n}\left[\frac{c^2}{n^2}t - \frac{c^4}{n^4}\right] \\ &\quad + \left\{\frac{2}{n\pi}[1-(-1)^n] + (-1)^{n+1}\frac{2c^4}{n^5}\right\}e^{\frac{-n^2 t}{c^2}} \end{aligned} \qquad (2.5.60)$$

with the solution being

$$w(x,t) = \sum_{n=1}^{\infty} w_n(t) \sin nx \tag{2.5.61}$$

where $w_n(t)$ is given above.

As a final example in this section consider Laplace's equation with more than one non-homogeneous boundary condition in the form

$$\frac{\partial^2 u}{\partial x^2} + \frac{\partial^2 u}{\partial y^2} = 0 \tag{2.5.62}$$

subject to

$$\begin{aligned}
&\text{(i)} \quad \alpha_1 u(0,y) + \beta_1 \frac{\partial u}{\partial x}(0,y) = f_1(y) \\
&\text{(ii)} \quad \alpha_2 u(l,y) + \beta_2 \frac{\partial u}{\partial x}(l,y) = f_2(y) \\
&\text{(iii)} \quad \alpha_3 u(x,0) + \beta_3 \frac{\partial u}{\partial y}(x,0) = g_1(x) \\
&\text{(iv)} \quad \alpha_4 u(x,m) + \beta_4 \frac{\partial u}{\partial y}(x,l) = g_2(x).
\end{aligned} \tag{2.5.63}$$

For this problem write

$$u(x,y) = u_1(x,y) + u_2(x,y) + u_3(x,y) + u_4(x,y) \tag{2.5.64}$$

where u_1, u_2, u_3 and u_4 each satisfy Laplace's equation and the boundary conditions are

$$\begin{aligned}
\text{(a)} \qquad & \alpha_1 u_1(0,y) + \beta_1 \frac{\partial u_1}{\partial x}(0,y) = f_1(y) \\
& \alpha_2 u_1(l,y) + \beta_2 \frac{\partial u_1}{\partial x}(l,y) = 0 \\
& \alpha_3 u_1(x,0) + \beta_3 \frac{\partial u_1}{\partial y}(x,0) = 0 \\
& \alpha_4 u_1(x,m) + \beta_4 \frac{\partial u_1}{\partial y}(x,m) = 0,
\end{aligned} \tag{2.5.65}$$

$$\begin{aligned}
\text{(b)} \qquad & \alpha_1 u_2(0,y) + \beta_1 \frac{\partial u_2}{\partial x}(0,y) = 0 \\
& \alpha_2 u_2(l,y) + \beta_2 \frac{\partial u_2}{\partial x}(l,y) = f_2(y) \\
& \alpha_3 u_2(x,0) + \beta_3 \frac{\partial u_2}{\partial y}(x,0) = 0 \\
& \alpha_4 u_2(x,m) + \beta_4 \frac{\partial u_2}{\partial y}(x,m) = 0,
\end{aligned} \tag{2.5.66}$$

$$\text{(c)}\qquad \begin{aligned} &\alpha_1 u_3(0,y) + \beta_1 \frac{\partial u_3}{\partial x}(0,y) = 0 \\ &\alpha_2 u_3(l,y) + \beta_2 \frac{\partial u_3}{\partial x}(l,y) = 0 \\ &\alpha_3 u_3(x,0) + \beta_3 \frac{\partial u_3}{\partial y}(x,0) = g_1(x) \\ &\alpha_4 u_3(x,m) + \beta_4 \frac{\partial u_3}{\partial y}(x,m) = 0, \end{aligned} \tag{2.5.67}$$

$$\text{(d)}\qquad \begin{aligned} &\alpha_1 u_4(0,y) + \beta_1 \frac{\partial u_4}{\partial x}(0,y) = 0 \\ &\alpha_2 u_4(l,y) + \beta_2 \frac{\partial u_4}{\partial x}(l,y) = 0 \\ &\alpha_3 u_4(x,0) + \beta_3 \frac{\partial u_4}{\partial y}(x,0) = 0 \\ &\alpha_4 u_4(x,m) + \beta_4 \frac{\partial u_4}{\partial y}(x,m) = g_2(x), \end{aligned} \tag{2.5.68}$$

so that each u_i satisfies Laplace's equation and one non-homogeneous boundary condition. It is left as an exercise to show that

$$u(x,y) = u_1(x,y) + u_2(x,y) + u_3(x,y) + u_4(x,y) \tag{2.5.69}$$

where the u_i satisfy the above conditions and the original equations. Problems 2.5.65–2.5.68 are each of precisely the type solved in section 2.3, the point being that one non-homogeneous term can be handled by using the three homogeneous conditions to obtain a Fourier series which can then be fitted to the non-homogeneous part.

EXERCISES

2.15 Confirm that the solution to Laplace's equation by splitting the problem into four as above in 2.5.64, indeed generates the required result.

2.16 Show that the solution of

$$\frac{\partial^2 u}{\partial x^2} = \frac{1}{c^2}\frac{\partial u}{\partial t}$$

which satisfies the conditions $u = 0$ at $x = 0$ for $t \geq 0$, $u = 1$ at $x = 1$ for $t \geq 0$ and $u = 0$ at $t = 0$ for $0 \leq x \leq 1$ is

$$u(x,t) = x + \frac{2}{\pi}\sum_{n=1}^{\infty} \frac{(-1)^n}{n} e^{-n^2\pi^2c^2t} \sin n\pi x.$$

2.17 Solve

$$\frac{\partial u}{\partial t} = \frac{\partial^2 u}{\partial x^2} + 2x$$

for $0 < x < 1$, $t > 0$,$u(0,t) = 0$, $u(1,t) = 0$, $u(x,0) = x - x^2$.

2.18 Solve

$$\frac{\partial u}{\partial t} = \frac{1}{4}\frac{\partial^2 u}{\partial x^2} + 1$$

for $0 < x < 10$, $t > 0$,$\frac{\partial u}{\partial x} = 0$ when $x = 0$, $t > 0$, $u(10,t) = 20$, $u(x,0) = 50$.

2.19 Solve

$$\frac{\partial u}{\partial t} = \frac{\partial^2 u}{\partial x^2}$$

for $0 < x < 1$, $t > 0$,$u(0,t) = 0$, $u(1,t) = t$, $u(x,0) = 0$.

2.20 Solve

$$\frac{\partial^2 u}{\partial x^2} + \frac{\partial^2 u}{\partial y^2} = 0$$

for $0 < x < 1$, $0 < y < 1$ with $u(x,0) = 0$, $u(0,y) = 0$, $u(x,1) = x$, $\frac{\partial u}{\partial x}(1,y) = 1$.

As an example which is in the way of a revision exercise consider the following.

2.21 Show that the transmission line equation

$$\frac{\partial^2 e}{\partial x^2} = LC\frac{\partial^2 e}{\partial t^2} + (RC + GL)\frac{\partial e}{\partial t} + RGe$$

with boundary conditions $e(0,t) = E_0 \cos wt$ and $e(x,t)$ bounded as $x \to \infty$, where $e(x,t)$ is the voltage at distance x and time t does not have a separable solution of the form

$$e(x,t) = X(x)T(t).$$

Further show that there exists a solution of the form

$$e(x,t) = E_0 e^{-ax}\cos(wt + bx)$$

provided that

$$a^2 - b^2 = RG - LCw^2, \qquad \text{and} \qquad 2ab = -(RC + GL)w$$

with the lesson that not every problem can be solved by separation of the variables.

At this stage, some of the simpler problems which are soluble by separation of the variables have been considered, including how the method may be modified to extend it to non-homogeneous problems.

2.6 Separation of variables in other coordinate systems

In the first part of this chapter, the powerful method of separation of variables was employed in Cartesian coordinates for the three standard canonical forms. The method can be used in different geometries obtained when the standard equations are transformed by change of variable into other coordinate systems, so opening up a large range of new solutions. The most common coordinate systems are cylindrical polar coordinates which are used on problems with cylindrical symmetry, and spherical polar coordinates for problems with spherical symmetry. Typical examples are waves on circular membranes, and the gravitational potential around a planet. These problems result in an extension of the use of orthogonal polynomials which have already been used to represent solutions in the Cartesian case. As in the earlier part of the chapter the various cases will be treated as worked examples, followed by a series of exercises.

Consider first problems with cylindrical symmetry, as shown in Figure 2.3.

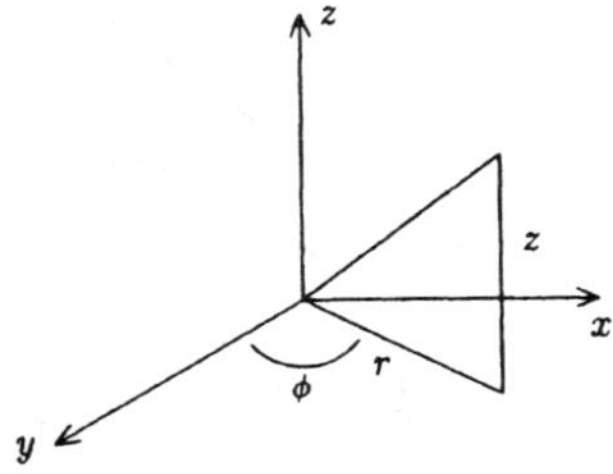

Fig. 2.3.

As a starting point in this work, Laplace's equation in cylindrical polar coordinates (r, ϕ, z) will be used in the form:

$$\nabla^2 u = \frac{1}{r}\frac{\partial}{\partial r}\left(r\frac{\partial u}{\partial r}\right) + \frac{1}{r^2}\frac{\partial^2 u}{\partial \phi^2} + \frac{\partial^2 u}{\partial z^2} = 0. \tag{2.6.1}$$

By restricting consideration to two dimensions so that there is no z dependence, and assuming circular symmetry so that the dependence on ϕ also vanishes, gives

$$\nabla^2 u = \frac{1}{r}\frac{\partial}{\partial r}\left(r\frac{\partial u}{\partial r}\right) = 0 \tag{2.6.2}$$

where $u = u(r)$ only. Hence

$$\frac{d}{dr}\left(r\frac{du}{dr}\right) = 0 \tag{2.6.3}$$

which solves as an ordinary differential equation to give

$$r\frac{du}{dr} = A \qquad \text{or} \qquad \frac{du}{dr} = \frac{A}{r} \tag{2.6.4}$$

with solution

$$u = A\ln r + B \tag{2.6.5}$$

where A and B are arbitrary constants. The solution is singular at both $r = 0$ and at infinity provided $A \neq 0$.

(Note: $v(r) = u(\frac{1}{r}) = A\ln(\frac{1}{r}) + B$ is also a solution since $A\ln(\frac{1}{r}) + B = -A\ln r + B = A'\ln r + B$.)

Another simple case is that of spherical symmetry for which Laplace's equation is expressed in spherical polar coordinates which are illustrated in Figure 2.4.

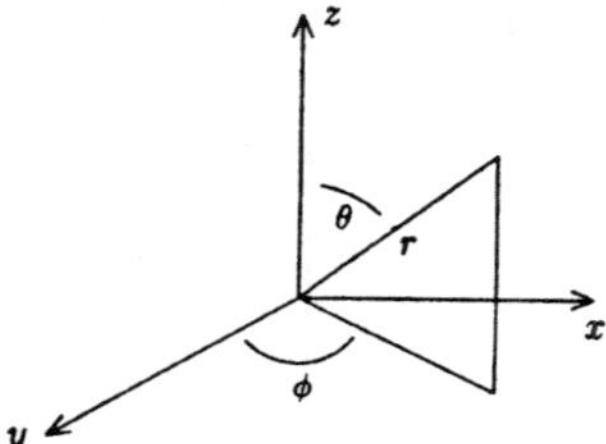

Fig. 2.4.

Laplace's equation in spherical polar coordinates (r, ϕ, θ) is

$$\nabla^2 u = \frac{1}{r^2}\left[\frac{\partial}{\partial r}\left(r^2\frac{\partial u}{\partial r}\right) + \frac{1}{\sin^2\theta}\frac{\partial^2 u}{\partial\phi^2} + \frac{1}{\sin\theta}\frac{\partial}{\partial\theta}\left(\sin\theta\frac{\partial u}{\partial\theta}\right)\right] = 0 \tag{2.6.6}$$

and by spherical symmetry, we mean that u is a function of r only giving

$$\frac{d}{dr}\left(r^2\frac{du}{dr}\right) = 0 \tag{2.6.7}$$

which reduces to

$$r^2\frac{du}{dr} = A \quad \text{or} \quad \frac{du}{dr} = \frac{A}{r^2} \tag{2.6.8}$$

with the solution

$$u = \frac{A'}{r} + B \tag{2.6.9}$$

where A, A' and B are arbitrary constants. The solution is singular at $r = 0$, provided $A' \neq 0$.

The next development is to move from single variable solutions, which reduce to ordinary differential equations, onto various two variable models for which separation of the variables becomes the main method. The first of these cases is Laplace's equation in plane polar coordinates.

The solution of $\nabla^2 u = 0$ within a finite region R with given values of u on the boundary of R is called the Dirichlet problem. The solution of this *interior* Dirichlet problem is unique. The exterior Dirichlet problem, consisting of solving $\nabla^2 u = 0$ in the infinite region exterior to R with given values of u on the boundary of R, can also be shown to have a unique solution in the two-dimensional case provided u remains bounded as $r \to \infty$.

Consider R to be a two-dimensional circular region of radius a, then on S, the boundary of R, let u be a known function

$$u(a, \phi) = f(\phi) \tag{2.6.10}$$

where f is a given function. Laplace's equation in plane polar coordinates is

$$\frac{1}{r}\frac{\partial}{\partial r}\left(r\frac{\partial u}{\partial r}\right) + \frac{1}{r^2}\frac{\partial^2 u}{\partial \phi^2} = 0. \tag{2.6.11}$$

Using separation of variables requires

$$u(r, \phi) = R(r)\Phi(\phi) \tag{2.6.12}$$

giving

$$\frac{\Phi}{r}\frac{d}{dr}\left(r\frac{dR}{dr}\right) + \frac{R}{r^2}\frac{d^2\Phi}{d\phi^2} = 0 \tag{2.6.13}$$

or

$$\frac{r}{R}\frac{d}{dr}\left(r\frac{dR}{dr}\right) + \frac{1}{\Phi}\frac{d^2\Phi}{d\phi^2} = 0. \tag{2.6.14}$$

As the two terms depend on different variables, the traditional split takes place to give

$$\frac{r}{R}\frac{d}{dr}\left(r\frac{dR}{dr}\right) = k \tag{2.6.15}$$

and

$$\frac{1}{\Phi}\frac{d^2\Phi}{d\phi^2} = -k. \tag{2.6.16}$$

Consider first 2.6.16 in the form

$$\frac{d^2\Phi}{d\phi^2} + k\Phi = 0. \tag{2.6.17}$$

Then, as $u(r, \phi)$ is the solution of a physical problem it must be single-valued requiring that $\Phi(\alpha) = \Phi(2\pi + \alpha)$ for all real α. If $k < 0$ or $k = -q^2$ the solution is

$$\Phi(\phi) = Ae^{q\phi} + Be^{-q\phi} \tag{2.6.18}$$

and this is multivalued in the sense that

$$\Phi(\phi + 2\pi) = Ae^{q(\phi+2\pi)} + Be^{-q(\phi+2\pi)} \neq A^{q\phi} + Be^{-q\phi} = \Phi(\phi). \tag{2.6.19}$$

On the other hand if $k = 0$, the solution is $\Phi = A\phi + B$ which is again multi-valued unless $A = 0$ since

$$\Phi(\phi + 2\pi) = \Phi(\phi) \Rightarrow A(\phi + 2\pi) + B = A\phi + B \Rightarrow A = 0 \tag{2.6.20}$$

to leave just

$$\Phi = \text{const.} \tag{2.6.21}$$

Finally, if $k > 0$ or $k = p^2$ the solution is

$$\Phi(\phi) = A\cos p\phi + B\sin p\phi \tag{2.6.22}$$

and now

$$\Phi(\phi + 2\pi) = \Phi(\phi) \tag{2.6.23}$$

implies

$$A\cos p(\phi + 2\pi) + B\sin p(\phi + 2\pi) = A\cos p\phi + B\sin p\phi \tag{2.6.24}$$

which is true if $p = n$ where n is an integer, to give,

$$\Phi_n(\phi) = A_n \cos n\phi + B_n \sin n\phi, \qquad n = 1, 2, 3, \ldots. \tag{2.6.25}$$

The equation for R becomes

$$r^2\frac{d^2R}{dr^2} + r\frac{dR}{dr} - n^2R = 0, \qquad n = 1, 2, 3, \ldots \tag{2.6.26}$$

and in the special case of $k = 0$,

$$\frac{d}{dr}\left(r\frac{dR}{dr}\right) = 0. \tag{2.6.27}$$

To solve the general equation 2.6.26 seek $R(r)$ in the form

$$R(r) = Ar^m \qquad \text{to get} \quad m(m-1)r^m + mr^m - n^2r^m = 0 \tag{2.6.28}$$

which reduces to

$$m^2 - n^2 = 0 \qquad \text{or} \qquad m = \pm n. \tag{2.6.29}$$

Thus the solution is

$$R_n(r) = \frac{C_n}{r^n} + D_n r^n \tag{2.6.30}$$

where C_n and D_n are arbitrary constants.

For the special case $k = 0$, equation 2.6.27 gives

$$\frac{d}{dr}\left(r\frac{dR}{dr}\right) = 0 \qquad \text{or} \qquad \frac{dR}{dr} = \frac{C_0}{r} \tag{2.6.31}$$

with solution

$$R(r) = C_0 \ln r + D_0. \tag{2.6.32}$$

Hence the solutions are

$$u_n(r,\phi) = \left(\frac{C_n}{r^n} + D_n r^n\right)(A_n \cos n\phi + B_n \sin n\phi) \tag{2.6.33}$$

and

$$u_0(r,\phi) = A_0(C_0 \ln r + D_0). \tag{2.6.34}$$

For the interior Dirichlet problem with $r \le a$, $C_n = 0$ (all n) to avoid a singularity at the origin, leaving the remaining terms

$$u_n(r,\phi) = r^n[E_n \cos n\phi + F_n \sin n\phi] \tag{2.6.35}$$

and

$$u_0(r,\phi) = E_0. \tag{2.6.36}$$

To satisfy 2.6.10 consider therefore the solution

$$u(r,\phi) = E_0 + \sum_{n=1}^{\infty} r^n(E_n \cos n\phi + F_n \sin n\phi) \tag{2.6.37}$$

and obtain

$$u(a,\phi) = E_0 + \sum_{n=1}^{\infty} a^n(E_n \cos n\phi + F_n \sin n\phi) = f(\phi) \qquad 0 \le \phi \le 2\pi. \tag{2.6.38}$$

Hence

$$E_0 = \frac{1}{2\pi}\int_0^{2\pi} f(\phi)\, d\phi \tag{2.6.39}$$

$$E_n = \frac{1}{a^n \pi}\int_0^{2\pi} f(\phi) \cos n\phi\, d\phi \tag{2.6.40}$$

and

$$F_n = \frac{1}{a^n \pi}\int_0^{2\pi} f(\phi) \sin n\phi\, d\phi. \tag{2.6.41}$$

For the exterior Dirichlet problem with $r \ge a$, D_n and C_0 must be zero to ensure that $u(r,\phi)$ remains bounded as $r \to \infty$. Hence in this case

$$u_n(r,\phi) = \frac{1}{r^n}(\overline{E}_n \cos n\phi + \overline{F}_n \sin n\phi). \tag{2.6.42}$$

Thus to satisfy $u(a,\phi) = f(\phi)$, consider

$$u(r,\phi) = \overline{E}_0 + \sum_{n=1}^{\infty} \frac{1}{r^n}(\overline{E}_n \cos n\phi + \overline{F}_n \sin n\phi) \tag{2.6.43}$$

which now gives

$$u(a,\phi) = \overline{E}_0 + \sum_{n=1}^{\infty} \frac{1}{a^n}[\overline{E}_n \cos n\phi + \overline{F}_n \sin n\phi] = f(\phi). \tag{2.6.44}$$

Hence the Fourier coefficients are given by

$$\overline{E}_0 = \frac{1}{2\pi}\int_0^{2\pi} f(\phi)\, d\phi \tag{2.6.45}$$

$$\overline{E}_n = \frac{a^n}{\pi}\int_0^{2\pi} f(\phi)\cos n\phi\, d\phi \tag{2.6.46}$$

and

$$\overline{F}_n = \frac{a^n}{\pi}\int_0^{2\pi} f(\phi)\sin n\phi\, d\phi. \tag{2.6.47}$$

Suppose for example that $f(\phi) = \phi$, $-\pi < \phi < \pi$, then by periodic extension $\int_0^{2\pi}$ becomes $\int_{-\pi}^{\pi}$. Then

$$\frac{1}{2\pi}\int_0^{2\pi} f(\phi)\, d\phi = \frac{1}{2\pi}\int_{-\pi}^{\pi} \phi\, d\phi = 0 \tag{2.6.48}$$

and

$$\frac{1}{\pi}\int_0^{2\pi} f(\phi)\cos n\phi\, d\phi = \frac{1}{\pi}\int_{-\pi}^{\pi} \phi\cos n\phi\, d\phi = 0 \tag{2.6.49}$$

and

$$\frac{1}{\pi}\int_0^{2\pi} f(\phi)\sin n\phi\, d\phi = \frac{1}{\pi}\int_{-\pi}^{\pi} \phi\sin n\phi\, d\phi = \frac{-2}{n}(-1)^n. \tag{2.6.50}$$

Hence the solution to the interior problem is

$$u(r,\phi) = -2\sum_{n=1}^{\infty}\left(\frac{r}{a}\right)^n \frac{(-1)^n}{n}\sin n\phi \tag{2.6.51}$$

and the solution to the exterior problem is

$$u(r,\phi) = -2\sum_{n=1}^{\infty}\left(\frac{a}{r}\right)^n \frac{(-1)^n}{n}\sin n\phi. \tag{2.6.52}$$

As a second example, consider finding the solution $u(r,\phi)$ of Laplace's equation $\nabla^2 u = 0$ in the section of a circle $r < a$, $0 < \phi < \frac{\pi}{2}$ which vanishes on $\phi = 0$ and $\phi = \frac{\pi}{2}$, and takes the value 1 on the curved boundary $r = a$. The equation being solved is

$$\nabla^2 u = \frac{1}{r}\frac{\partial}{\partial r}\left(r\frac{\partial u}{\partial r}\right) + \frac{1}{r^2}\frac{\partial^2 u}{\partial \phi^2} = 0. \tag{2.6.53}$$

Use separation of the variables to make the assumption

$$u(r,\phi) = R(r)\Phi(\phi) \tag{2.6.54}$$

to give

$$\frac{r}{R}\frac{d}{dr}\left(r\frac{dR}{dr}\right) + \frac{1}{\Phi}\frac{d^2\Phi}{d\phi^2} = 0. \tag{2.6.55}$$

As each term is a function of a different variable, the separate equations are

$$\frac{r}{R}\frac{d}{dr}\left(r\frac{dR}{dr}\right) = k \tag{2.6.56}$$

and

$$\frac{1}{\Phi}\frac{d^2\Phi}{d\phi^2} = -k. \tag{2.6.57}$$

The given conditions then give

$$\begin{aligned} u(r,0) = 0 &\Rightarrow R(r)\Phi(0) = 0 \Rightarrow \Phi(0) = 0 \\ u\left(r,\tfrac{\pi}{2}\right) = 0 &\Rightarrow R(r)\Phi\left(\tfrac{\pi}{2}\right) = 0 \Rightarrow \Phi\left(\tfrac{\pi}{2}\right) = 0 \end{aligned} \tag{2.6.58}$$

for non-trivial solutions.

Consider first 2.6.57 in the form

$$\frac{d^2\Phi}{d\phi^2} + k\Phi = 0 \tag{2.6.59}$$

subject to

$$\Phi(0) = \Phi\left(\frac{\pi}{2}\right) = 0.$$

As in the Cartesian case, three possibilities arise and the non-trivial solution arises when $k > 0$ or $k = p^2$. The equation to be solved is

$$\frac{d^2\Phi}{d\phi^2} + p^2\Phi = 0 \tag{2.6.60}$$

with solution

$$\Phi(\phi) = A\cos p\phi + B\sin p\phi. \tag{2.6.61}$$

The boundary conditions give $\Phi(0) = 0$ from which $A = 0$ and $\Phi(\pi/2) = 0$ which implies

$$0 = B\sin p\frac{\pi}{2}.$$

Hence there is a non-trivial solution when

$$\frac{p\pi}{2} = n\pi \quad \text{or} \quad p = 2n, \qquad n = 1, 2, 3, \dots . \tag{2.6.62}$$

Hence if $k = 4n^2$, the solution is

$$\Phi_n(\phi) = B_n \sin 2n\phi, \qquad n = 1, 2, 3, \dots . \tag{2.6.63}$$

Hence for the R solution, the equation is

$$r^2\frac{d^2R}{dr^2} + r\frac{dR}{dr} - 4n^2R = 0. \tag{2.6.64}$$

Equation 2.6.64 is solved by letting

$R(r) = r^m$ to give $m(m-1)+m-4n^2 = 0$ or $m^2 = 4n^2$ or $m = \pm 2n$.

Hence

$$R_n(r) = \frac{C_n}{r^{2n}} + D_n r^{2n}. \tag{2.6.65}$$

Thus u_n has the form

$$u_n(r,\phi) = \frac{E_n}{r^{2n}}\sin 2n\phi + F_n r^{2n}\sin 2n\phi, \qquad n = 1, 2, 3, \dots \tag{2.6.66}$$

with

$$E_n = B_n C_N, \qquad F_n = B_n D_n,$$

so the solution can be written as

$$u(r,\phi) = \sum_{n=1}^{\infty}\left(\frac{E_n}{r^{2n}} + F_n r^{2n}\right)\sin 2n\phi. \tag{2.6.67}$$

However, $u(r,\phi)$ is finite when $r = 0$, and hence $E_n = 0$, to leave

$$u(r,\phi) = \sum_{n=1}^{\infty} F_n r^{2n}\sin 2n\phi. \tag{2.6.68}$$

When $r = a$,

$$u(a,\phi) = \sum_{n=1}^{\infty} F_n a^{2n}\sin 2n\phi = 1 \tag{2.6.69}$$

and again the Fourier coefficients are:

$$a^{2n}F_n = -\frac{2}{n\pi}[(-1)^n - 1] \tag{2.6.70}$$

and the final solution is

$$u(r,\phi) = \frac{2}{\pi}\sum_{n=1}^{\infty}\frac{(1-(-1)^n)}{n}\left(\frac{r}{a}\right)^{2n}\sin 2n\phi. \tag{2.6.71}$$

So far solutions of Laplace's equation have been considered in some non-Cartesian coordinate systems. Attention will now move to the wave equation and in particular the vibrations of a circular membrane of radius R for which the wave equation in polar coordinates becomes:

$$\frac{\partial^2 u}{\partial t^2} = c^2\left(\frac{\partial^2 u}{\partial r^2} + \frac{1}{r}\frac{\partial u}{\partial r} + \frac{1}{r^2}\frac{\partial^2 u}{\partial \theta^2}\right). \tag{2.6.72}$$

Let the membrane be fixed along the boundary $r = R$, giving

$$u(R,\theta,t) = 0 \qquad \text{for} \quad t \geq 0,\ 0 \leq \theta \leq 2\pi.$$

Then the separation of variables assumption gives

$$u(r,\theta,t) = W(r)Q(\theta)T(t), \tag{2.6.73}$$

$$\frac{T''}{c^2T} = \frac{W''}{W} + \frac{1}{r}\frac{W'}{W} + \frac{1}{r^2}\frac{Q''}{Q}. \tag{2.6.74}$$

In the usual way, the independence of each side gives

$$\frac{T''}{c^2T} = k \tag{2.6.75}$$

and

$$\frac{W''}{W} + \frac{1}{r}\frac{W'}{W} + \frac{1}{r^2}\frac{Q''}{Q} = k. \tag{2.6.76}$$

The second equation can itself now be split to give

$$r^2\frac{W''}{W} + r\frac{W'}{W} - r^2k = -\frac{Q''}{Q} \tag{2.6.77}$$

yielding the two equations

$$r^2\frac{W''}{W} + r\frac{W'}{W} - r^2k = \lambda \tag{2.6.78}$$

and

$$-\frac{Q''}{Q} = \lambda. \tag{2.6.79}$$

The conditions on the solution $u(r,\theta,t)$ are:

(i) It is single valued;

(ii) It is bounded for $0 \le r \le R$;

(iii) $u(R, \theta, t) = 0$.

To impose these conditions, consider first 2.6.79, for which the non-trivial solution arises when $\lambda \geq 0$. If $\lambda > 0$ or $\lambda = p^2$, the solution is

$$Q(\theta) = A\cos p\theta + B\sin p\theta \tag{2.6.80}$$

which is single valued when $p = n, \quad n = 1, 2, 3, \ldots$ and for $\lambda = 0$, the solution $Q(\theta) = A$ is single-valued. Hence λ must be $0, 1^2, 2^2, \ldots, n^2, \ldots$.

It can be shown that k must be negative to satisfy the other conditions. Thus, writing $k = -\alpha^2$ and $\lambda = n^2$, equation 2.6.78 takes the form

$$r^2W'' + rW' + (\alpha^2r^2 - n^2)W = 0. \tag{2.6.81}$$

This equation is Bessel's equation whose solutions are a further transcendental function called Bessel's functions. Like the trigonometric functions, these are tabulated and can be found in computer libraries, though they are not so common in calculators. These functions are treated in Chapter 1.

The next variation to be considered is the use of spherical polar coordinates defined by the variables (r, ϕ, θ). Consider the solution of Laplace's equation

$$\frac{1}{r^2}\frac{\partial}{\partial r}\left(r^2\frac{\partial u}{\partial r}\right) + \frac{1}{r^2\sin^2\theta}\frac{\partial^2 u}{\partial\phi^2} + \frac{1}{r^2\sin\theta}\frac{\partial}{\partial\theta}\left(\sin\theta\frac{\partial u}{\partial\theta}\right) = 0 \tag{2.6.82}$$

in these coordinates. Assume that the line $\theta = 0$ is an axis of symmetry and that u is finite in the solution domain which is a sphere of radius R.

The symmetry assumption implies that there is no dependence on ϕ and hence

$$\frac{\partial}{\partial r}\left(r^2\frac{\partial u}{\partial r}\right) + \frac{1}{\sin\theta}\frac{\partial}{\partial\theta}\left(\sin\theta\frac{\partial u}{\partial\theta}\right) = 0. \tag{2.6.83}$$

The separation assumption is

$$u(r, \theta) = R(r)\Theta(\theta) \tag{2.6.84}$$

which gives

$$\frac{1}{R}\frac{d}{dr}\left(r^2\frac{dR}{dr}\right) + \frac{1}{\Theta\sin\theta}\frac{d}{d\theta}\left(\sin\theta\frac{d\Theta}{d\theta}\right) = 0 \tag{2.6.85}$$

and the separated equations are

$$\frac{1}{R}\frac{d}{dr}\left(r^2\frac{dR}{dr}\right) = \lambda \tag{2.6.86}$$

and

$$\lambda + \frac{1}{\Theta\sin\theta}\frac{d}{d\theta}\left(\sin\theta\frac{d\Theta}{d\theta}\right) = 0. \tag{2.6.87}$$

Clearly λ can be replaced by $s(s+1)$ without any loss of generality to give

$$\frac{1}{\sin\theta}\frac{d}{d\theta}\left(\sin\theta\frac{d\Theta}{d\theta}\right)+s(s+1)\Theta=0 \tag{2.6.88}$$

for the equation for Θ. To obtain the standard form for this equation let $\cos\theta = w$, then

$$\frac{d}{d\theta}=\frac{d}{dw}\cdot\frac{dw}{d\theta}=-\sin\theta\frac{d}{dw}$$

and

$$\sin^2\theta = 1-w^2$$

to give

$$-\frac{d}{dw}\left(-(1-w^2)\frac{d\Theta}{dw}\right)+s(s+1)\Theta=0 \tag{2.6.89}$$

or

$$\frac{d}{dw}\left((1-w^2)\frac{d\Theta}{dw}\right)+s(s+1)\Theta=0 \tag{2.6.90}$$

or

$$(1-w^2)\frac{d^2\Theta}{dw^2}-2w\frac{d\Theta}{dw}+s(s+1)\Theta=0. \tag{2.6.91}$$

This is Legendre's equation. Now $0\le\theta\le\pi$ and so $-1\le w\le 1$ and the only solutions which satisfy the requirements of being bounded and continuous are the Legendre polynomials which were first introduced in Chapter 1.

This chapter is concluded with two examples in which the special functions of Legendre and Bessel are employed. The defining equations were introduced in Chapter 1. For the first of these, consider the vibration of a circular membrane of radius R with radial symmetry, that is no θ dependence. The wave equation is then

$$\frac{\partial^2 u}{\partial t^2}=c^2\left(\frac{\partial^2 u}{\partial r^2}+\frac{1}{r}\frac{\partial u}{\partial r}\right). \tag{2.6.92}$$

The membrane is fixed along the boundary $r=R$ so that $u(R,t)=0$ for all $t\ge 0$. Solutions not depending on θ will occur if the initial conditions do not depend on θ, that is, if they are of the form $u(r,0)=f(r)$, the initial deflection, and

$$\left.\frac{\partial u}{\partial t}\right|_{t=0}=g(r)$$

the initial velocity.

Separation of the variables requires

$$u(r,t)=W(r)T(t) \tag{2.6.93}$$

which results in

$$T''-c^2kT=0 \tag{2.6.94}$$

and

$$W'' + \frac{1}{r}W' - kW = 0. \tag{2.6.95}$$

Now the membrane is fixed at $r = R$ so that $u(R,t) = 0$ gives $W(R)T(t) = 0$, and hence $W(R) = 0$ for non-trivial solutions. Also $u(r,t)$ must be finite, hence $W(r)$ and $T(t)$ must each be finite. To solve 2.6.95, the non-trivial solution arises when $k < 0$, hence letting $k = -q^2$ gives

$$W'' + \frac{1}{r}W' + q^2W = 0. \tag{2.6.96}$$

Let $s = qr$ then the equation becomes

$$\frac{d^2W}{ds^2} + \frac{1}{s}\frac{dW}{ds} + W = 0. \tag{2.6.97}$$

This is Bessel's equation of order zero and has a general solution

$$W(s) = AJ_0(s) + BY_0(s) \tag{2.6.98}$$

where J_0 and Y_0 are the Bessel functions of the first and second kind of order zero. Y_0 is singular when $s = 0$, and therefore we must have $B = 0$ to leave

$$W(r) = AJ_0(qr) \tag{2.6.99}$$

and

$$W(R) = 0 \qquad \text{implies} \quad AJ_0(qR) = 0.$$

Thus for a non-trivial solution, we require $J_0(qR) = 0$. The Bessel function J_0 has infinitely many real zeros. Let us denote the positive zeros of $J_0(s)$ by $\alpha_1, \alpha_2, \ldots$.

For $J_0(qR) = 0$, we must have $qR = \alpha_m$ or

$$q_m = \frac{\alpha_m}{R}, \qquad m = 1, 2, \ldots \tag{2.6.100}$$

Hence

$$W_m(r) = J_0(q_m r) = J_0\left(\frac{\alpha_m}{R}r\right), \qquad m = 1, 2, \ldots \tag{2.6.101}$$

are the solutions that vanish at $r = R$.

Now turn to the T equation and seek solutions of

$$T'' = -c^2q^2T \qquad \text{with} \quad q = \frac{\alpha_m}{R}. \tag{2.6.102}$$

These are given by

$$T(t) = C_m \cos\left(\frac{c\alpha_m t}{R}\right) + D_m \sin\left(\frac{c\alpha_m t}{R}\right) \tag{2.6.103}$$

so that the basic solution of 2.6.92 has the form

$$u_m(r,t) = \left[C_m \cos\left(\frac{c\alpha_m t}{R}\right) + D_m \sin\left(\frac{c\alpha_m t}{R}\right)\right] J_0\left(\frac{\alpha_m}{R} r\right). \tag{2.6.104}$$

To satisfy the initial conditions, the solutions are summed over m to give

$$\begin{aligned} u(r,t) &= \sum_{m=1}^{\infty} u_m(r,t) \\ &= \sum_{m=1}^{\infty} \left[C_m \cos\left(\frac{c\alpha_m t}{R}\right) + D_m \sin\left(\frac{c\alpha_m t}{R}\right)\right] J_0\left(\frac{\alpha_m}{R} r\right). \end{aligned} \tag{2.6.105}$$

At $t = 0$, $u(r,0) = f(r)$ and hence

$$f(r) = \sum_{m=1}^{\infty} C_m J_0\left(\frac{\alpha_m}{R} r\right). \tag{2.6.106}$$

However,

$$J_0\left(\frac{\alpha_m}{R} r\right), \qquad m = 1, 2, 3, \dots, \tag{2.6.107}$$

form a complete set of orthogonal functions with respect to the weight r, on the interval $0 \leq r \leq R$ and

$$\begin{aligned} \left\| J_0\left(\frac{\alpha_m}{R} r\right)\right\|^2 &= \int_0^R r\left(J_0\left(\frac{\alpha_m}{R} r\right)\right)^2 dr \\ &= \frac{R^2}{2}[J_1(\alpha_m)]^2. \end{aligned} \tag{2.6.108}$$

Hence

$$C_m = \frac{2}{R^2[J_1(\alpha_m)]^2} \int_0^R r f(r) J_0\left(\frac{\alpha_m}{R} r\right) dr, \qquad m = 1, 2, \dots. \tag{2.6.109}$$

(Differentiability of $f(r)$ on $0 \leq r \leq R$ is sufficient for the existence of the series development).

Also

$$\begin{aligned} \frac{\partial u}{\partial t} &= \sum_{m=1}^{\infty} \left[-C_m \frac{c\alpha_m}{R} \sin\left(\frac{c\alpha_m t}{R}\right)\right. \\ &\quad \left. + D_m \frac{c\alpha_m}{R} \cos\left(\frac{c\alpha_m t}{R}\right)\right] J_0\left(\frac{\alpha_m}{R} r\right) \end{aligned} \tag{2.6.110}$$

and the second boundary condition gives

$$\frac{\partial u}{\partial t}(r,0) = g(r) = \sum_{m=1}^{\infty} \frac{c\alpha_m}{R} D_m J_0\left(\frac{\alpha_m}{R} r\right). \tag{2.6.111}$$

Hence

$$D_m = \frac{2}{Rc\alpha_m [J_1(\alpha_m)]^2} \int_0^R r g(r) J_0\left(\frac{\alpha_m}{R} r\right) dr. \tag{2.6.112}$$

For the final example using these new transcendental functions, consider the solution of Laplace's equation in spherical polar coordinates with axial-symmetry. Let the spherical polar coordinates be (r, ϕ, θ), and Laplace's equation in these coordinates is

$$\frac{1}{r^2}\frac{\partial}{\partial r}\left(r^2 \frac{\partial u}{\partial r}\right) + \frac{1}{r^2 \sin^2\theta}\frac{\partial^2 u}{\partial \phi^2} + \frac{1}{r^2 \sin\theta}\frac{\partial}{\partial \theta}\left(\sin\theta \frac{\partial u}{\partial \theta}\right) = 0. \tag{2.6.113}$$

Consider a spherical domain of radius a, subject to the Dirichlet boundary conditions

$$u(a,\theta) = \begin{cases} u_1(\text{const}) & 0 \le \theta \le \frac{\pi}{2} \\ u_2(\text{const}) & \frac{\pi}{2} \le \theta \le \pi, \end{cases}$$

with u_1 and u_2 being two constants, where the line $\theta = 0$ is taken as the axis of symmetry, and u is finite in the domain. Hence Laplace's equation is being solved within a sphere, the upper hemisphere surface having $u = u_1$, and the lower hemisphere surface having $u = u_2$. By the assumption that there is no dependence on ϕ, the equation becomes:

$$\frac{\partial}{\partial r}\left(r^2 \frac{\partial u}{\partial r}\right) + \frac{1}{\sin\theta}\frac{\partial}{\partial \theta}\left(\sin\theta \frac{\partial u}{\partial \theta}\right) = 0. \tag{2.6.114}$$

Writing $u(r,\theta) = R(r)\Theta(\theta)$ to separate the variables gives

$$\frac{1}{R}\frac{d}{dr}\left(r^2 \frac{dR}{dr}\right) + \frac{1}{\Theta \sin\theta}\frac{d}{d\theta}\left(\sin\theta \frac{d\Theta}{d\theta}\right) = 0. \tag{2.6.115}$$

As each term is a function of a different variable, two ordinary differential equations arise

$$\frac{1}{R}\frac{d}{dr}\left(r^2 \frac{dR}{dr}\right) = \lambda \tag{2.6.116}$$

and

$$\lambda + \frac{1}{\Theta \sin\theta}\frac{d}{d\theta}\left(\sin\theta \frac{d\Theta}{d\theta}\right) = 0. \tag{2.6.117}$$

We can replace λ by $s(s+1)$ without any loss of generality. The equation for Θ then becomes

$$\frac{1}{\sin\theta}\frac{d}{d\theta}\left(\sin\theta \frac{d\Theta}{d\theta}\right) + s(s+1)\Theta = 0. \tag{2.6.118}$$

Let

$$\cos\theta = w, \qquad \sin^2\theta = 1 - w^2, \qquad \frac{d}{d\theta} = -\sin\theta\frac{d}{dw} \tag{2.6.119}$$

so reducing 2.6.118 to

$$\frac{d}{dw}\left((1-w^2)\frac{d\Theta}{dw}\right) + s(s+1)\Theta = 0, \tag{2.6.120}$$

which is Legendre's equation of Chapter 1. Now $0 \le \theta \le \pi$ and so $-1 \le w \le 1$ and the only solutions of Legendre's equation which are bounded and continuous, together with their derivatives, in the required interval occur when $s = n \quad (n = 0, 1, 2, \ldots)$. Hence s is restricted to non-negative integers and n will be used for these. Hence the finite solutions in $0 \le \theta \le \pi$ $(-1 \le w \le 1)$ are the Legendre polynomials, $P_n(w)$ or $P_n(\cos\theta)$ so that

$$\Theta_n(\theta) = P_n(\cos\theta) \tag{2.6.121}$$

with $\lambda = n(n+1), n = 0, 1, 2, \ldots$. The equation for R is

$$r^2R'' + 2rR' - n(n+1)R = 0 \tag{2.6.122}$$

which gives the solution

$$R_n(r) = A_n r^n + \frac{B_n}{r^{n+1}}. \tag{2.6.123}$$

Thus the basic solution of 2.6.114 is

$$u_n(r,\theta) = R_n(r)\Theta_n(\theta) = \left(A_n r^n + \frac{B_n}{r^{n+1}}\right)P_n(\cos\theta). \tag{2.6.124}$$

Now the solution must be finite in the ball of radius a, hence $B_n = 0$ and so

$$u_n(r,\theta) = A_n r^n P_n(\cos\theta) \tag{2.6.125}$$

and to fit the boundary conditions sum over n to give

$$u(r,\theta) = \sum_{n=0}^{\infty} A_n r^n P_n(\cos\theta). \tag{2.6.126}$$

The boundary conditions give

$$\sum_{n=0}^{\infty} A_n a^n P_n(\cos\theta) = f(\theta) \tag{2.6.127}$$

or writing $w = \cos\theta$ gives

$$\sum_{n=0}^{\infty} A_n a^n P_n(w) = f^*(w) \tag{2.6.128}$$

where $f^*(w) = f(\theta)$. Now $P_n(w)$, $n = 0, 1, 2, \ldots$, form an orthogonal system of functions on $-1 \le w \le 1$ with $||P_n||^2 = \frac{2}{2n+1}$. Hence

$$A_n = \frac{2n+1}{2a^n} \int_{-1}^{1} f^*(w) P_n(w)\, dw. \qquad (2.6.129)$$

Since $w = \cos\theta$, the range $-1 \le w \le 0$ corresponds to $\pi \ge \theta \ge \frac{\pi}{2}$ and $0 \le w \le 1$ to $\frac{\pi}{2} \ge \theta \ge 0$, hence we have

$$A_n = \frac{(2n+1)}{2a^n} \int_{-1}^{0} u_2 P_n(w) dw + \frac{(2n+1)}{2a^n} \int_{0}^{1} u_1 P_n(w) dw \qquad (2.6.130)$$

on substituting the boundary conditions. Hence

$$A_0 = \frac{1}{2} u_2 \int_{-1}^{0} dw + \frac{u}{2} \int_{0}^{1} dw = \frac{1}{2}(u_2 - u_1) \qquad (2.6.131)$$

$$A_1 = \frac{3u_2}{2a} \int_{-1}^{0} w dw + \frac{3u_1}{2a} \int_{0}^{1} w dw = -\frac{3u_2}{4a} + \frac{3u_1}{4a} \qquad (2.6.132)$$

and

$$A_2 = \frac{5u_2}{2a} \int_{-1}^{0} \frac{1}{2}(3w^2 - 1) dw + \frac{5u_1}{2a} \int_{0}^{1} \frac{1}{2}(3w^2 - 1) dw = 0. \qquad (2.6.133)$$

Hence the full solution is

$$\begin{aligned} u(r,\theta) &= \frac{1}{2}(u_1 + u_2) - \frac{3}{4}(u_2 - u_1)\frac{r}{a} P_1(\cos\theta) \\ &\quad + \frac{7}{16}(u_2 - u_1)\left(\frac{r}{a}\right)^3 P_3(\cos\theta) + \ldots. \end{aligned} \qquad (2.6.134)$$

In summary, the method of separating the variables is a powerful device in the solution of partial differential equations and is one of the the most widely applied techniques for their solution. Some criticism may be made of the usefulness of the resulting series from a practical point of view. The one aim of a solution is to be able to compute values in a real problem, though the analytic solution will give insight into the properties of the physical problem modelled by the partial differential equation. The series obtained by separating the variables are of a Fourier type and may display very slow convergence. Hence, their use in a numerical solution can be frustrating. In the companion volume, direct numerical methods are given which avoid these issues. Nevertheless, the analytic information yielded by these series solutions is valuable in determining the nature of the solutions. The chapter is concluded with a further set of exercises on the use of non-Cartesian coordinates.

EXERCISES

2.22 Solve

$$\frac{\partial^2 v}{\partial r^2} + \frac{1}{r}\frac{\partial v}{\partial r} + \frac{1}{r^2}\frac{\partial^2 v}{\partial \theta^2} = 0$$

in the quadrant $0 \le r \le 1$, $0 \le \theta \le \frac{\pi}{2}$ subject to:

(i) $v = 0$ at $\theta = 0$ for $0 \le r \le 1$

(ii) $\frac{\partial v}{\partial \theta} = 0$ at $\theta = \frac{\pi}{2}$ for $0 \le r \le 1$

(iii) $v = 2\theta$ at $r = 1$ for $0 \le \theta \le \frac{\pi}{2}$.

2.23 Solve

$$\frac{\partial}{\partial r}\left(r\frac{\partial u}{\partial r}\right) + \frac{1}{r}\frac{\partial^2 u}{\partial \phi^2} = 0$$

in the quadrant $0 < r < a$, $0 < \phi < \frac{\pi}{2}$ with

$$\frac{\partial u}{\partial \phi}(r,0) = \frac{\partial u}{\partial \phi}\left(r, \frac{\pi}{2}\right) = 0, \qquad u(a,\phi) = \phi.$$

2.24 Show that the solution $u(r,\phi)$ of Laplace's equation $\nabla^2 u = 0$ in the semicircular region $r < a$, $0 < \phi < \pi$ which vanishes at $\phi = 0$ and takes the value A (cons) on $\theta = \pi$ and on the curved boundary $r = a$ is

$$u(r,\phi) = \frac{A}{\pi}\left[\phi + 2\sum_{n=1}^{\infty}\left(\frac{r}{a}\right)^n \frac{\sin n\phi}{n}\right].$$

(Note: the boundary conditions are non-homogeneous.)

2.25 Given that $f(x) = 1$, $(0 < x < a)$ has the expansion

$$f(x) = \frac{2}{a}\sum_{n=1}^{\infty}\frac{J_0(\lambda_n x)}{\lambda_n J_1(\lambda_n a)}$$

where λ_n are the positive roots of the equation $J_0(\lambda a) = 0$, show that solution of the equation

$$\frac{1}{r}\frac{\partial}{\partial r}\left(r\frac{\partial u}{\partial r}\right) = \frac{1}{k}\frac{\partial u}{\partial t}$$

for $0 \le r \le a$, $t \ge 0$ (where k and a are constants) subject to the boundary conditions

$$u(r,0) = \tau_0 \quad (= \text{const}), \qquad u(a,t) = 0$$

for all t is

$$u(r,t) = \frac{2\tau_0}{a}\sum_{n=1}^{\infty}\frac{J_0(\lambda_n r)}{\lambda_n J_1(\lambda_n a)}e^{-k\lambda_n^2 t}.$$

2.26 Find $u(r, \theta)$ outside the sphere for the worked example defined in the text after equation 2.6.113.

3

First-order Equations and Hyperbolic Second-order Equations

3.1 Introduction

The concept of a characteristic curve for a second-order equation was introduced in Chapter 1, and led to a classification of these equations. When the characteristics are real as in the hyberbolic case, they can be used to solve partial differential equations directly. Along the characteristic curve, an ordinary differential equation holds, and in both an analytic and numerical context, can be used to solve the original partial differential equation. This chapter will begin with a consideration of first-order partial differential equations from an analytic point of view, where characteristics play an important role. This will be followed by the specific consideration of the wave equation and a characteristic based method known as d'Alembert's method.

3.2 First-order equations

In section 1.2, a quasilinear second-order equation was introduced. The corresponding first-order equation will have the form

$$Pp + Qq = R \tag{3.2.1}$$

where, as in Chapter 1, $p = \frac{\partial z}{\partial x}$, $q = \frac{\partial z}{\partial y}$ and the functions P, Q and R are functions of x, y and z. It was this equation which was first solved by Lagrange in 1772.

We use the earlier definition of the characteristic curve as the curve along which the highest derivatives are not uniquely defined. In addition to 3.2.1 we have the relation

$$p\,dx + q\,dy = dz. \tag{3.2.2}$$

The highest derivatives are now simply p and q, and hence 3.2.1 and 3.2.2 give two linear algebraic equations for these terms. Following the same lines as in Chapter 1, there will be no unique solution when the determinant

$$\begin{vmatrix} P & Q \\ dx & dy \end{vmatrix} = 0 \tag{3.2.3}$$

and as before for a non-unique solution to exist

$$\begin{vmatrix} P & R \\ dx & dz \end{vmatrix} = 0. \tag{3.2.4}$$

Hence the equation of the characteristic curves is

$$\frac{dx}{P} = \frac{dy}{Q} \tag{3.2.5}$$

and along these curves the condition

$$\frac{dx}{P} = \frac{dz}{R} \tag{3.2.6}$$

must hold.

The important theorem for the general solution of 3.2.1 is as follows.

Theorem 3.1

The general solution of 3.2.1 is

$$F(\xi, \eta) = 0$$

where F is an arbitrary differentiable function and $\xi(x, y, z) = c_1$ and $\eta(x, y, z) = c_2$ are the solution of

$$\frac{dx}{P} = \frac{dy}{Q} = \frac{dz}{R} \tag{3.2.7}$$

which will establish that the characteristic curves and the equation along these curves yields the required solution.

Proof

The function ξ must satisfy

$$\xi_x dx + \xi_y dy + \xi_z dz = 0 \tag{3.2.8}$$

and hence

$$P\xi_x + Q\xi_y + R\xi_z = 0. \tag{3.2.9}$$

Similarly for η

$$P\eta_x + Q\eta_y + R\eta_z = 0 \tag{3.2.10}$$

and equations 3.2.9 and 3.2.10 solve to give

$$\frac{P}{\partial(\xi,\eta)/\partial(y,z)} = \frac{Q}{\partial(\xi,\eta)/\partial(z,x)} = \frac{P}{\partial(\xi,\eta)/\partial(x,y)} \tag{3.2.11}$$

where

$$\frac{\partial(\xi,\eta)}{\partial(x,y)} = \begin{vmatrix} \frac{\partial\xi}{\partial x} & \frac{\partial\xi}{\partial y} \\ \frac{\partial\eta}{\partial x} & \frac{\partial\eta}{\partial y} \end{vmatrix}. \tag{3.2.12}$$

However, from $F(\xi,\eta)=0$, differentiating w.r.t. x and y gives

$$\begin{aligned} \frac{\partial F}{\partial\xi}\left[\frac{\partial\xi}{\partial x} + \frac{\partial\xi}{\partial z}p\right] + \frac{\partial F}{\partial\eta}\left[\frac{\partial\eta}{\partial x} + \frac{\partial\eta}{\partial z}p\right] &= 0 \\ \frac{\partial F}{\partial\xi}\left[\frac{\partial\xi}{\partial y} + \frac{\partial\xi}{\partial z}q\right] + \frac{\partial F}{\partial\eta}\left[\frac{\partial\eta}{\partial y} + \frac{\partial\eta}{\partial z}q\right] &= 0 \end{aligned}$$

and now eliminating $\partial F/\partial\xi$ and $\partial F/\partial\eta$ yields

$$p\frac{\partial(\xi,\eta)}{\partial(y,z)} + q\frac{\partial(\xi,\eta)}{\partial(z,x)} = \frac{\partial(\xi,\eta)}{\partial(x,y)} \tag{3.2.13}$$

which with 3.2.11 gives 3.2.1 as required.

As an example, consider the first-order equation

$$(x+z)\frac{\partial z}{\partial x} + y\frac{\partial z}{\partial y} = z + y^2$$

for which the solution of the ordinary differential equations

$$\frac{dx}{x+z} = \frac{dy}{y} = \frac{dz}{z+y^2}$$

is required for the functions ξ and η. The second equation has the standard form

$$\frac{dz}{dy} - \frac{z}{y} = y$$

with integral $z = c_1 y + y^2$ and the first equation then gives $x = c_1 y \ln y + c_2 y + y^2$. Hence

$$\begin{aligned} \xi(x,y,z) &= \frac{z-y^2}{y} = c_1 \\ \eta(x,y,z) &= \frac{x - y^2 - (z-y^2)\ln y}{y} = c_2. \end{aligned}$$

First-order equations which are not necessarily linear in p and q can also be solved in some cases by generalising the above approach. These generalised characteristic equations were first used by Paul Charpit in 1784 (the year of his death), and are known as Charpit's equations. It seems that Lagrange also found these equations in 1779, but Charpit makes no reference to the earlier work. The more general problem is

$$F(x, y, z, p, q) = 0 \tag{3.2.14}$$

and in the setting up of the determinants 3.2.3 and 3.2.4, equation 3.2.1 is replaced by

$$\frac{\partial F}{\partial p} dp + \frac{\partial F}{\partial q} dq = 0 \tag{3.2.15}$$

which yields the characteristic equations:

$$\frac{dx}{F_p} = \frac{dy}{F_q} = \frac{dz}{pF_p + qF_q}. \tag{3.2.16}$$

These equations reduce to the quasilinear case 3.2.7 correctly when 3.2.14 has the form 3.2.1. Now introduce an independent parameter t such that

$$\frac{dx}{dt} = F_p \quad \text{and} \quad \frac{dy}{dt} = F_q \tag{3.2.17}$$

which gives

$$\frac{dz}{dt} = pF_p + qF_q. \tag{3.2.18}$$

In addition

$$\begin{aligned} \frac{dp}{dt} &= \frac{\partial p}{\partial x}\frac{dx}{dt} + \frac{\partial p}{\partial y}\frac{dy}{dt} \\ &= \frac{\partial p}{\partial x}\frac{\partial F}{\partial p} + \frac{\partial p}{\partial y}\frac{\partial F}{\partial q} \\ &= \frac{\partial p}{\partial x}\frac{\partial F}{\partial p} + \frac{\partial q}{\partial x}\frac{\partial F}{\partial q} \end{aligned} \tag{3.2.19}$$

and differentiating 3.2.14 with respect to x gives

$$\frac{\partial F}{\partial x} + \frac{\partial F}{\partial z} p + \frac{\partial F}{\partial p}\frac{\partial p}{\partial x} + \frac{\partial F}{\partial q}\frac{\partial q}{\partial x} = 0 \tag{3.2.20}$$

to yield with 3.2.19 the equation

$$\frac{dp}{dt} = -(F_x + pF_z) \tag{3.2.21}$$

and in a similar fashion

$$\frac{dq}{dt} = -(F_y + qF_z). \tag{3.2.22}$$

Equations 3.2.17, 3.2.18, 3.2.21 and 3.2.22 are the characteristic equations for the non-linear problem 3.2.14.

The general solution, which in the quasilinear case was expressed in terms of a general function of the two determinable forms ξ and η, can be alternatively expressed as a two-parameter family of surfaces known as the *complete solution* or *complete integral*. Hence the two-parameter family

$$G(x, y, z, \alpha, \beta) = 0 \tag{3.2.23}$$

for a known function G can be differentiated with respect to both x and y to yield two equations. By eliminating α and β from these equations, a first-order partial differential equation remains. Hence, for example if

$$G(x, y, z, \alpha, \beta) = z - \alpha x - \beta y = 0 \tag{3.2.24}$$

then differentiating w.r.t. x and y gives

$$\frac{\partial z}{\partial x} - \alpha = 0 \qquad \frac{\partial z}{\partial y} - \beta = 0.$$

Eliminating α and β demonstrates that 3.2.24 is the complete integral for

$$x\frac{\partial z}{\partial x} + y\frac{\partial z}{\partial y} = u.$$

Suppose a complete integral can be found, then for the solution to contain a curve such as

$$x = x_0(s), \qquad y = y_0(s), \qquad z = z_0(s), \tag{3.2.25}$$

it is required that

$$G(x_0(s), y_0(s), z_0(s), \alpha, \beta) = 0 \tag{3.2.26}$$

and therefore that α is a function of β, say $\alpha = \phi(\beta)$. The solution then has the form

$$G(x, y, z, \alpha(x, y, z), \phi(\alpha(x, y, z))) = 0 \tag{3.2.27}$$

which has recovered the general solution with an arbitrary function ϕ, remembering that G is known.

In order to complete the initial conditions for the solution of the characteristic equations which contains the curve 3.2.25, initial conditions for p_0 and q_0 are required. These follow by using $dz = p\,dx + q\,dy$ and the original equation to give the pair of equations:

$$z_0'(s) = p_0(s)x_0'(s) + q_0(s)y_0'(s)$$
$$F(x_0(s), y_0(s), z_0(s), p_0(s), q_0(s)) = 0$$

which can then be solved in principle for p_0 and q_0.

To illustrate this work, consider solving

$$p^2 + q^2 = z$$

for those solutions for which $z = x^2/4 + 1$ when $y = 0$. Then the characteristic equations are

$$\frac{dx}{dt} = 2p, \qquad \frac{dy}{dt} = 2q, \qquad \frac{dz}{dt} = 2p^2 + 2q^2,$$

$$\frac{dp}{dt} = p, \qquad \frac{dq}{dt} = q,$$

and $p = a_1 e^t$ and $q = a_2 e^t$ immediately. Hence $x = 2p + a_3$, $y = 2q + a_4$ and $z = p^2 + q^2 + a_5$ where $(a_i, \quad i = 1, \ldots, 5)$ are the constants of integration. To force $z = x^2/4 + 1$ when $y = 0$ makes $a_3 = 0$, $a_4 = 2$ and $a_5 = 0$ to leave the solution

$$z = \frac{x^2}{4} + \frac{(y-2)^2}{4}.$$

The following exercises may now be considered.

EXERCISES

3.1 Find the general solution of the quasilinear first-order equation

$$\frac{\partial z}{\partial x} + c\frac{\partial z}{\partial y} = 0$$

with c a constant. This is the equation of unidirectional waves.

3.2 Find the general solution of the quasilinear equation

$$[y(x+y) + z]\frac{\partial z}{\partial x} + [x(x+y) - z]\frac{\partial z}{\partial y} = z(x+y).$$

3.3 Find the general solution of the quasilinear problem

$$(y+z)\frac{\partial z}{\partial x} + (z+x)\frac{\partial z}{\partial y} = x + y.$$

3.4 Solve the nonlinear equation

$$\frac{\partial z}{\partial x}\frac{\partial z}{\partial y} = 1$$

with initial condition $z(x, 0) = x$.

3.5 Solve the nonlinear unidirectional wave equation

$$\frac{\partial z}{\partial t} + \left(\frac{\partial z}{\partial x}\right)^2 = 0$$

subject to $z(x, 0) = xd$ where d is a constant.

3.3 Introduction to d'Alembert's Method

The previous chapter has covered the process of separation of the variables for the three standard problems of heat flow, waves and the potential problem of Laplace. This is a convenient place to make a diversion into another method of solution for the wave equation which in fact yields the full general solution in terms of arbitrary functions. Much of this section is then devoted to how the boundary conditions can be utilised to fix these arbitrary functions. In the process, the concept of characteristics will arise again. Consider the wave equation in the form

$$\frac{\partial^2 u}{\partial x^2} - \frac{1}{c^2}\frac{\partial^2 u}{\partial t^2} = 0 \tag{3.3.1}$$

and consider the change of variables

$$\zeta = x - ct \qquad \text{and} \qquad \phi = x + ct. \tag{3.3.2}$$

Then in terms of ζ and η,

$$\begin{aligned}\frac{\partial u}{\partial x} &= \frac{\partial u}{\partial \zeta}\frac{\partial \zeta}{\partial x} + \frac{\partial u}{\partial \phi}\frac{\partial \phi}{\partial x} \\ &= \frac{\partial u}{\partial \zeta} + \frac{\partial u}{\partial \phi}\end{aligned} \tag{3.3.3}$$

and similarly

$$\frac{\partial u}{\partial t} = -c\frac{\partial u}{\partial \zeta} + c\frac{\partial u}{\partial \phi}. \tag{3.3.4}$$

Differentiating again gives:

$$\frac{\partial^2 u}{\partial x^2} = \frac{\partial^2 u}{\partial \zeta^2} + 2\frac{\partial^2 u}{\partial \zeta \partial \phi} + \frac{\partial^2 u}{\partial \phi^2} \tag{3.3.5}$$

and

$$\frac{\partial^2 u}{\partial t^2} = c^2\frac{\partial^2 u}{\partial \zeta^2} - 2c^2\frac{\partial^2 u}{\partial \zeta \partial \phi} + c^2\frac{\partial^2 u}{\partial \phi^2} \tag{3.3.6}$$

and hence 3.3.1 yields only the term

$$4\frac{\partial^2 u}{\partial \zeta \partial \phi} = 0. \tag{3.3.7}$$

In this form, the equation can be integrated in general. The coefficient 4 may be ignored and then integrating w.r.t. ζ gives

$$\frac{\partial u}{\partial \phi} = \hat{f}(\phi) \tag{3.3.8}$$

where $\hat{f}$ is an arbitrary function of ϕ. This step can be confirmed by partially differentiating with respect to ζ to recover 3.3.7. The integration w.r.t. ϕ can now be performed to give:

$$u = f(\phi) + g(\zeta) \tag{3.3.9}$$

where f is the integral of $\hat{f}$, and g is the arbitrary function of integration. Again differentiation w.r.t. ϕ will recover 3.3.8. Hence the general solution to 3.3.1 is

$$u(x,t) = f(x-ct) + g(x+ct) \tag{3.3.10}$$

where f and g are arbitrary functions.

The equations 3.3.2 are called the characteristic equations for the partial differential equation 3.3.1, and because the roots of the associated quadratic are real and distinct the partial differential equation is said to be hyperbolic. In the sister volume to this book on numerical solutions, numerical methods based on the properties of these real characteristics are pursued at length.

In order to apply boundary conditions to the general solution, some initial observations will be made. Firstly, what are the properties of the general function $f(x-ct)$? To answer this question let

$$f(\zeta) = \begin{cases} \sin \pi\zeta & 0 \le \zeta \le 1 \\ 0 & \text{otherwise} \end{cases} \tag{3.3.11}$$

then

$$f(x-ct) = \begin{cases} \sin \pi(x-ct) & 0 \le (x-ct) \le 1 \\ 0 & \text{otherwise} \end{cases}. \tag{3.3.12}$$

What does this function look like?

If $t=0$, $f(x-ct)$ becomes

$$f(x) = \begin{cases} \sin \pi x & 0 \le x \le 1 \\ 0 & \text{otherwise} \end{cases}. \tag{3.3.13}$$

If $t=1$, $f(x-ct)$ becomes

$$f(x-c) = \begin{cases} \sin \pi(x-c) & 0 \le x-c \le 1 \\ 0 & \text{otherwise} \end{cases} \tag{3.3.14}$$

and so on for increasing t.

Putting these curves together in a three-dimensional plot gives the diagram in Figure 3.1.

The graph for $t=0$ is sliding at a constant rate along the x-axis without changing shape, and the rate of sliding is c. This translation is due to the argument $x-ct$. Hence $f(x-ct)$ represents a wave propagating to the right.

Similarly $g(x+ct)$ represents a wave propagation to the left. As the left- and right-propagating features of

$$u(x,t) = f(x-ct) + g(x+ct) \tag{3.3.15}$$

depend solely on the arguments $x+ct$ and $x-ct$ then the general solution of the one-dimensional wave equation represents a combination of left- and right-propagating waves.

To see an example of these waves, consider the simple example

$$u(x,t) = \sin(x-ct) + \sin(x+ct). \tag{3.3.16}$$

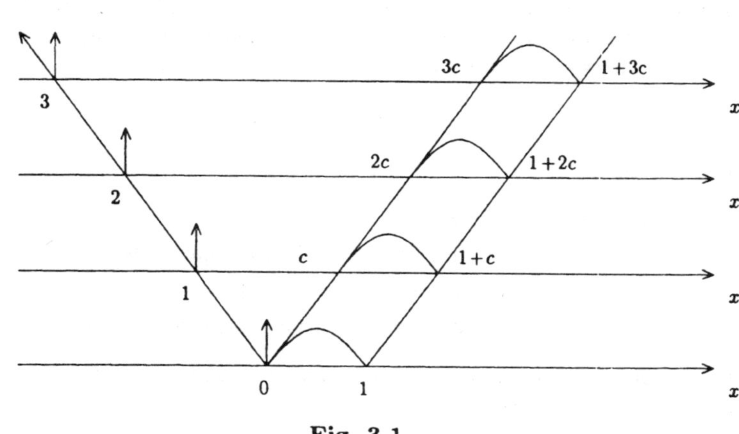

Fig. 3.1.

This is a solution of the wave equation and is the sum of a left-propagating and right-propagating wave. However

$$\sin(x - ct) + \sin(x + ct) = 2 \sin x \cos ct \tag{3.3.17}$$

and drawing the function for $t = 0, \frac{\pi}{3c}, \frac{\pi}{2c}, \frac{2\pi}{3c}, \frac{\pi}{c}, \ldots$ gives the forms in Figure 3.2.

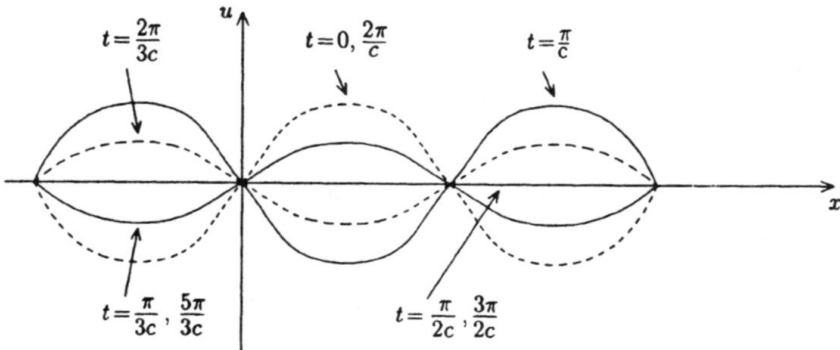

Fig. 3.2.

The shapes of all the graphs are $\sin x$ with a y-scaling by $2 \cos ct$. In particular note that the points satisfying $\sin x = 0$ are always at rest. Such a wave is a standing wave, and $\sin x = 0$ at $x = \pm n\pi, \quad n = 0, 1, 2, \ldots$ which are the nodes for this wave. Thus the wave equation possesses solutions that can be described as (a) left- or right-propagating waves of constant speed, or, (b) standing waves.

Moreover, the combination of certain $f(x - ct)$ and $g(x + ct)$ functions can give rise to solutions which are not solely left-propagating waves, right-propagating waves or standing waves. For example consider

$$u(x,t) = \sin(x - ct) + 2\sin(x + ct) \tag{3.3.18}$$

which can be rewritten as

$$\begin{aligned} u(x,t) &= 2\sin x \cos ct + \sin(x + ct) \\ &\quad \text{standing wave + left-propagating wave} \\ &= 4\sin x \cos ct - \sin(x - ct) \\ &\quad \text{standing wave + right-propagating wave.} \end{aligned}$$

As an example of d'Alembert's method, which is based on the general solution, consider the solution of the equation:

$$\frac{\partial^2 u}{\partial x^2} = \frac{1}{c^2}\frac{\partial^2 u}{\partial t^2} \qquad (-\infty < x < \infty,\ t > 0) \tag{3.3.19}$$

subject to the conditions

$$\left.\begin{aligned} u(x,0) &= 2\sin x & (-\infty < x < \infty) \\ \frac{\partial u}{\partial t}(x,0) &= 0 & (-\infty < x < \infty) \end{aligned}\right\} \text{infinite string.}$$

The general solution is

$$u(x,t) = f(x - ct) + g(x + ct) \tag{3.3.20}$$

and the condition

$$u(x,0) = 2\sin x \tag{3.3.21}$$

implies

$$2\sin x = f(x) + g(x). \tag{3.3.22}$$

The derivative condition leads to

$$\frac{\partial u}{\partial t}(x,t) = -cf'(x - ct) + cg'(x + ct)$$

so that

$$\frac{\partial u}{\partial t}(x,0) = 0$$

implies that

$$0 = -cf'(x) + cg'(x) \qquad \text{or} \qquad f'(x) = g'(x). \tag{3.3.23}$$

Differentiating 3.3.22 gives

$$2\cos x = f'(x) + g'(x) = 2g'(x) \tag{3.3.24}$$

and from 3.3.23

$$g'(x) = \cos(x) \quad \text{and} \quad g(x) = \sin x + A.$$

Equation 3.3.22 then gives

$$f(x) = \sin x - A$$

and the full solution is

$$\begin{aligned} u(x,t) &= f(x-ct) + g(x+ct) \\ &= \sin(x-ct) - A + \sin(x+ct) + A \\ &= \sin(x-ct) + \sin(x+ct). \end{aligned} \tag{3.3.25}$$

This solution is a combination of a left- and right-propagating wave and is in fact a standing wave.

A few exercises on this approach are now presented.

EXERCISES

3.6 Let g be defined by

$$g(\zeta) = \begin{cases} \zeta(\zeta - 1) & 0 \le \zeta \le 1 \\ 0 & \text{otherwise.} \end{cases}$$

Then

$$g(x+ct) = \begin{cases} (x+ct)(x+ct-1) & 0 \le x+ct \le 1 \\ 0 & \text{otherwise.} \end{cases}$$

Plot the graphs of $y = g(x+ct)$ for $t = 0, 1, 2, 3, \ldots$ assuming $c = \frac{1}{2}$. Hence identify $g(x+ct)$ as a left-propagating wave.

3.7 Find the solution of the wave equation

$$\frac{\partial^2 u}{\partial x^2} = \frac{1}{c^2}\frac{\partial^2 u}{\partial t^2} \qquad (-\infty < x < \infty,\, t > 0)$$

subject to

$$\begin{aligned} &u(x,0) = 0 && (-\infty < x < \infty) \\ &\frac{\partial u}{\partial t}(x,0) = \cos x && (-\infty < x < \infty). \end{aligned}$$

3.4 d'Alembert's General Solution

Having seen a worked example of a specific case, it is feasible to write a fairly general solution of the wave equation 3.3.1 subject to the general initial conditions

$$u(x,0) = \phi(x), \qquad \frac{\partial u}{\partial t}(x,0) = q(x) \qquad (-\infty < x < \infty) \tag{3.4.1}$$

where ϕ and q are known functions.

The general solution is as before

$$u(x,t) = f(x - ct) + g(x + ct) \tag{3.4.2}$$

and

$$\frac{\partial u}{\partial t}(x,t) = -cf'(x - ct) + cg'(x + ct). \tag{3.4.3}$$

Hence

$$u(x,0) = \phi(x) \qquad \text{implies} \quad f(x) + g(x) = \phi(x) \tag{3.4.4}$$

and

$$\frac{\partial u}{\partial t}(x,0) = q(x) \quad \text{implies} \qquad -cf'(x) + cg'(x) = q(x). \tag{3.4.5}$$

Differentiating 3.4.4 gives

$$f'(x) + g'(x) = \phi'(x). \tag{3.4.6}$$

Hence 3.4.5 and 3.4.6 give

$$2cg'(x) = q(x) + c\phi'(x) \tag{3.4.7}$$

or

$$g'(x) = \frac{1}{2c}\left[q(x) + c\phi'(x)\right]. \tag{3.4.8}$$

Hence the solution is

$$g'(\zeta) = \frac{1}{2c}\left[q(\zeta) + c\phi'(\zeta)\right] \tag{3.4.9}$$

which integrates to give

$$g(\zeta) = \frac{1}{2c}\left[Q(\zeta) + c\phi(\zeta)\right] + A \tag{3.4.10}$$

where $Q'(\zeta) = q(\zeta)$ and A is an arbitrary constant.

From 3.4.4

$$f(\zeta) = \phi(\zeta) - g(\zeta) = \phi(\zeta) - \frac{1}{2c}\left[Q(\zeta) + c\phi(\zeta)\right] - A \tag{3.4.11}$$

or

$$f(\zeta) = \frac{1}{2c}\left[c\phi(\zeta) - Q(\zeta)\right] - A. \tag{3.4.12}$$

Hence the full solution in terms of the initial functions ϕ and Q is

$$\begin{aligned} u(x,t) &= f(x-ct)+g(x+ct) \\ &= \frac{1}{2}\phi(x-ct)-\frac{1}{2c}Q(x-ct)-A \\ &\quad +\frac{1}{2}\phi(x+ct)+\frac{1}{2c}Q(x+ct)+A \\ &= \frac{1}{2}\left[\phi(x-ct)+\phi(x+ct)\right]+\frac{1}{2c}\left[Q(x+ct)-Q(x-ct)\right]. \quad (3.4.13) \end{aligned}$$

Now, since $Q' = q$, it follows that

$$\begin{aligned} Q(x+ct)-Q(x-ct) &= [Q(\zeta)]_{x-ct}^{x+ct} \\ &= \int_{x-ct}^{x+ct} Q'(\zeta)d\zeta = \int_{x-ct}^{x+ct} q(\zeta)d\zeta. \quad (3.4.14) \end{aligned}$$

Hence

$$u(x,t)=\frac{1}{2}\left[\phi(x-ct)+\phi(x+ct)\right]+\frac{1}{2c}\int_{x-ct}^{x+ct} q(\zeta)d\zeta. \quad (3.4.15)$$

Equation 3.4.15 is known as d'Alembert's solution of the wave equation, although the result as it stands, is of limited use in that the range of x is unbounded. Hence the solution applies to an "infinite" string.

Consider an example of the d'Alembert solution. The motion of an infinite string is modelled by the equation

$$\frac{\partial^2 u}{\partial x^2}=\frac{1}{c^2}\frac{\partial^2 u}{\partial t^2} \quad (3.4.16)$$

with the initial conditions

$$u(x,0)=0, \qquad \frac{\partial u}{\partial t}(x,0)=xe^{-x^2}.$$

Apply d'Alembert's solution to find $u(x,t)$.

d'Alembert's solution is

$$u(x,t)=\frac{1}{2}\left[\phi(x-ct)+\phi(x+ct)\right]+\frac{1}{2c}\int_{x-ct}^{x+ct} q(\zeta)d\zeta \quad (3.4.17)$$

with the functions ϕ and q being:

$$\phi(\zeta)=0, \qquad \text{and} \qquad q(\zeta)=\zeta e^{-\zeta^2} \quad (3.4.18)$$

yielding the general solution

$$\begin{aligned} u(x,t) &= 0 + \frac{1}{2c}\int_{x-ct}^{x+ct} \zeta e^{-\zeta^2}\, d\zeta \\ &= \frac{1}{4c}\left[e^{-(x-ct)^2} - e^{-(x+ct)^2}\right]. \end{aligned} \tag{3.4.19}$$

There are now some further exercises on this technique.

EXERCISES

3.8 Find the solution of the wave equation

$$\frac{\partial^2 u}{\partial x^2} = 4\frac{\partial^2 u}{\partial t^2} \qquad (-\infty < x < \infty,\ t > 0)$$

subject to

$$\begin{aligned} &u(x,0) = e^{-x^2} && (-\infty < x < \infty) \\ &\frac{\partial u}{\partial t}(x,0) = xe^{-x^2} && (-\infty < x < \infty). \end{aligned}$$

Sketch on the same diagram the graphs of $y = u(x,t)$ for $t = 0, 2, 4$.

3.9 Using d'Alembert's solution, solve the wave equation

$$\frac{\partial^2 u}{\partial x^2} = \frac{\partial^2 u}{\partial t^2}$$

with the initial conditions

$$u(x,0) = \sin x, \qquad \frac{\partial u}{\partial t}(x,0) = \cos x \qquad (-\infty < x < \infty).$$

3.5 Characteristics

In the introductory discussion in Chapter 1, the concept of characteristics was introduced. This concept is a very important part of the whole study of partial differential equations, and will play a significant role in the numerical solution of hyperbolic equations in the companion volume. Some of the relevant ideas concerning characteristics can be sensibly introduced here.

The wave equation

$$\frac{\partial^2 u}{\partial x^2} = \frac{1}{c^2}\frac{\partial^2 u}{\partial t^2} \tag{3.5.1}$$

has characteristics

$$\zeta = x - ct \qquad \text{and} \qquad \phi = x + ct. \tag{3.5.2}$$

The more general definition of characteristics was presented in Chapter 1, and now the discussion will be centred on the specific characteristics for 3.5.1. The first observation is that through every point $P(x_0, t_0)$ in the xt-plane pass two characteristics, which are

$$x - ct = x_0 - ct_0 \qquad \text{and} \qquad x + ct = x_0 + ct_0. \tag{3.5.3}$$

These characteristics intersect the x-axis at $A(x_0 - ct_0, 0)$ and $B(x_0 + ct_0, 0)$ respectively and this is shown in Figure 3.3.

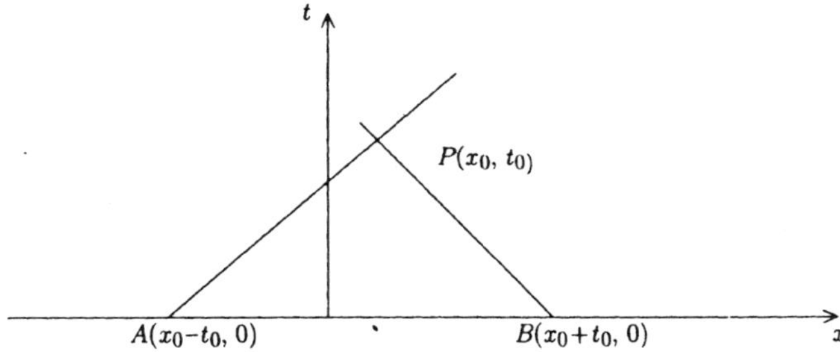

Fig. 3.3.

However, the solution of the wave equation at (x_0, t_0) is

$$u(x_0, t_0) = \frac{1}{2}\left[\phi(x_0 - ct_0) + \phi(x_0 + ct_0)\right] + \frac{1}{2c}\int_{x_0 - ct_0}^{x_0 + ct_0} q(\zeta)d\zeta \tag{3.5.4}$$

and hence the value of u at (x_0, t_0) depends upon:

(a) the mean of the values of $\phi(= u(x, 0))$ at points $A(x_0 - ct_0, 0)$ and $B(x_0 + ct_0, 0)$

(b) the integral of $q\left(= \frac{\partial u}{\partial t}(x, 0)\right)$ along the segment AB.

Segment AB (or that part of the axis enclosed by the characteristics through P) is known as the *interval of dependence* of P for $t = 0$. From this it can be seen that the solution at (x_0, t_0) depends upon $u(x, 0)$ and $\frac{\partial u}{\partial t}(x, 0)$ for only a limited range of x.

For example, suppose it is given that

$$\phi(x) = u(x, 0) = \begin{cases} 1 & -1 \leq x \leq 0 \\ 0 & \text{otherwise} \end{cases} \tag{3.5.5}$$

and

$$g(x) = \frac{\partial u}{\partial t}(x,0) = \begin{cases} 1 & 0 \le x \le 1 \\ 0 & \text{otherwise.} \end{cases} \tag{3.5.6}$$

If in addition $c = 1$, find the interval of dependence of each of the points $R(\frac{1}{2},1), S(-2,\frac{1}{2})$ in the (x,t) plane. Sketch these intervals of dependence and find $u(\frac{1}{2},1)$, $u(-2,\frac{1}{2})$.

The interval of dependence for $R(\frac{1}{2},1)$ is from $(\frac{1}{2},0)$ to $(\frac{3}{2},0)$, whereas, the interval of dependence for $S(-2,\frac{1}{2})$ is from $(-\frac{5}{2},0)$ to $(-\frac{3}{2},0)$. The sketch is shown in Figure 3.4.

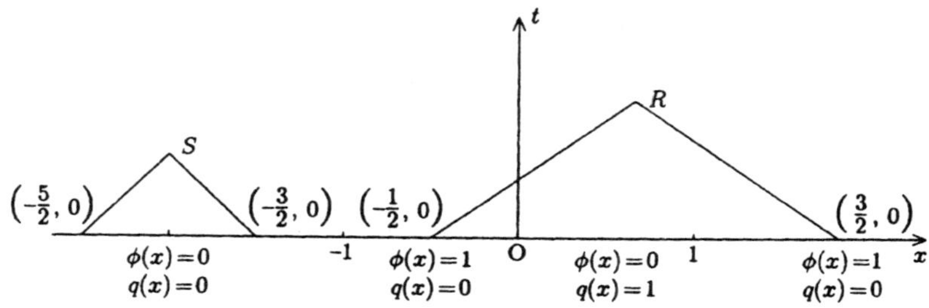

Fig. 3.4.

The particular solution values at $(\frac{1}{2},1)$ and $(-2,\frac{1}{2})$ can then be found from:

$$\begin{aligned} u\left(\frac{1}{2},1\right) &= \frac{1}{2}\left[\phi\left(-\frac{1}{2}\right)+\phi\left(\frac{3}{2}\right)\right] + \frac{1}{2}\int_{-1/2}^{3/2} q(\zeta)\,d\zeta \\ &= \frac{1}{2}[1+0] + \frac{1}{2}\left[\int_{-1/2}^{0} 0\,d\zeta + \int_{0}^{1} 1\,d\zeta + \int_{1}^{3/2} 0\,d\zeta\right] = 1 \end{aligned} \tag{3.5.7}$$

and

$$\begin{aligned} u\left(-2,\frac{1}{2}\right) &= \frac{1}{2}\left[\phi\left(-\frac{5}{2}\right)+\phi\left(-\frac{3}{2}\right)\right] + \frac{1}{2}\int_{-5/2}^{-3/2} q(\zeta)\,d\zeta \\ &= \frac{1}{2}[0+0] + \frac{1}{2}\int_{-5/2}^{-3/2} 0\,d\zeta = 0. \end{aligned} \tag{3.5.8}$$

This method can be applied, in principle at least, to find the solution at every point in the (x,t) plane.

Consider the wave equation

$$\frac{\partial^2 u}{\partial x^2} = \frac{1}{c^2}\frac{\partial^2 u}{\partial t^2} \tag{3.5.9}$$

subject to the initial conditions

$$u(x,0) = \begin{cases} 1 & x \geq 0 \\ 0 & x < 0 \end{cases} \quad \text{and} \quad \frac{\partial u}{\partial t}(x,0) = 0. \tag{3.5.10}$$

Find the solution $u(x,t)$.

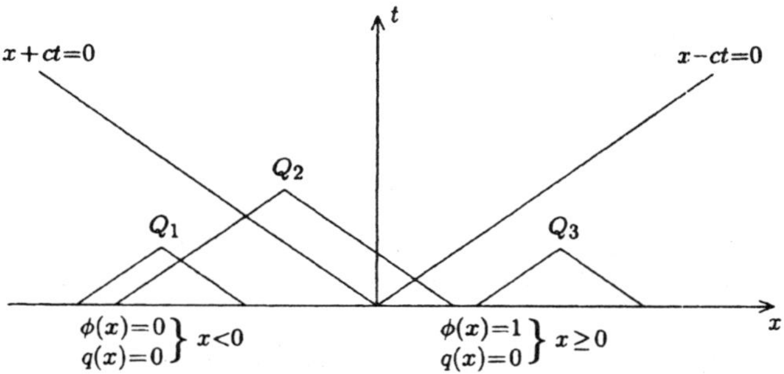

Fig. 3.5.

Figure 3.5 shows the ϕ characteristics through the origin because $x = 0$ is the point at which the rule for $\phi(x) = u(x,0)$ changes. Consider $Q_1(x_1,t_1), Q_2(x_2,t_2)$ and $Q_3(x_3,t_3)$ as typical points, one in each of the three regions into which the (x,t) plane is divided. These are also shown in Figure 3.5.

(1) $Q_1(x_1,t_1)$: The characteristic through Q_1 intersects the negative x-axis where $\phi(x) = u(x,0) = 0$ and

$$u(x_1,t_1) = \frac{1}{2}[0+0] + \frac{1}{2c}\int_{x_1-ct_1}^{x_1+ct_1} 0\, d\zeta = 0.$$

(2) $Q_2(x_2,t_2)$: One characteristic through Q_2 intersects the negative x-axis where $\phi(x) = u(x,0) = 0$, the other intersects the positive x-axis, at a point where $\phi(x) = u(x,0) = 1$. Hence

$$u(x_2,t_2) = \frac{1}{2}[0+1] + \frac{1}{2c}\int_{x_2-ct_2}^{x_2+ct_2} 0\, d\zeta = \frac{1}{2}.$$

(3) $Q_3(x_3, t_3)$: Both characteristics through Q_3 intersect the positive x-axis where $\phi(x) = u(x, 0) = 1$ and so

$$u(x_3, t_3) = \frac{1}{2}[1+1] + \frac{1}{2c}\int_{x_3-ct_3}^{x_3+ct_3} 0\, d\zeta = 1.$$

Since the values of u do not depend on the choice within a region of the specific values of (x_1, t_1), (x_2, t_2) and (x_3, t_3), it follows that within a region $u(x, t)$ is constant, the constant being 0, $\frac{1}{2}$ or 1.

The boundaries of the regions are the x-axis, and the characteristics $x+ct = 0$, and $x - ct = 0$.

(1) On the x-axis ($t = 0$) and

$$u(x, 0) = \begin{cases} 1 & x \geq 0 \\ 0 & x < 0 \end{cases}$$

(2) At all points (except the origin) on the characteristic $x + ct = 0$, we have $u(x, t) = \frac{1}{2}$ since $\phi(0) = 1$ and the value of ϕ at all negative values of x is 0.

(3) At all points (except the origin) on the characteristic $x - ct = 0$ we have $u(x, t) = 1$ since $\phi(0) = 1$ and the value of ϕ at all positive values of x is 1.

The string profiles can be plotted for given values of t, in particular values of $u(x, t_0)$ can be found for $t_0 = 0, \frac{1}{2}, 1, 2$ to give:

(i) $t = 0$

$$u(x, 0) = \begin{cases} 1 & x \geq 0 \\ 0 & x < 0. \end{cases}$$

(ii) $t = \frac{1}{2}$. The line $t = \frac{1}{2}$ cuts $x + ct = 0$ at $x = \frac{-c}{2}$ and $x - ct = 0$ at $x = \frac{c}{2}$ and

$$u\left(x, \frac{1}{2}\right) = \begin{cases} 0 & x < \frac{-c}{2} \\ \frac{1}{2} & -\frac{c}{2} \leq x < \frac{c}{2} \\ 1 & \frac{c}{2} \leq x. \end{cases}$$

(iii) $t = 1$. The line $t = 1$ intersects $x + ct = 0$ and $x - ct = 0$ at $x = -c$ and $x = c$ respectively, and

$$u(x, 1) = \begin{cases} 0 & x < -c \\ \frac{1}{2} & -c \leq x < c \\ 1 & c \leq x. \end{cases}$$

(iv) $t = 2$ gives

$$u(x, 2) = \begin{cases} 0 & x < -2c \\ \frac{1}{2} & -2c \leq x < 2c \\ 1 & 2c \leq x. \end{cases}$$

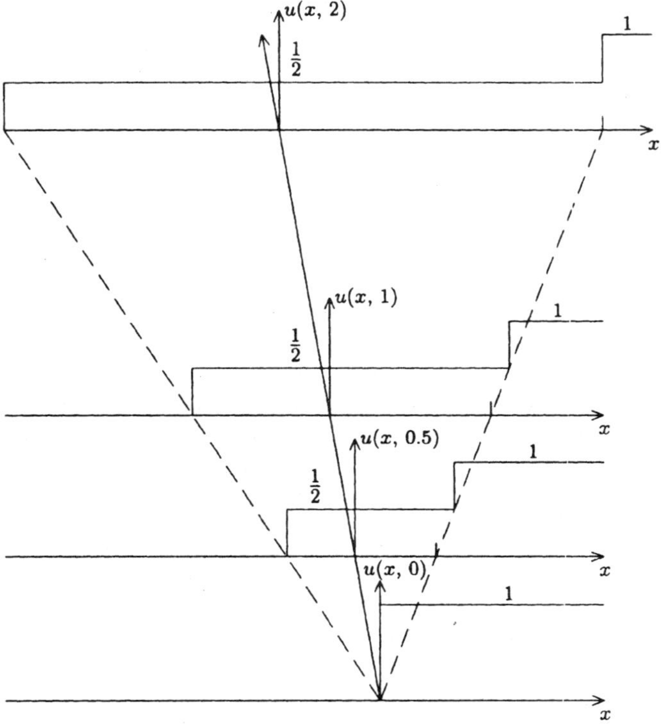

Fig. 3.6.

These results are shown together in Figure 3.6.

It is clear that the initial disturbance (at $t = 0$) is propagated through the system both to the left and right as waves. The wavefronts may be supposed to propagate along the characteristics in the xt-plane, shown in the diagram as dotted lines.

In solving wave problems, one important feature is the propagation of discontinuities. The solution of the wave equation at ϕ depends only on the values of $\phi = u(x,0)$, $q(x) = \frac{\partial u}{\partial t}(x,0)$ on AB (that part of the x-axis cut off by the characteristics through P). This knowledge of $\phi(x)$ and $q(x)$ on AB also determines the solution at any point (x,t) in triangle APB. Hence, changes ϕ and q outside AB do not affect the values of the solution at points inside the triangle APB, as shown in Figure 3.7.

Hence, if ϕ and/or q possess discontinuities at A and/or B such discontinuities will not affect the solutions at points within triangle APB. (If ϕ and q are continuous on AB, then the solution is continuous in APB.) However, such discontinuities in the initial conditions at A and/or B will affect the solution on

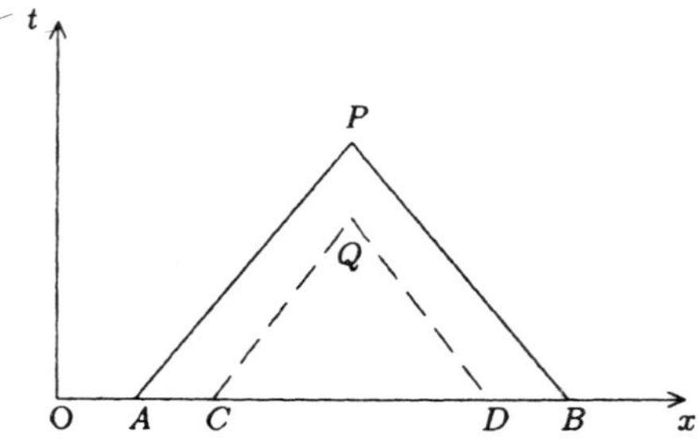

Fig. 3.7.

and outside PAB. In fact, *any discontinuities in ϕ and q or their derivatives will be transmitted along the characteristics in the xt-plane.* This is an important property of characteristics from a physical point of view.

In the last example, a discontinuity in $\phi(x)$ resulted in discontinuities in $u(x,t)$ along the characteristic passing through $(0,0)$, the point of discontinuity of $\phi(x)$. Hence the concept of a classical solution in which $u(x,t)$ is twice differentiable with respect to t and x needs replacing with the concept of a weak generalized solution and this is beyond the scope of the immediate discussion here and may be seen in Zauderer (1983).

As an example to illustrate these effects, solve the wave equation

$$\frac{\partial^2 u}{\partial x^2} = \frac{1}{c^2}\frac{\partial^2 u}{\partial t^2} \tag{3.5.11}$$

subject to the initial conditions

$$u(x,0) = 0 \qquad (-\infty < x < \infty) \tag{3.5.12}$$

and

$$\frac{\partial u}{\partial t}(x,0) = \begin{cases} 1 & x \geq 0 \\ 0 & x < 0. \end{cases} \tag{3.5.13}$$

Sketch the xt-plane and the characteristics passing through the point $(0,0)$. Is $\frac{\partial u}{\partial t}$ continuous across the pair of characteristics drawn?

The situation is shown in Figure 3.8.

The characteristics divide the xt-plane into three regions. Now as $\phi(x) = 0$, d'Alembert's solution reduces to

$$u(x,t) = \frac{1}{2c}\int_{x-ct}^{x+ct} q(\zeta)\,d\zeta. \tag{3.5.14}$$

Consider first the point P: both characteristics through P intersect the negative x-axis and so for $P(x,t)$, $x - ct < 0$ and $x + ct < 0$, and $q(\zeta) = 0$ on $x - ct \leq$

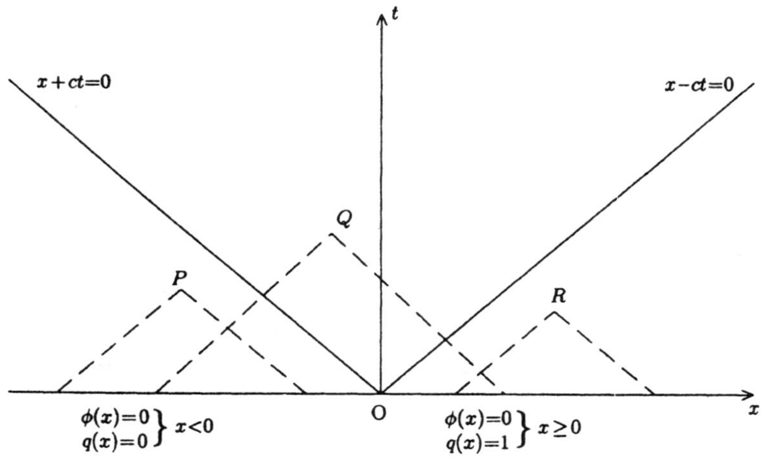

Fig. 3.8.

$\zeta \leq x + ct < 0$. Hence

$$u(x,t) = \frac{1}{2c} \int_{x-ct}^{x+ct} 0 d\zeta = 0. \tag{3.5.15}$$

For point Q, $x - ct < 0$ and $x + ct > 0$ so

$$u(x,t) = \frac{1}{2c} \int_{x-ct}^{x+ct} q(s)\, ds = \frac{1}{2c}(x + ct).$$

Similarly for point R: here $x - ct > 0$ and $x + ct > 0$ so

$$\begin{aligned} u(x,t) &= \frac{1}{2c} \int_{x-ct}^{x+ct} q(\zeta)\, d\zeta = \frac{1}{2c} \int_{x-ct}^{x+ct} 1\, d\zeta \\ &= \frac{1}{2c}[x + ct - (x - ct)] = t. \end{aligned} \tag{3.5.16}$$

In summary:

at P	$u(x,t) = 0,$	$\frac{\partial u}{\partial t} = 0$
at Q	$u(x,t) = \frac{1}{2c}(x + ct),$	$\frac{\partial u}{\partial t} = \frac{1}{2}$
and		
at R	$u(x,t) = t,$	$\frac{\partial u}{\partial t} = 1.$

On the characteristic $x + ct = 0$, we have $u(x,t) = \frac{1}{2c}(x + ct) = 0$ and on the characteristic $x - ct = 0$ we have $u(x,t) = t$. The derivative $\frac{\partial u}{\partial t}$ is not defined at

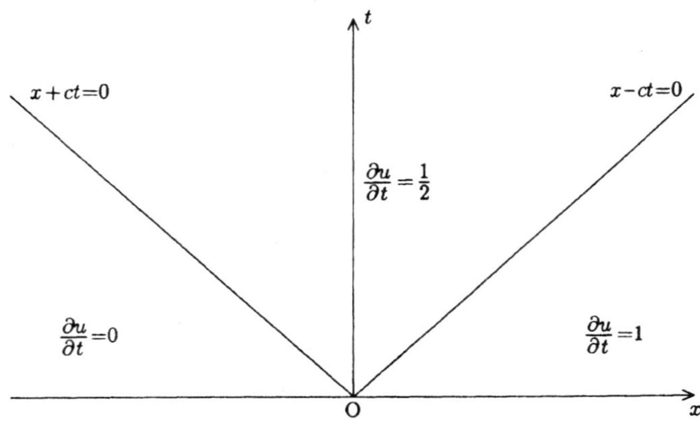

Fig. 3.9.

points on these characteristics, hence $\frac{\partial u}{\partial t}$ is discontinuous there. These features are shown in Figure 3.9.

So far the only problem being considered by d'Alembert's method has been the infinite string. Attention now turns to the semi-infinite string, and again an example with quite general boundary conditions will be used to illustrate the solution.

First there are some further exercises on this section.

EXERCISES

3.10 Consider the wave equation

$$\frac{\partial^2 u}{\partial x^2} = \frac{\partial^2 u}{\partial t^2}.$$

Initially the system is at rest

$$\frac{\partial u}{\partial t}(x,0) = 0$$

and is undisplaced except in the region of the point $x = 2$. The displacement is modelled by

$$u(x,0) = \begin{cases} 0 & x \neq 2 \\ a & x = 2. \end{cases}$$

Draw the (x,t) diagram and the characteristics passing through $x = 2$, $t = 0$. Show that for $t > 0$, $u(x,t) = 0$ or $\frac{a}{2}$. Sketch the wave profile for various values of t on a three-dimensional diagram.

3.11 Solve the wave equation

$$\frac{\partial^2 u}{\partial x^2} = \frac{1}{c^2}\frac{\partial^2 u}{\partial t^2}$$

subject to the initial conditions

$$u(x,0) = \begin{cases} x & x \geq 0 \\ 0 & x < 0 \end{cases}$$

and

$$\frac{\partial u}{\partial t}(x,0) = 0 \qquad (-\infty < x < \infty)$$

and show that $\frac{\partial u}{\partial x}$ is discontinuous over a pair of characteristics.

3.6 Semi-infinite Strings

Consider a string stretching along the positive x-axis with its end at $x = 0$ fixed. Suppose the initial conditions are

$$\begin{aligned} u(x,0) &= \phi_1(x), \qquad & 0 \leq x < \infty \\ \tfrac{\partial u}{\partial t}(x,0) &= q_1(x), \qquad & 0 \leq x < \infty. \end{aligned} \tag{3.6.1}$$

The condition at $x = 0$ imposes a boundary condition of $u(0,t) = 0, \quad 0 \leq t$, and it implies that $\phi_1(0) = q_1(0) = 0$.

d'Alembert's solution gives formally

$$u = \frac{1}{2}\left[\phi_1(x-ct) + \phi_1(x+ct)\right] + \frac{1}{2c}\int_{x-ct}^{x+ct} q_1(\zeta)\,d\zeta \tag{3.6.2}$$

which satisfies the initial conditions. Now for $0 \leq x \leq ct$, $x - ct$ is negative and so this solution cannot be evaluated as $\phi_1(\zeta)$ and $q_1(\zeta)$ are not defined for ζ negative.

The question to be asked is whether this motion of the semi-infinite string may be just the right hand part, corresponding to $x \geq 0$, of some infinite string. The problem would be solved if an infinite string with initial displacement and velocity functions $\phi(x)$ and $q(x)$ can be found so that for $x \geq 0$ they give the initial displacement and velocity functions for the semi-infinite string and also satisfy the condition $u(0,t) = 0$.

We are therefore seeking

$$\begin{aligned} \phi(x) &= \phi_1(x), \qquad & x \geq 0 \\ q(x) &= q_1(x), \qquad & x \geq 0. \end{aligned} \tag{3.6.3}$$

Then

$$u = \frac{1}{2}\left[\phi(x-ct) + \phi(x+ct)\right] + \frac{1}{2c}\int_{x-ct}^{x+ct} q(\zeta)\,d\zeta \tag{3.6.4}$$

will satisfy the wave equation and also the initial conditions. It will also satisfy

$$u(0,t) = 0 \qquad \text{for all} \quad t \geq 0$$

if

$$\frac{1}{2}\left[\phi(-ct) + \phi(+ct)\right] + \frac{1}{2c}\int_{x-ct}^{x+ct} q(\zeta)\,d\zeta = 0 \tag{3.6.5}$$

for all $t \geq 0$.

To achieve this end choose

$$\phi(-ct) = -\phi(ct) = -\phi_1(ct) \qquad \text{for all} \quad ct \geq 0 \tag{3.6.6}$$

and

$$q(-\zeta) = -q(\zeta) = -q_1(\zeta) \qquad \text{for all} \quad \zeta \geq 0. \tag{3.6.7}$$

The functions $\phi(x)$ and $q(x)$ are already defined for $x > 0$, and 3.6.7 and 3.6.7 now define $\phi(x)$ and $q(x)$ for negative x as

$$\begin{aligned}
\phi(-x) &= -\phi_1(x) && \text{for} \quad x > 0 \\
\text{so } \phi(x) &= -\phi_1(-x) && \text{for} \quad x < 0, \\
q(-x) &= -q_1(x) && \text{for} \quad x > 0 \\
\text{so } q(x) &= -q_1(-x) && \text{for} \quad x < 0.
\end{aligned} \tag{3.6.8}$$

Hence the infinite string, whose motion for $x \geq 0$ correctly gives the motion of the semi-infinite string, has to have initial displacement and velocity functions defined by

$$\begin{aligned}
\phi(x) &= \begin{cases} \phi_1(x) & \text{if } x \geq 0 \\ -\phi_1(-x) & \text{if } x < 0, \end{cases} \\
q(x) &= \begin{cases} q_1(x) & \text{if } x \geq 0 \\ -q_1(-x) & \text{if } x < 0. \end{cases}
\end{aligned} \tag{3.6.9}$$

The functions $\phi(x)$ and $q(x)$ are the **odd extensions** of $\phi_1(x)$ and $q_1(x)$.

To consider this approach in practice, take the example:

$$\frac{\partial^2 u}{\partial x^2} = \frac{1}{c^2}\frac{\partial^2 u}{\partial t^2} \tag{3.6.10}$$

for $x \geq 0$, $t \geq 0$ subject to

(i) $u(0,t) = 0$ for $t \geq 0$

(ii) $u(x,0) = \sin^2 x$ for $x \geq 0$

(iii) $\frac{\partial u}{\partial t}(x,0)$ for $x \geq 0$.

This is the problem of a semi-infinite string with $\phi_1(x) = \sin^2 x$ and $q_1(x) = 0$. The initial displacements and velocity functions of the infinite string whose motion for $x \geq 0$ gives the correct motion for this semi-infinite string are

$$\phi(x) = \begin{cases} \phi_1(x) & x \geq 0 \\ -\phi_1(-x) & x < 0 \end{cases}$$

or

$$\phi(x) = \begin{cases} \sin^2 x & x \geq 0 \\ -\sin^2 x & x < 0 \end{cases}$$

and

$$q(x) = \begin{cases} q_1(x) & x \geq 0 \\ -q_1(-x) & x < 0 \end{cases}$$

or $q(x) = 0$ for all x.

From d'Alembert's solution

$$\begin{aligned} u(x,t) &= \frac{1}{2}[\phi(x-ct) + \phi(x+ct)] + \frac{1}{2c}\int_{x-ct}^{x+ct} 0\, d\zeta \\ &= \frac{1}{2}[\phi(x-ct) + \phi(x+ct)] \end{aligned} \tag{3.6.11}$$

and

$$\phi(x-ct) = \sin^2(x-ct) \quad \text{if} \quad x-ct \geq 0 \quad (x \geq ct) \tag{3.6.12}$$

$$\phi(x-ct) = -\sin^2(x-ct) \quad \text{if} \quad x-ct < 0 \quad (x < ct) \tag{3.6.13}$$

$$\phi(x+ct) = \sin^2(x+ct) \quad \text{if} \quad x+ct \geq 0 \quad (x \geq -ct) \tag{3.6.14}$$

$$\phi(x+ct) = -\sin^2(x+ct) \quad \text{if} \quad x+ct < 0 \quad (x < -ct). \tag{3.6.15}$$

Hence for a fixed value of t, the rule for u in terms of x will change at $x = -ct$ and $x = ct$. However, we are only interested in $x \geq 0$ for the semi-infinite string, and so we will get just two rules for u, namely:

(i) for $0 < x < ct$ (x lies between $-ct$ and ct)

$$u = \frac{1}{2}[-\sin^2(x-ct) + \sin^2(x+ct)]$$

(ii) for $x \geq ct$

$$u = \frac{1}{2}[\sin^2(x-ct) + \sin^2(x+ct)].$$

The above solution deals with a semi-infinite string where one end is fixed. If the end at $x = 0$ is free then the free end boundary condition is $\frac{\partial u}{\partial x}(0,t) = 0 \quad (t \geq 0)$. In this case, the appropriate procedure for the solution of the infinite problem is to extend the initial conditions to form *even* functions, and hence find p and q such that

$$\phi(x) = \begin{cases} \phi_1(x), & x \geq 0 \\ \phi_1(-x), & x < 0, \end{cases} \quad q(x) = \begin{cases} q_1(x), & x \geq 0 \\ q_1(-x), & x < 0. \end{cases} \tag{3.6.16}$$

To illustrate this case with an example, consider

$$\frac{\partial^2 u}{\partial x^2} = \frac{\partial^2 u}{\partial t^2} \qquad (0 < x < \infty,\, t > 0) \tag{3.6.17}$$

subject to

$$u(x,0) = e^{-x^2} \qquad (0 < x < \infty)$$

and

$$\frac{\partial u}{\partial t}(x,0) = 0 \qquad (0 < x < \infty)$$

when there is a free end at $x = 0$.

Now e^{-x^2} is defined over $-\infty < x < \infty$ and is even about $x = 0$. Thus, the extension for $u(x,0)$ is

$$\phi(x) = e^{-x^2} \quad (-\infty < x < \infty) \tag{3.6.18}$$

and the extension for $\frac{\partial u}{\partial t}(x,0)$ is

$$q(x) = 0 \quad (-\infty < x < \infty). \tag{3.6.19}$$

Then d'Alembert's solution is

$$u(x,t) = \frac{1}{2}\left[e^{-(x-t)^2} + e^{-(x+t)^2}\right]. \tag{3.6.20}$$

Finally in this section, the case of a finite string with fixed ends is considered. Let the string lie along the x-axis between $x = 0$ and $x = l$ with its ends fixed and set in motion with

(i) $u(x,0) = \phi_1(x), \quad 0 \leq x \leq l$

(ii) $\frac{\partial u}{\partial t}(x,0) = q_1(x), \quad 0 \leq x \leq l$

(iii) $u(0,t) = 0 \quad t \geq 0 \quad (t \geq 0),$ fixed end

(iv) $u(l,t) = 0 \quad t \geq 0 \quad (t \geq 0),$ fixed end.

An infinite string is required whose motion for the portion from $x = 0$ to $x = l$ gives the correct motion of the finite string. To do this, the initial displacements and velocity functions for the infinite string are defined to be odd periodic extensions of period $2l$ of $\phi_1(x)$ and $q_1(x)$. Hence define

$$\begin{aligned} \phi(x) &= \phi_1(x), \qquad 0 < x \le l && (3.6.21)\\ \phi(x) &= -\phi_1(-x), \qquad -l \le x \le 0 && (3.6.22)\\ \phi(x+2l) &= \phi(x) && (3.6.23) \end{aligned}$$

and similarly for $q(x)$.

If the ends are not fixed, other periodic extensions may be used, but usually the problems for finite strings are solved using separation of variables as in Chapter 2. It can be seen from the solutions generated here that the method of d'Alembert is of only limited use and even with ingenuity can only be extended marginally.

The following set of exercises may now be attempted.

EXERCISES

3.12 Solve

$$\frac{\partial^2 u}{\partial x^2} = \frac{1}{c^2}\frac{\partial^2 u}{\partial t^2}$$

for $t > 0, \quad 0 < x < \infty$ with

$$\begin{aligned} u(0,t) &= 0, \qquad (t \ge 0),\\ u(x,0) &= \sin x \qquad (x > 0) \end{aligned}$$

and

$$\frac{\partial u}{\partial t}(x,0) = 0 \qquad (x > 0).$$

3.13 Solve

$$\frac{\partial^2 u}{\partial x^2} = \frac{1}{c^2}\frac{\partial^2 u}{\partial t^2}$$

$(0 < x < \infty, t > 0)$ subject to

$$\begin{aligned} u(x,0) &= 0 \qquad (0 < x < \infty)\\ \frac{\partial u}{\partial t}(x,0) &= xe^{-x^2} \qquad (0 < x < \infty) \end{aligned}$$

and

$$u(0,t) = 0 \qquad (t \ge 0).$$

3.14 Solve

$$\frac{\partial^2 u}{\partial x^2} = \frac{\partial^2 u}{\partial t^2}$$

$(0 < x < \infty,\ t > 0)$ subject to

$$\begin{aligned} u(x,0) &= 0 \qquad (0 < x < \infty) \\ \frac{\partial u}{\partial t}(x,0) &= \sin x \qquad (0 < x < \infty) \end{aligned}$$

and

$$u(0,t) = 0 \qquad (t \geq 0).$$

4
Integral Transforms

4.1 Introduction

The material in this chapter was first presented by Fourier, Cauchy and Poisson at more or less the same time. It is difficult to determine who played the major role, as they presented their work to the Paris Academy of Sciences orally in the first instance. Fourier submitted a paper in 1811 for an Academy prize, and Cauchy submitted "Théorie de la propagation des ondes" similarly in 1816. Poisson was not able to compete for a prize as he was a member of the Academy, but published his work in "Mémoire sur la théorie des ondes".

To introduce this transform method, the method of separation of variables will be reconsidered here, with the heat equation in the form

$$\frac{\partial u}{\partial t} = c^2 \frac{\partial^2 u}{\partial x^2} \tag{4.1.1}$$

being the case in question. Let a bar extend to infinity on both sides and be laterally insulated. In this case, there are no boundary conditions (except that in physical problems u must remain finite), but there is the initial condition

$$u(x, 0) = f(x), \qquad -\infty < x < \infty, \tag{4.1.2}$$

where $f(x)$ is the initial temperature of the bar.

To solve the equation, by separation of the variables set

$$u(x, t) = X(x)T(t) \tag{4.1.3}$$

to give in the usual way

$$X'' = kX, \qquad T' = c^2 kT. \tag{4.1.4}$$

For the solution to remain bounded as $t \to \infty$, take $k = -p^2$ with $p \geq 0$ so that the solution of $T' = c^2 kT$ is not a positive exponential.

Thus

$$X(x) = A\cos px + B\sin px, \qquad T(t) = e^{-c^2p^2t} \tag{4.1.5}$$

where A and B are arbitrary constants. Therefore

$$u(x,t;p) = (A\cos px + B\sin px)e^{-c^2p^2t} \tag{4.1.6}$$

is a solution of 4.1.1. The usual Fourier series expansion cannot be used as p does not necessarily take on only integer multipliers of a fixed number. (Any series formed by taking p as multiples of a fixed number would give a function which is periodic in x at $t = 0$ and we are not told that $f(x)$ is periodic.)

Now A and B are arbitrary, but may depend on p. Hence, assume that $A = A(p)$ and $B = B(p)$. Then as the heat equation is linear and homogeneous

$$u(x,t) = \int_0^\infty u(x,t;p)\,dp = \int_0^\infty (A(p)\cos px + B(p)\sin px)e^{-c^2p^2t}\,dp \tag{4.1.7}$$

is a formal solution of 4.1.1, assuming the existence of the integrals, as $p \geq 0$. To satisfy the initial conditions,

$$u(x,0) = \int_0^\infty (A(p)\cos px + B(p)\sin px)\,dp \tag{4.1.8}$$

which leaves the problem of finding $A(p)$ and $B(p)$. This leads to the use of Fourier integrals.

4.2 Fourier Integrals

Let $f_T(t)$ be a periodic function with period T. Then $f_T(t)$ may be expressed formally in a Fourier trigonometric series in terms of $\cos(2n\pi t/T)$ and $\sin(2n\pi t/T)$ in the form

$$f_T(t) = \frac{a_0}{2} + \sum_{n=1}^{\infty}\left[a_n\cos\left(\frac{2n\pi t}{T}\right) + b_n\sin\left(\frac{2n\pi t}{T}\right)\right] \tag{4.2.1}$$

where

$$a_n = \frac{2}{T}\int_{\frac{-T}{2}}^{\frac{T}{2}} f_T(t)\cos\left(\frac{2n\pi t}{T}\right)dt \tag{4.2.2}$$

and

$$b_n = \frac{2}{T} \int_{\frac{-T}{2}}^{\frac{T}{2}} f_T(t) \sin\left(\frac{2n\pi t}{T}\right) dt. \tag{4.2.3}$$

The convergence of the series in 4.2.1 is considered in Apostol (1974), and results in assumptions on $f_T(t)$. A Fourier representation for the function $f_T(t)$ is now required as the period T is allowed to become infinite.

Substituting for a_n and b_n in 4.2.1 and writing w_n for $\frac{2n\pi}{T}$ gives

$$\begin{aligned} f_T(t) &= \frac{1}{T} \int_{\frac{-T}{2}}^{\frac{T}{2}} f(t)\, dt \\ &+ \frac{2}{T} \sum_{n=1}^{\infty} \left[\cos w_n t \int_{\frac{-T}{2}}^{\frac{T}{2}} f_T(v) \cos w_n v \, dv + \sin w_n t \int_{\frac{-T}{2}}^{\frac{T}{2}} f_T(v) \sin w_n v \, dv \right]. \end{aligned} \tag{4.2.4}$$

Now

$$w_{n+1} - w_n = \frac{2(n+1)\pi}{T} - \frac{2n\pi}{T} = \frac{2\pi}{T} \tag{4.2.5}$$

and define

$$\Delta w = w_{n+1} - w_n = \frac{2\pi}{T} \tag{4.2.6}$$

which implies that

$$\frac{2}{T} = \frac{\Delta w}{\pi} \tag{4.2.7}$$

and the Fourier series may be written as

$$\begin{aligned} f_T(t) &= \frac{1}{T} \int_{\frac{-T}{2}}^{\frac{T}{2}} f(t)\, dt \\ &+ \frac{1}{\pi} \left[\sum_{n=1}^{\infty} \left\{ \cos w_n t \int_{\frac{-T}{2}}^{\frac{T}{2}} f_T(v) \cos w_n v \, dv \Delta w \right.\right. \\ &\left.\left. + \sin w_n t \int_{\frac{-T}{2}}^{\frac{T}{2}} f_T(v) \sin w_n v \, dv \Delta w \right\} \right]. \end{aligned} \tag{4.2.8}$$

This representation is valid for any fixed T, arbitrary large, but finite. Now let $T \to \infty$ and assume that the resulting non-periodic function

$$f(t) = \lim_{T \to \infty} f_T(t) \tag{4.2.9}$$

is absolutely integrable on the t-axis, namely

$$\int_{-\infty}^{\infty} |f(t)|dt \tag{4.2.10}$$

exists; then $\frac{1}{T} \to 0$ and the value of the first term on the right-hand side approaches zero. Furthermore

$$\Delta w = \frac{2\pi}{T} \to 0 \tag{4.2.11}$$

and it seems plausible that the infinite series in 4.2.9 becomes an integral from 0 to ∞. ($w_n = n\Delta w$ tending to the general frequency variable w which is continuous.) Then

$$\begin{aligned} f(t) &= \frac{1}{\pi}\int_0^{\infty}\left[\cos wt \int_{-\infty}^{\infty} f(v)\cos wv\,dv \right. \\ &\qquad \left. + \sin wt \int_{-\infty}^{\infty} f(v)\sin wv\,dv\right] dw. \end{aligned} \tag{4.2.12}$$

Now define

$$A(w) = \frac{1}{\pi}\int_{-\infty}^{\infty} f(v)\cos wv\,dv \tag{4.2.13}$$

and

$$B(w) = \frac{1}{\pi}\int_{-\infty}^{\infty} f(v)\sin wv\,dv \tag{4.2.14}$$

giving

$$f(t) = \int_0^{\infty} [A(w)\cos wt + B(w)\sin wt]\,dw. \tag{4.2.15}$$

This is the Fourier integral representation of $f(t)$.

It is clear that this naive approach suggests the representation but by no means establishes it. The limit of the series as $\Delta w \to 0$ is not the definition of the integral. Sufficient conditions for the validity are as follows.

Theorem 4.1

If $f(t)$ is piecewise continuous in every finite interval and has a right- and left-hand derivative at every point, and the integral

$$\int_{-\infty}^{\infty} |f(t)|\,dt \tag{4.2.16}$$

exists then $f(t)$ can be represented by a Fourier integral. At a point where $f(t)$ is discontinuous the value of the Fourier integral equals the average of the left- and right-hand limits of $f(t)$ at that point. (For a proof see Apostol (1974).)

If $f(t)$ is an even function, namely $f(t) = f(-t)$, then

$$B(w) = 0 \qquad \text{and} \qquad A(w) = \frac{2}{\pi}\int_0^\infty f(u)\cos wu\,du \tag{4.2.17}$$

and

$$f(t) = \int_0^\infty A(w)\cos wt\,dw, \tag{4.2.18}$$

which is the Fourier cosine integral.

Similarly if $f(t)$ is an odd function, namely $f(-t) = -f(t)$, then

$$A(w) = 0 \quad \text{and} \quad B(w) = \frac{2}{\pi}\int_0^\infty f(u)\sin wu\,du \tag{4.2.19}$$

and now

$$f(t) = \int_0^\infty B(w)\sin wt\,dw \tag{4.2.20}$$

which is the Fourier sine integral.

As a worked example, find the Fourier integral representation of

$$f(t) = e^{-kt} \tag{4.2.21}$$

when $t > 0$ and $f(-t) = f(t)$, $k > 0$, and hence we find

$$\int_0^\infty \frac{\cos wt}{k^2 + w^2}\,dw. \tag{4.2.22}$$

Since $f(t)$ is even, $B(w) = 0$ and

$$\begin{aligned} A(w) &= \frac{2}{\pi}\int_0^\infty f(t)\cos wt\,dt = Re\left[\frac{2}{\pi}\int_0^\infty e^{-kt}e^{iwt}\,dt\right] \\ &= \frac{2k}{\pi(k^2 + w^2)} \end{aligned} \tag{4.2.23}$$

to give

$$f(t) = e^{-kt} = \frac{2k}{\pi}\int_0^\infty \frac{\cos wt}{k^2 + w^2}\,dw \tag{4.2.24}$$

for $t \geq 0$ and $k > 0$. Hence

$$\int_0^\infty \frac{\cos wt}{k^2 + w^2}\, dw = \frac{\pi}{2k} e^{-kt}. \tag{4.2.25}$$

As the function is continuous at $t = 0$ the integral representation holds for $t = 0$.

If we had required the Fourier integral of

$$f(t) = e^{-kt}, \qquad t > 0 \qquad \text{and} \qquad f(-t) = -f(t) \tag{4.2.26}$$

then

$$A(w) = 0, \quad \text{and} \qquad B(w) = \frac{2w}{\pi(k^2 + w^2)} \tag{4.2.27}$$

which gives

$$e^{-kt} = \frac{2}{\pi} \int_0^\infty \frac{w \sin wt}{k^2 + w^2}\, dw \tag{4.2.28}$$

for $t > 0$ and $k > 0$. The integral on the right-hand side of 4.2.28 does not exist for $t = 0$, as it is divergent.

Some exercises are now presented to illustrate the work of the previous section.

EXERCISES

Using the Fourier integral representation show that:

4.1

$$\int_0^\infty \frac{\cos xw + w \sin xw}{1 + w^2}\, dw = \begin{cases} 0 & x < 0 \\ \frac{\pi}{2} & x = 0 \\ \pi e^{-x} & x > 0. \end{cases}$$

4.2

$$\int_0^\infty \frac{w^3 \sin xw}{w^4 + 4}\, dw = \frac{\pi}{2} e^{-x} \cos x$$

when $x > 0$. (Use $f(x)$ odd.)

4.3

$$\int_0^\infty \frac{\sin \pi w \sin xw}{1 - w^2}\, dw = \begin{cases} \frac{\pi}{2} \sin x & 0 \leq x \leq \pi \\ 0 & x > \pi. \end{cases}$$

(Use $f(x)$ odd.)

4.4

$$\int_0^\infty \frac{1-\cos \pi w}{w} \sin xw\, dw = \begin{cases} \frac{\pi}{2} & 0 \le x \le \pi \\ 0 & x > \pi. \end{cases}$$

(Use $f(x)$ odd.)

4.5

$$\int_0^\infty \frac{\cos xw}{1+w^2}\, dw = \frac{\pi}{2} e^{-x} \qquad x > 0.$$

(Use $f(x)$ even.)

4.6

$$\int_0^\infty \frac{\sin w \cos xw}{w}\, dw = \begin{cases} \frac{\pi}{2} & 0 \le x \le 1 \\ \frac{\pi}{4} & x = 1 \\ 0 & x > 1. \end{cases}$$

(Use $f(x)$ even.)

4.7

$$\int_0^\infty \frac{\cos \frac{\pi w}{2} \cos xw}{1-w^2}\, dw = \begin{cases} \frac{\pi}{2}\cos x & |x| < \frac{\pi}{2} \\ 0 & |x| > \frac{\pi}{2}. \end{cases}$$

(Use $f(x)$ even.)

Represent the following functions $f(x)$ in the form

$$f(x) = \frac{1}{\pi} \int_0^\infty A(w) \cos wx\, dw$$

4.8

$$f(x) = \begin{cases} 1 & 0 < x < a \\ 0 & x > a. \end{cases}$$

4.9

$$f(x) = \begin{cases} x^2 & 0 < x < a \\ 0 & x > a. \end{cases}$$

4.10 If $f(x)$ has the representation

$$f(x) = \frac{1}{\pi} \int_0^\infty A(w) \cos wx\, dx$$

show that

$$f(ax) = \frac{1}{\pi a} \int_0^\infty A\left(\frac{w}{a}\right) \cos xw\, dw, \qquad a > 0.$$

4.11 Show that $f(x) = 1$ $(0 < x < \infty)$ cannot be represented by a Fourier integral.

4.3 Application to the Heat Equation

Returning to equation 4.1.7 gives:

$$u(x,t) = \int_0^\infty [A(p)\cos px + B(p)\sin px]e^{-c^2p^2t}\,dp \tag{4.3.1}$$

giving

$$u(x,0) = f(x) = \int_0^\infty [A(p)\cos px + B(p)\sin px]\,dp. \tag{4.3.2}$$

From the Fourier integral results of the previous section,

$$A(p) = \frac{1}{\pi}\int_{-\infty}^{\infty} f(v)\cos pv\,dv \tag{4.3.3}$$

and

$$B(p) = \frac{1}{\pi}\int_{-\infty}^{\infty} f(v)\sin pv\,dv \tag{4.3.4}$$

which leads to

$$\begin{aligned} u(x,t) &= \frac{1}{\pi}\int_0^\infty \left\{\left[\int_{-\infty}^{\infty} f(v)\cos pv\,dv\right]\cos px \right.\\ &\quad \left. + \left[\int_{-\infty}^{\infty} f(v)\sin pv\,dv\right]\sin px\right\} e^{-c^2p^2t}dp \\ &= \frac{1}{\pi}\int_0^\infty \left[\int_{-\infty}^{\infty} f(v)(\cos pv\cos px + \sin pv\sin px)dv\right] e^{-c^2p^2t}dp \\ &= \frac{1}{\pi}\int_0^\infty \int_{-\infty}^{\infty} f(v)\cos(p(v-x))dve^{-c^2p^2t}dp. \end{aligned} \tag{4.3.5}$$

Assuming the order of integration may be changed,

$$u(x,t) = \frac{1}{\pi}\int_{-\infty}^{\infty} f(v)\int_0^\infty \cos p(x-v)e^{-c^2p^2t}\,dp\,dv. \tag{4.3.6}$$

Now

$$\int_0^\infty e^{-s^2} \cos 2bs\, ds = \frac{\sqrt{\pi}}{2} e^{-b^2}, \tag{4.3.7}$$

then by setting

$$s = cp\sqrt{t} \tag{4.3.8}$$

and choosing

$$b = \frac{x - v}{2c\sqrt{t}} \tag{4.3.9}$$

gives

$$\int_0^\infty e^{-c^2p^2t} \cos p(x - v) dp = \frac{\sqrt{\pi}}{2} \frac{1}{c\sqrt{t}} e^{\frac{-(x-v)^2}{4c^2t}}. \tag{4.3.10}$$

Hence

$$u(x,t) = \frac{1}{2c\sqrt{\pi t}} \int_{-\infty}^{\infty} f(v) \exp\left[\frac{-(x - v)^2}{4c^2t}\right] dv. \tag{4.3.11}$$

Putting

$$w = \frac{v - x}{2c\sqrt{t}} \tag{4.3.12}$$

yields the result

$$u(x,t) = \frac{1}{\sqrt{\pi}} \int_{-\infty}^{\infty} f(x + 2cw\sqrt{t}) e^{-w^2} dw. \tag{4.3.13}$$

Some further exercises are now presented.

EXERCISES

4.12 Show that the solution of Laplace's equation $\nabla^2 u = 0$ in the region $0 < y < a, -\infty < x < \infty$ satisfying

$$u(x,0) = f(x) \qquad \text{and} \qquad u(x,a) = 0$$

(where $f(x)$ is a given function and a is a constant) is

$$u(x,y) = \frac{1}{\pi} \int_0^\infty \left[\int_{-\infty}^{\infty} \frac{\sinh[\lambda(a - y)]}{\sinh(\lambda a)} f(\zeta) \cos \lambda(\zeta - x) d\zeta \right] d\lambda.$$

4.13 Show that the bounded solution $u(x,y)$ of Laplace's equation $\nabla^2 u = 0$ in the quadrant $x > 0, y > 0$ which satisfies $u(0,y) = 0$ (for $y > 0$) and $u(x,0) = f(x)$ (for $x > 0$) is

$$u(x,y) = \frac{2}{\pi}\int_0^\infty\int_0^\infty e^{-\lambda y} f(\zeta)\sin\lambda x \sin\lambda\zeta\, d\zeta\, d\lambda.$$

By integrating with respect to λ transform this solution into

$$u(x,y) = \frac{y}{\pi}\int_0^\infty f(\zeta)\left[\frac{1}{y^2+(\zeta-x)^2} - \frac{1}{y^2+(\zeta+x)^2}\right] d\zeta.$$

4.14 Show that the solution $u(r,t)$ of the heat equation

$$\frac{1}{r}\frac{\partial}{\partial r}\left(r\frac{\partial u}{\partial r}\right) = \frac{1}{k}\frac{\partial u}{\partial t}$$

$(0 < r < \infty, t > 0)$ which remains finite at $r = 0$ and for which $u(r,0) = f(r)$ (where $f(r)$ is assumed known) may be written as

$$u(r,t) = \int_0^\infty A(\lambda) e^{-\lambda^2 kt} J_0(\lambda r)\, d\lambda$$

where

$$f(r) = \int_0^\infty A(\lambda) J_0(\lambda r)\, d\lambda.$$

4.15 Show that the solution $u(r,t)$ of

$$\frac{\partial^2 u}{\partial x^2} = \frac{1}{k}\frac{\partial u}{\partial t}$$

subject to the boundary conditions

(i) $u(0,t) = 0, \quad t > 0$
(ii) $u(x,0) = f(x), \quad x \geq 0$
(iii) $u(x,t)$ is bounded in the interval $0 < x < \infty$

is

$$u(x,t) = \frac{2}{\pi}\int_0^\infty e^{-\lambda^2 kt}\sin\lambda x\left[\int_0^\infty f(\zeta)\sin\lambda\zeta\, d\zeta\right] d\lambda.$$

4.16 Show that a bounded solution of Laplace's equation in the semi-infinite region $x \geq 0, 0 \leq y \leq 1$ subject to the boundary conditions

$$\left(\frac{\partial u}{\partial x}\right)_{x=0} = 0, \qquad \left(\frac{\partial u}{\partial y}\right)_{y=0} = 0$$

and $u(x,1) = f(x)$ is

$$u(x,y) = \frac{2}{\pi}\int_0^\infty \frac{\cos\lambda x \cosh\lambda y}{\cosh\lambda}\left[\int_0^\infty f(\zeta)\cos\lambda\zeta d\zeta\right] d\lambda.$$

If

$$f(x) = \begin{cases} 1 & 0 \leq x \leq 1 \\ 0 & x > 1 \end{cases}$$

show that

$$u(x,y) = \frac{2}{\pi}\int_0^\infty \frac{\cos\lambda x \cosh\lambda y \sin\lambda}{\lambda\cosh\lambda} d\lambda.$$

4.4 Fourier Sine and Cosine Transforms

The work of the previous sections motivates the following approach. Suppose that $u(x,t)$ is a function defined for $0 < x < \infty, t > 0$. Then the Fourier sine transform of $u(x,t)$ with respect to x is defined by

$$\mathcal{F}_s(u(x,t)) = \mathcal{U}_s(p,t) = \int_0^\infty u(x,t)\sin px\, dx \tag{4.4.1}$$

with inverse transform

$$\mathcal{F}_s^{-1}(\mathcal{U}_s(p,t)) = u(x,t) = \frac{2}{\pi}\int_0^\infty \mathcal{U}_s(p,t)\sin px\, dp. \tag{4.4.2}$$

Similarly the Fourier cosine transform of $u(x,t)$ with respect to x is

$$\mathcal{F}_c(u(x,t)) = \mathcal{U}_c(p,t) = \int_0^\infty u(x,t)\cos px\, dx \tag{4.4.3}$$

with inverse transform

$$\mathcal{F}_c^{-1}(\mathcal{U}_c(p,t)) = u(x,t) = \frac{2}{\pi}\int_0^\infty \mathcal{U}_c(p,t)\cos px\, dp. \tag{4.4.4}$$

The sine transform of $\frac{\partial u}{\partial x}$ w.r.t. x is given by

$$\begin{aligned}\mathcal{F}_s\left(\frac{\partial u}{\partial x}\right) &= \int_0^\infty \frac{\partial u}{\partial x}\sin px\,dx \\ &= [u\sin px]_0^\infty - p\int_0^\infty u(x,t)\cos p\,dx \qquad (4.4.5)\end{aligned}$$

and provided $u(x,t)\to 0$ as $x\to\infty$ (which is often the case in physical problems)

$$\mathcal{F}_s\left(\frac{\partial u}{\partial x}\right) = -p\mathcal{U}_c(p,t). \qquad (4.4.6)$$

Similarly the cosine transform of $\frac{\partial u}{\partial x}$ with respect to x is

$$\begin{aligned}\mathcal{F}_c\left(\frac{\partial u}{\partial x}\right) &= \int_0^\infty \frac{\partial u}{\partial x}\cos px\,dx \\ &= [u\cos px]_0^\infty + p\int_0^\infty u(x,t)\sin px\,dx \\ &= p\mathcal{U}_s(p,t) - u(0,t), \qquad (4.4.7)\end{aligned}$$

again provided $u(x,t)\to 0$ as $x\to\infty$.

The transforms of $\frac{\partial^2 u}{\partial x^2}$ are

$$\begin{aligned}\mathcal{F}_s\left(\frac{\partial^2 u}{\partial x^2}\right) &= -p\mathcal{F}_c\left(\frac{\partial u}{\partial x}\right) \\ &= -p[p\mathcal{U}_s(p,t) - u(0,t)] = -p^2\mathcal{U}_s(p,t) + pu(0,t) \quad (4.4.8)\end{aligned}$$

and

$$\begin{aligned}\mathcal{F}_c\left(\frac{\partial^2 u}{\partial x^2}\right) &= p\mathcal{F}_s\left(\frac{\partial u}{\partial x}\right) - \frac{\partial u}{\partial x}(0,t) \\ &= -p^2\mathcal{U}_c(p,t) - \left.\frac{\partial u}{\partial x}\right|_{x=0} \qquad (4.4.9)\end{aligned}$$

where, in addition to assuming that $u(x,t)\to 0$ as $x\to\infty$, it is also assumed that $\frac{\partial u}{\partial x}\to 0$.

The choice between using a sine transform or a cosine transform for the solution of a second-order partial differential equation will depend on the boundary conditions, since the sine transform of $\frac{\partial^2 u}{\partial x^2}$ requires a knowledge of $u(0,t)$, whilst the cosine transform of $\frac{\partial^2 u}{\partial x^2}$ requires $\frac{\partial u}{\partial x}$ to be given at $x=0$. These transforms are used when $0<x<\infty$ is the required range for x.

The transforms of $\frac{\partial^2 u}{\partial t^2}$ and $\frac{\partial u}{\partial t}$ are (w.r.t. x)

$$\mathcal{F}_s\left(\frac{\partial u}{\partial t}\right) = \int_0^\infty \frac{\partial u}{\partial t}\sin px\,dx$$

$$= \frac{d}{dt}\int_0^\infty u\sin px\,dx = \frac{d}{dt}\mathcal{U}_s. \tag{4.4.10}$$

Hence

$$\mathcal{F}_s\left(\frac{\partial^2 u}{\partial t^2}\right) = \frac{d^2}{dt^2}\mathcal{U}_s \tag{4.4.11}$$

$$\mathcal{F}_c\left(\frac{\partial u}{\partial t}\right) = \frac{d}{dt}\mathcal{U}_s \tag{4.4.12}$$

and

$$\mathcal{F}_c\left(\frac{\partial^2 u}{\partial t^2}\right) = \frac{d^2}{dt^2}\mathcal{U}_s. \tag{4.4.13}$$

To see how this works in practice consider the example of the heat equation

$$\frac{\partial^2 u}{\partial x^2} = \frac{1}{k}\frac{\partial u}{\partial t}, \tag{4.4.14}$$

given

$$u(0,t) = u_0 = \text{const}, \qquad \text{and} \qquad u(x,0) = 0, \qquad 0 < x < \infty. \tag{4.4.15}$$

Since $u(0,t)$ is given, the sine transform is appropriate to give

$$\frac{d\mathcal{U}_s}{dt} + kp^2\mathcal{U}_s = pku_0. \tag{4.4.16}$$

The integrating factor is

$$e^{\int kp^2 dt} = e^{kp^2 t} \tag{4.4.17}$$

which gives

$$\mathcal{U}_s e^{kp^2 t} = \int pku_0 e^{kp^2 t}dt = \frac{u_0}{p}e^{kp^2 t} + A. \tag{4.4.18}$$

Now $u(x,0) = 0$ which implies that

$$\mathcal{F}_s(u(x,0)) = \mathcal{U}_s(p,0) = \int_0^\infty 0\sin px\,dx = 0 \tag{4.4.19}$$

which gives $A = -u_0/p$ and therefore

$$\mathcal{U}_s = \frac{u_0}{p}\left(1 - e^{-kp^2 t}\right). \tag{4.4.20}$$

The inversion for the sine transform gives

$$\begin{aligned} u(x,t) &= \frac{2}{\pi}\int_0^\infty \frac{u_0}{p}\left(1-e^{-kp^2t}\right)\sin px\,dp \\ &= \frac{2u_0}{\pi}\left[\int_0^\infty \frac{\sin px}{p}\,dp - \int_0^\infty e^{-pk^2t}\sin px\,dp\right]. \end{aligned} \tag{4.4.21}$$

Using the result

$$\int_0^\infty \frac{\sin px}{p}\,dp = \frac{\pi}{2}, \quad x>0, \tag{4.4.22}$$

gives

$$u(x,t) = u_0\left(1-\frac{2}{\pi}\int_0^\infty e^{-kp^2t}\frac{\sin px}{p}\,dp\right). \tag{4.4.23}$$

The exercises for this section are closely related to those for the next and will appear at that stage.

4.5 General Fourier Transforms

The sine and cosine transformations can be combined to form the general Fourier transform. Consider

$$\begin{aligned} f(t) &= \frac{1}{\pi}\int_0^\infty \left[\cos wt \int_{-\infty}^\infty f(v)\cos wv\,dv \right. \\ &\qquad \left. + \sin wt \int_{-\infty}^\infty f(v)\sin wv\,dv\right] dw \\ &= \frac{1}{\pi}\int_0^\infty\int_{-\infty}^\infty f(v)[\cos wt\cos wv + \sin wt\sin wv]\,dv\,dw \\ &= \frac{1}{\pi}\int_0^\infty\int_{-\infty}^\infty f(v)\cos w(v-t)\,dv\,dw. \end{aligned} \tag{4.5.1}$$

Now $\cos w(v-t)$ is an even function of w and so we can write

$$f(t) = \frac{1}{2\pi}\int_{-\infty}^\infty\int_{-\infty}^\infty f(v)\cos w(v-t)\,dv\,dw. \tag{4.5.2}$$

Also as $\sin w(v-t)$ is an odd function of w

$$0 = \frac{i}{2\pi}\int_{-\infty}^{\infty}\int_{-\infty}^{\infty} f(v)\sin w(v-t)\,dv\,dw. \tag{4.5.3}$$

Adding these equations gives

$$\begin{aligned} f(t) &= \frac{1}{2\pi}\int_{-\infty}^{\infty}\int_{-\infty}^{\infty} f(v)[\cos w(v-t) + i\sin w(v-t)]\,dv\,dw \\ &= \frac{1}{2\pi}\int_{-\infty}^{\infty}\int_{-\infty}^{\infty} f(v)e^{i(v-t)w}\,dv\,dw \\ &= \frac{1}{2\pi}\int_{-\infty}^{\infty} e^{-itw}\left[\int_{-\infty}^{\infty} \cos f(v)e^{ivw}\,dv\right] dw. \end{aligned} \tag{4.5.4}$$

Let

$$F(w) = \int_{-\infty}^{\infty} f(v)e^{iwv}dv \tag{4.5.5}$$

then

$$f(t) = \frac{1}{2\pi}\int_{-\infty}^{\infty} F(w)e^{-itw}dw. \tag{4.5.6}$$

The Fourier transform of $f(t)$ is

$$\mathcal{F}(f) = F(w) = \int_{-\infty}^{\infty} f(t)e^{iwt}dt \tag{4.5.7}$$

and the inverse Fourier transform is

$$f(t) = \mathcal{F}^{-1}(f) = \frac{1}{2\pi}\int_{-\infty}^{\infty} F(w)e^{-iwt}dw. \tag{4.5.8}$$

The use of this general Fourier transform is for ranges of the variable, w.r.t. which the transform is taken, from $-\infty$ to ∞.

Consider now how this transform affects derivatives. Let x be the variable w.r.t. which the transform is taken. Then $\frac{d^n f}{dx^n}$ has Fourier transform

$$(-iw)^n F(w) \qquad \text{for} \quad n = 1, 2, \ldots \tag{4.5.9}$$

provided that

$$\frac{d^m f}{dx^m} \to 0 \qquad \text{as} \qquad x \to \pm\infty \quad \text{for} \quad m = 0, 1, 2, \ldots, (n-1),$$

as

$$\mathcal{F}\left(\frac{d^n f}{dx^n}\right) = \int_{-\infty}^{\infty} e^{iwx}\frac{d^n f}{dx^n}\,dx$$
$$= (-iw)\mathcal{F}\left(\frac{d^{n-1} f}{dx^{n-1}}\right) \tag{4.5.10}$$

by integration by parts and the result follows.

One of the many important formulae which is used in this field is given in the convolution theorem which is stated below in 4.5.14. The convolution $f * g$ of two functions f and g is defined by

$$(f * g)(x) = \int_{-\infty}^{\infty} f(u)g(x-u)\,du$$
$$= \int_{-\infty}^{\infty} f(x-u)g(u)\,du. \tag{4.5.11}$$

Now

$$\mathcal{F}(f * g) = \int_{-\infty}^{\infty} e^{iwx} \int_{-\infty}^{\infty} f(u)g(x-u)\,dudx$$
$$= \int_{-\infty}^{\infty} f(u) \int_{-\infty}^{\infty} g(x-u)e^{iwx}\,dxdu. \tag{4.5.12}$$

Let $v = x - u$ and $t = u$ or $x = t + v$ and $u = t$ then

$$\frac{\partial(x,u)}{\partial(t,v)} = \begin{vmatrix} 1 & 1 \\ 1 & 0 \end{vmatrix} = -1. \tag{4.5.13}$$

After applying this change of variables in 4.5.12, we deduce the convolution theorem which states that

$$\mathcal{F}(f * g) = \int_{-\infty}^{\infty} f(t) \int_{-\infty}^{\infty} g(v)e^{iw(t+v)}\,dvdt$$
$$= \int_{-\infty}^{\infty} f(t)e^{iwt}dt \int_{-\infty}^{\infty} g(v)e^{iwv}dv$$
$$= \mathcal{F}(f)\mathcal{F}(g) \tag{4.5.14}$$

and

$$\mathcal{F}^{-1}(\mathcal{F}(f)\mathcal{F}(g)) = f * g = \int_{-\infty}^{\infty} f(u)g(x-u)\,du. \tag{4.5.15}$$

The Fourier transform is now used in a worked example in which Laplace's equation in two dimensions is solved. Hence consider

$$\frac{\partial^2 u}{\partial x^2}+\frac{\partial^2 u}{\partial y^2}=0 \tag{4.5.16}$$

in the half plane $y \geq 0$ subject to the boundary condition $u(x,0)=f(x)$ $(-\infty < x < \infty)$ and the condition $u(x,t) \to 0$ as $\sqrt{x^2+y^2} \to \infty$.

Using a Fourier transform with respect to x,

$$\mathcal{F}(u(x,y))=\mathcal{U}(w,y)=\int_{-\infty}^{\infty} u(x,y)e^{iwx}dx \tag{4.5.17}$$

and

$$\mathcal{F}\left(\frac{\partial^2 u}{\partial y^2}\right)=\frac{d^2\mathcal{U}}{dy^2}, \qquad \mathcal{F}\left(\frac{\partial^2 u}{\partial x^2}\right)=(-iw)^2\mathcal{U} \tag{4.5.18}$$

which implies

$$\frac{d^2\mathcal{U}}{dy^2}-w^2\mathcal{U}=0 \tag{4.5.19}$$

with the solution

$$\mathcal{U}=Ae^{wy}+Be^{-wy}. \tag{4.5.20}$$

The boundary conditions give

$$\mathcal{U}(w,0)=\mathcal{F}(f)=F(w) \tag{4.5.21}$$

and

$$\mathcal{U}(w,y)\to 0 \qquad \text{as} \qquad y\to\infty. \tag{4.5.22}$$

If

$$\left.\begin{array}{lll} w>0, & \text{we must have} & A=0, \quad B=F(w) \\ w<0, & \text{we must have} & B=0, \quad A=F(w) \end{array}\right\}$$

which gives

$$\mathcal{U}(x,y)=F(w)e^{-|w|y}. \tag{4.5.23}$$

Now

$$\mathcal{F}^{-1}\left(e^{-|w|y}\right)=\frac{1}{\pi}\frac{y}{x^2+y^2} \tag{4.5.24}$$

so the Convolution Theorem yields

$$\begin{aligned} u(x,y) &= \int_{-\infty}^{\infty} f(x-u)\frac{1}{\pi}\frac{y}{u^2+y^2}\,du \\ &= \frac{y}{\pi}\int_{-\infty}^{\infty}\frac{f(x-u)}{u^2+y^2}\,du \\ &= \frac{y}{\pi}\int_{-\infty}^{\infty}\frac{f(u)}{(u-x)^2+y^2}\,du. \end{aligned} \tag{4.5.25}$$

A further set of exercises is presented at this point.

EXERCISES

4.17 Solve

$$\frac{\partial^2 u}{\partial x^2} = \frac{1}{k}\frac{\partial u}{\partial t}$$

given

$$\left(\frac{\partial u}{\partial x}\right)_{x=0} = -v \quad \text{(const)}$$

$u(x,0) = 0$ for $0 < x < \infty$, $t > 0$. (Use Fourier cosine transforms.)

4.18 Using Fourier sine transforms with respect to x, show that the solution to

$$\frac{\partial u}{\partial t} = \frac{\partial^2 u}{\partial x^2} \qquad x > 0, t > 0$$

subject to the conditions

$$u(0,t) = 0, \qquad u(x,0) = \begin{cases} 1 & 0 < x < 1 \\ 0 & 1 \le x \end{cases}$$

with $u(x,t)$ bounded, is

$$u(x,t) = \frac{2}{\pi}\int_0^\infty \frac{1-\cos p}{p} e^{-p^2 t} \sin px \, dp.$$

4.19 Using Fourier transforms solve

$$\frac{\partial u}{\partial t} = 2\frac{\partial^2 u}{\partial x^2} \qquad x > 0,\ t > 0$$

if $u(0,t) = 0$, $u(x,0) = e^{-x}$ and $u(x,t)$ is bounded.

4.20 Solve

$$\frac{\partial u}{\partial t} = \frac{\partial^2 u}{\partial x^2}$$

$$\frac{\partial u}{\partial x}(0,t) = 0 \qquad u(x,0) = \begin{cases} x & 0 \le x \le 1 \\ 0 & 1 < x \end{cases}$$

and $u(x,t)$ is bounded where $x > 0, t > 0$.

4.21 Using the Fourier transform solve

$$a^2\frac{\partial^4 u}{\partial x^4} + \frac{\partial^2 u}{\partial t^2} = 0$$

$0 \le t \le \infty$, $-\infty \le x \le \infty$, subject to the conditions $u(x,0) = f(x)$, $\frac{\partial u}{\partial t}(x,0) = 0$ and u and its partial derivatives tend to 0 as $x \to \pm\infty$. You may assume that

$$\mathcal{F}^{-1}(\cos ap^2 t) = \frac{1}{\sqrt{4\pi at}} \cos\left(\frac{x^2}{4at} - \frac{\pi}{4}\right).$$

Deduce that the solution is

$$u(x,t) = \frac{1}{\sqrt{4\pi at}} \int_{-\infty}^{\infty} f(x-y) \cos\left(\frac{y^2}{4at} - \frac{\pi}{4}\right) dy.$$

4.22 Use the cosine transform to show that the solution to

$$\frac{\partial^2 u}{\partial x^2} + \frac{\partial^2 u}{\partial y^2} = 0, \qquad y > 0, x > 0,$$

where

$$u(x,0) = \begin{cases} 1 & 0 < x < a \\ 0 & a < x, \end{cases} \qquad \left.\frac{\partial u}{\partial x}\right|_{x=0} = 0$$

and

$$u(x,y) \to 0 \qquad \text{as} \quad \sqrt{x^2+y^2} \to \infty$$

is

$$u(x,y) = \frac{1}{\pi}\left[\tan^{-1}\left(\frac{a+x}{y}\right) + \tan^{-1}\left(\frac{a-x}{y}\right)\right].$$

You may assume

$$\int_0^\infty e^{-sx} x^{-1} \sin rx \, dx = \tan^{-1}\left(\frac{r}{s}\right), \qquad r, s > 0.$$

4.23 Show that the solution of Laplace's equation $\nabla^2 v = 0$ for v inside the semi-infinite strip $x > 0, 0 < y < b$ subject to the boundary conditions

(i)	$v = f(x)$	on $y = 0,\ 0 < x < \infty$
(ii)	$v = 0$	when $y = b,\ 0 < x < \infty$
(iii)	$v = 0$	when $x = 0,\ 0 < y < b$

is given by

$$v = \frac{2}{\pi} \int_0^\infty f(u) \int_0^\infty \frac{\sinh p(b-y)}{\sinh pb} \sin xp \sin up \, dp \, du.$$

4.6 Laplace transform

The Laplace transform of $f(x)$ is defined as

$$\mathcal{L}(f) = F(p) = \int_0^\infty f(x)e^{-px}\,dx \tag{4.6.1}$$

where $f(x)$ is defined for $x \geq 0$ and is integrable over any finite positive interval, and where $f(x)$ is such that $e^{-kx}|f(x)|$ is integrable over the interval $0 < x < \infty$ for some real k. For convergence of 4.6.1, we require $Re(p) > k$. For many purposes, it is sufficient to think of p as being a real number which is large enough to make the integral converge, that is if $f(x)$ behaves like e^{qx} for large x, then p must be greater than q.

In order to use Laplace transforms in practice, a formula is required to transform back from $F(p)$ to $f(x)$. Such a formula is the inversion formula. However, to prove the inversion formula, it is convenient to suppose that p is complex and write $p = \gamma + i\beta$. We assume that $\gamma = Re(p)$ is large enough for the convergence of the integral. From 4.6.1,

$$\int_{\gamma-iw}^{\gamma+iw} e^{xp}F(p)\,dp = \int_{\gamma-iw}^{\gamma+iw} e^{xp}\int_0^\infty f(u)e^{-pu}du dp \tag{4.6.2}$$

where γ is constant. Writing $p = \gamma + i\beta$ gives the right-hand side:

$$\int_{-w}^{w} e^{x(\gamma+i\beta)}\int_0^\infty f(u)e^{-(\gamma+i\beta)u}\,du\,id\beta$$

$$= ie^{\gamma x}\int_{-w}^{w} e^{ix\beta}\int_0^\infty e^{-i\beta u}\left[e^{-\gamma u}f(u)\right]\,du\,d\beta. \tag{4.6.3}$$

If we now assume $f(u) = 0$ for $u < 0$, so that the integral $\int_0^\infty$ over u becomes $\int_{-\infty}^\infty$ and let $w \to \infty$, then

$$\int_{\gamma-i\infty}^{\gamma+i\infty} e^{xp}F(p)dp = ie^{\gamma x}\int_{-\infty}^{\infty} e^{ix\beta}\int_{-\infty}^{\infty} e^{-i\beta u}[e^{-\gamma u}f(u)]dud\beta. \tag{4.6.4}$$

Using 4.5.4 gives

$$2\pi g(x) = \int_{-\infty}^{\infty} e^{-ixw}\int_{-\infty}^{\infty} g(u)e^{iuw}\,du\,dw. \tag{4.6.5}$$

Let $\beta = -w$ then

$$2\pi g(x) = \int_{-\infty}^{\infty} e^{ix\beta} \int_{-\infty}^{\infty} e^{-iu\beta} g(u)\, du\, d\beta \tag{4.6.6}$$

and from 4.6.4 it follows that

$$\int_{\gamma-i\infty}^{\gamma+i\infty} e^{xp} F(p) dp = ie^{\gamma x} 2\pi [e^{-\gamma x} f(x)]. \tag{4.6.7}$$

Therefore,

$$2\pi i f(x) = \int_{\gamma-i\infty}^{\gamma+i\infty} e^{xp} F(p)\, dp. \tag{4.6.8}$$

It can be shown that in order for this argument to be meaningful $\gamma =$ (Real part of p) must be greater than the largest of the real parts of all the possible singularities of $F(p)$.

Consider $f(x) = e^{-x}$ as an example, then

$$\begin{aligned} F(p) &= \int_0^{\infty} e^{-x} e^{-px}\, dx = \int_0^{\infty} e^{-(1+p)x}\, dx \\ &= \left. \frac{-e^{-(1+p)x}}{1+p} \right|_0^{\infty} = \frac{1}{1+p}, \qquad Re(p) > -1. \end{aligned} \tag{4.6.9}$$

From the inversion formula

$$\begin{aligned} 2\pi i f(x) &= \int_{\gamma-i\infty}^{\gamma+i\infty} e^{xp} F(p)\, dp, \\ &= \int_{\gamma-i\infty}^{\gamma+i\infty} \frac{e^{xp}}{1+p}\, dp, \qquad \gamma > -1 \end{aligned} \tag{4.6.10}$$

(the real part of the singularity of $F(p)$ is -1).

The line of integration is parallel to the imaginary axis in the p-plane and must lie on the right of the singularities of the integrand, in this case, a simple pole at $p = -1$.

This integral is most easily evaluated by taking $\gamma = 0$ and completing the path by a semicircle as illustrated in Figure 4.1. On the circular arc $p = Re^{i\theta}$, $\frac{\pi}{2} < \theta < \frac{3\pi}{2}$ and it can be shown that the integral along the arc tends to zero as $R \to \infty$ (Jordan's lemma, see Wunsch (1994)). From Cauchy's residue theorem

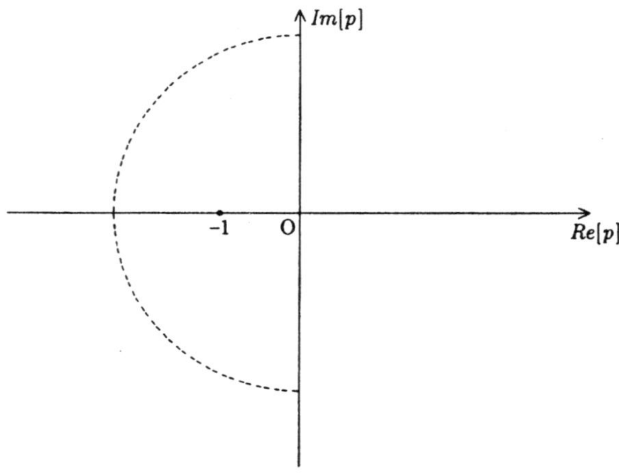

Fig. 4.1.

(see Section 1.7), we have that the line integral is equal to $2\pi i\times$ residue at the pole $p = -1$.

$$\operatorname*{Res}_{p=-1} \frac{e^{xp}}{1+p} = \lim_{p\to -1} \frac{(1+p)e^{xp}}{1+p} = e^{-x}. \tag{4.6.11}$$

Hence

$$2\pi i f(x) = 2\pi i e^{-x}$$

gives $f(x) = e^{-x}$.

4.7 Inverting Laplace Transforms

It is clear from the discussion in the previous section that the main problem using Laplace transforms will arise at the inversion stage. Some discussion has been already been given in Chapter 1 on residues and the use of residues for complex integration. This work will be revisited with the view to inverting Laplace transforms.

In complex contour integration, the interest is with line integrals in the complex plane, of the type shown in Figure 4.2.

Cauchy's residue theorem for closed contours states that if $f(z)$ is a function which is analytic inside and on a simple closed path C except for finitely many isolated singular points $a_1, a_2, \ldots, a_n$ inside C, then

$$\int_C f(z)\,dz = 2\pi i \sum_{j=1}^{n} \operatorname*{Res}_{z=a_j} f(z), \tag{4.7.1}$$

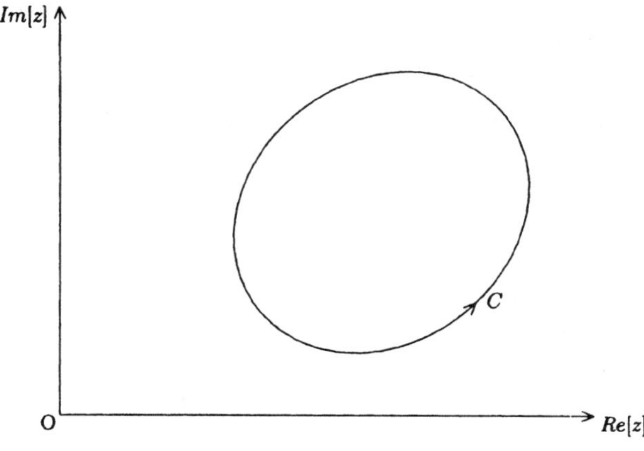

Fig. 4.2.

the integral being taken in the anticlockwise sense around C. The only type of singularity that will be considered are poles, which may be defined by considering $f(z) = P(z)/Q(z)$ where $P(z)$ and $Q(z)$ are both analytic functions of z. Suppose that $P(z)$ is not zero where $Q(z)$ is zero, then the zeros of $Q(z) = 0$ are the poles of $f(z)$. Suppose that $z = a$ is a pole of $f(z)$, then if $(z - a)$ occurs to the first power as a factor of $Q(z)$, the pole is said to be the first order (or a simple pole), if $(z - a)$ occurs to the second power as a factor of $Q(z)$ the pole is said to be of order 2, and so on to higher orders.

Hence:

(a)
$$\frac{1}{z(z-2)^2}$$
has a simple pole $z = 0$, and a double pole $z = 2$.

(b)
$$\frac{z^3}{(z-i)(z+i)}$$
has simple poles at $z = i$ and $z = -i$.

(c) Consider the function
$$\frac{e^{az}}{1+e^z}.$$
This has poles when $1 + e^z = 0$, that is $e^z = -1 = e^{(2n\pi+\pi)i}$ or when
$$z = (2n+1)\pi i, \qquad n = 0, \pm 1, \pm 2, \dots.$$
Each of these is a simple root of the equation $e^z = -1$, and therefore gives a simple pole.

To use Cauchy's residue theorem, we need to find the residues at poles. If $f(z)$ has a simple pole at $z = a$, then the residue there is given by

$$(z-a)f(z)\Big|_{z=a} \qquad \text{or} \qquad \lim_{z\to a}(z-a)f(z) \tag{4.7.2}$$

or letting $f(z) = P(z)/Q(z)$ then

$$\operatorname*{Res}_{z=a} f(z) = \frac{P(a)}{Q'(a)}. \tag{4.7.3}$$

If $f(z)$ has a pole of order $m > 1$ at $z = a$ then the residue there is given by

$$\frac{1}{(m-1)!}\lim_{z\to a}\left[\frac{d^{m-1}}{dz^{m-1}}\left[(z-a)^m f(z)\right]\right]. \tag{4.7.4}$$

Suppose that we want to find the residues for

$$(a) \quad \frac{1}{z(z-2)^2} \qquad (b) \quad \frac{e^{az}}{1+e^z}.$$

Consider

(a)

$$\frac{1}{z(z-2)^2}.$$

This has a simple pole at $z = 0$, and a double pole at $z = 2$. Hence

$$\operatorname*{Res}_{z=0} f(z) = \lim_{z\to 0} z\frac{1}{z(z-2)^2} = \frac{1}{4}$$

and

$$\begin{aligned}\operatorname*{Res}_{z=2} f(z) &= \frac{1}{1!}\lim_{z\to 2}\left[\frac{d}{dz}\left(\frac{(z-2)^2}{z(z-2)^2}\right)\right] \\ &= \frac{1}{1!}\lim_{z\to 2}\frac{d}{dz}\left(\frac{1}{z}\right) = \lim_{z\to 2}\left(-\frac{1}{z^2}\right) = -\frac{1}{4}.\end{aligned}$$

(b) The function $\frac{e^{az}}{1+e^z}$ has simple poles at $z = (2n+1)\pi i$, $(n = 0, \pm 1, \pm 2, \ldots)$, and hence

$$\begin{aligned}\operatorname*{Res}_{z=(2n+1)\pi i} f(z) &= \frac{e^{az}}{e^z}\bigg|_{z=(2n+1)\pi i} = \frac{e^{a(2n+1)\pi i}}{e^{(2n+1)\pi i}} \\ &= -e^{a(2n+1)\pi i}.\end{aligned}$$

To find the inverse Laplace transform using residues, the integral

$$\lim_{w\to\infty} \int_{\gamma-iw}^{\gamma+iw} e^{xp} F(p)\, dp \tag{4.7.5}$$

needs to be evaluated. The line of integration is parallel to the imaginary axis in the p plane, and must lie to the right of all the singularities of the integrand. Complete the semi-circle as illustrated in Figure 4.3.

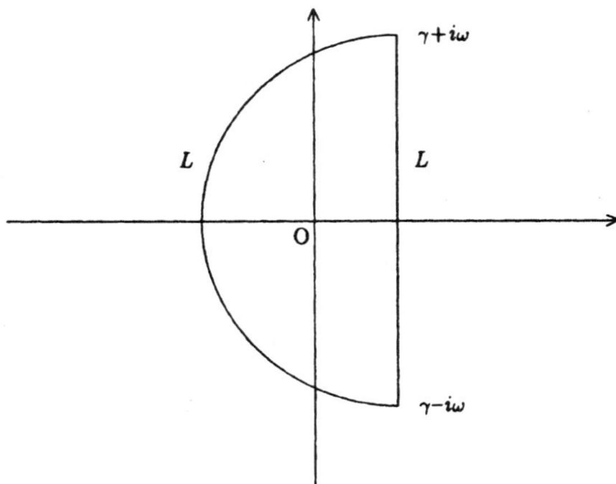

Fig. 4.3.

Then the integral around the contour $l + L$ has the value

$$\int_{l+L} e^{xp} F(p)\, dp = 2\pi i \sum \text{Residues enclosed in } l + L \tag{4.7.6}$$

where $l + L$ encloses all the singularities of the integrand.

Now (without getting into a technical discussion about the properties that F has to satisfy),

$$\int_{L} e^{xp} F(p)\, dp \to 0 \qquad \text{as} \quad w \to \infty \tag{4.7.7}$$

for most functions $F(p)$. Hence

$$\begin{aligned} \lim_{w\to\infty} \int_{l} e^{xp} F(p)\, dp &= \int_{\gamma-i\infty}^{\gamma+i\infty} e^{xp} F(p)\, dp \\ &= 2\pi i \sum \text{Residues of } e^{xp} F(p). \end{aligned} \tag{4.7.8}$$

However

$$\int_{\gamma-i\infty}^{\gamma+i\infty} e^{xp}F(p)\,dp = 2\pi i f(x) \tag{4.7.9}$$

and therefore

$$f(x) = \sum \text{Residues of } e^{xp}F(p) \tag{4.7.10}$$

which holds if the only singularities of the integrand are poles.

The above result is Heaviside's inversion formula which states that, the inverse Laplace transform of $F(p)$ is

$$f(x) = \sum \text{ Residues of } F(p)e^{px} \tag{4.7.11}$$

provided the only singularities of $F(p)$ are poles.

As examples, consider finding the inverse Laplace transforms of

$$(a)\; F(p) = \frac{1}{p^2+2p+2} \qquad (b)\; G(p) = \frac{1}{(p+1)(p-2)^2}$$

$$(c)\; H(p) = \frac{1}{p}\tanh\frac{p}{2}.$$

(a) Let us factorise $F(p)$ as follows:

$$F(p) = \frac{1}{p^2+2p+2} = \frac{1}{(p+1+i)(p+1-i)}.$$

For the simple poles at $p = -1-i$ and $p = -1+i$

$$\begin{aligned}\operatorname*{Res}_{p=-1-i} \frac{1}{p^2+2p+2}e^{px} &= \lim_{p\to-1-i} \frac{(p+1+i)e^{px}}{(p+1+i)(p+1-i)} \\ &= \frac{e^{(-1-i)x}}{-2i},\end{aligned}$$

$$\begin{aligned}\operatorname*{Res}_{p=-1+i} \frac{1}{p^2+2p+2}e^{px} &= \lim_{p\to-1+i} \frac{(p+1-i)e^{px}}{(p+1+i)(p+1-i)} \\ &= \frac{e^{(-1+i)x}}{2i}.\end{aligned}$$

Thus

$$\begin{aligned} f(x) &= \sum \text{Residues} = \frac{e^{(-1+i)x}}{2i} - \frac{e^{(-1-i)x}}{2i} \\ &= e^{-x}\frac{e^{ix}-e^{-x}}{2i} = e^{-x}\sin x. \end{aligned}$$

(b) Note that

$$G(p) = \frac{1}{(p+1)(p-2)^2}$$

has a simple pole $p = -1$, and a double pole $p = 2$:

$$\begin{aligned}
\operatorname*{Res}_{p=-1} \frac{1}{(p+1)(p-2)^2} e^{px} &= \lim_{p\to -1} \frac{(p+1)e^{px}}{(p+1)(p-2)^2} = \frac{1}{9}e^{-x}, \\
\operatorname*{Res}_{p=2} \frac{1}{(p+1)(p-2)^2} e^{px} &= \frac{1}{1!}\lim_{p\to 2} \frac{d}{dp}\left[(p-2)^2 \frac{e^{px}}{(p+1)(p-2)^2}\right] \\
&= \lim_{p\to 2} \frac{d}{dx}\left[\frac{e^{px}}{p+1}\right] \\
&= \frac{3xe^{2x} - e^{2x}}{9}.
\end{aligned}$$

Hence

$$g(x) = \frac{1}{9}e^{-x} + \frac{1}{3}xe^{2x} - \frac{1}{9}e^{2x}.$$

(c) The function

$$H(p) = \frac{1}{p}\tanh\frac{p}{2} = \frac{1}{p}\frac{\sinh\frac{p}{2}}{\cosh\frac{p}{2}}$$

has singularities when $\cosh\frac{p}{2} = 0$ since

$$\frac{\sinh\frac{p}{2}}{p} \qquad \text{is finite as} \quad p \to 0.$$

Hence singularities arise when

$$\frac{e^{\frac{p}{2}} + e^{-\frac{p}{2}}}{2} = 0 \qquad \text{or} \qquad e^p = -1 = e^{(2n+1)\pi i},$$

$$p = (2n+1)\pi i \qquad n = 0, \pm 1, \pm 2, \ldots.$$

The singularities of $H(p)$ are simple poles, hence

$$\begin{aligned}
\operatorname*{Res}_{p=(2n+1)\pi i} H(p)e^{px} &= \left.\frac{\sinh\frac{p}{2}e^{px}}{\frac{p}{2}\sinh\frac{p}{2} + \cosh\frac{p}{2}}\right|_{p=(2n+1)\pi i} \\
&= \frac{\sinh\left(\frac{(2n+1)\pi i}{2}\right) e^{(2n-1)\pi i x}}{(2n+1)\frac{\pi}{2} i \sinh\left(\frac{(2n+1)\pi i}{2}\right)} \\
&= -\frac{2i}{(2n+1)\pi} e^{i(2n+1)\pi x},
\end{aligned}$$

and for the pole at $-(2n+1)\pi i$ the residue is:

$$\begin{aligned}\operatorname*{Res}_{p=-(2n+1)\pi i} H(p)e^{px} &= \left.\frac{\sinh\frac{p}{2}e^{px}}{\frac{p}{2}\sinh\frac{p}{2}+\cosh\frac{p}{2}}\right|_{p=-(2n+1)\pi i} \\ &= \frac{2i}{(2n+1)\pi}e^{-i(2n+1)\pi x},\end{aligned}$$

and hence

$$\begin{aligned}h(x) &= \sum_{n=0}^{\infty}\left(-\frac{2i}{\pi}\right)\left\{\frac{e^{i(2n+1)\pi x}}{2n+1}-\frac{e^{-i(2n+1)\pi x}}{2n+1}\right\} \\ &= \sum_{n=0}^{\infty}\frac{4}{\pi}\frac{\sin(2n+1)\pi x}{2n+1}.\end{aligned}$$

Some exercises on the inversion of Laplace transforms using contour methods are provided next.

EXERCISES

4.24 Use the complex inversion formula to evaluate

$$\begin{aligned}&(a)\quad \mathcal{L}^{-1}\left[\frac{p}{p^2+a^2}\right] \\ &(b)\quad \mathcal{L}^{-1}\left[\frac{1}{p^2+a^2}\right] \\ &(c)\quad \mathcal{L}^{-1}\left[\frac{1}{(p+1)(p^2+1)}\right] \\ &(d)\quad \mathcal{L}^{-1}\left[\frac{1}{(1+p)^2}\right] \\ &(e)\quad \mathcal{L}^{-1}\left[\frac{1}{p^3(p^2+1)}\right].\end{aligned}$$

4.25 Find

$$\mathcal{L}^{-1}\left[\frac{1}{p(e^p+1)}\right]$$

using the complex inversion formula.

4.26 Prove that

$$\mathcal{L}^{-1}\left[\frac{1}{p\cosh p}\right]=1-\frac{4}{\pi}\left[\cos\frac{\pi x}{2}-\frac{1}{3}\cos\frac{3\pi x}{2}+\ldots\right].$$

4.27 Find

$$\mathcal{L}^{-1}\left[\frac{1}{p^2\sinh p}\right].$$

4.28 By using the inversion formula, prove that

$$\mathcal{L}^{-1}\left[\frac{1}{p^3\sinh ap}\right]=\frac{x(x^2-a^2)}{6a}-\frac{2a^2}{\pi^3}\sum_{n=1}^{\infty}\frac{(-1)^n}{n^3}\sin\frac{\pi nx}{a}.$$

4.8 Standard Transforms

From the definition of the Laplace transform, a table of standard transforms can be constructed, from which many common inversions can be made without direct recourse to contour methods. Some standard results are shown in Table 4.1.

Table 4.1.

$f(x)$	$F(p)$	$f(x)$	$F(p)$
a	$\frac{a}{p}$	$\sin\omega x$	$\frac{\omega}{p^2+\omega^2}$
e^{ax}	$\frac{1}{p-a}$	$\cos\omega x$	$\frac{p}{p^2+\omega^2}$
x^n	$\frac{n!}{p^{n+1}}$	$\delta(x-T)$	e^{-pT}
$\sinh\beta x$	$\frac{\beta}{p^2-\beta^2}$	$\cosh\beta x$	$\frac{p}{p^2-\beta^2}$

There are also some useful theorems which allow inverse transforms to be constructed from known results.

Theorem 4.2 (First shift theorem)

If $\mathcal{L}(f)=F(p)$ then

$$\mathcal{L}\left(e^{ax}f(x)\right)=F(p-a). \tag{4.8.1}$$

That is, the substitution of $p-a$ for p in the transform corresponds to the multiplication of the original function by e^{ax}. Direct application of this theorem gives the transforms shown in Table 4.2.

The convolution theorem of 4.5.11 has an equivalent form in terms of Laplace transforms:

Table 4.2.

$f(x)$	$F(p)$
$e^{ax}x^n$	$\frac{n!}{(p-a)^{n+1}}$
$e^{ax}\sin wx$	$\frac{w}{(p-a)^2+w^2}$
$e^{ax}\cos wx$	$\frac{p-a}{(p-a)^2+w^2}$

Theorem 4.3 (Convolution theorem)

Let $f(x)$ and $g(x)$ be two arbitrary functions each possessing a Laplace transform, $\mathcal{L}(f(x)) = F(p)$, $\mathcal{L}(g(x)) = G(p)$, then

$$\mathcal{L}\left[\int_0^x f(u)g(x-u)\,du\right] = F(p)G(p) \tag{4.8.2}$$

and

$$\mathcal{L}^{-1}[F(p)G(p)] = \int_0^x f(u)g(x-u)\,du. \tag{4.8.3}$$

As an application of this theorem, consider

$$\mathcal{L}^{-1}\left[\frac{1}{p^2(p+1)^2}\right]. \tag{4.8.4}$$

From the table of transforms

$$\mathcal{L}^{-1}\left(\frac{1}{p^2}\right) = x \qquad \text{and} \qquad \mathcal{L}^{-1}\left[\frac{1}{(p+1)^2}\right] = xe^{-x}.$$

Hence using the convolution theorem with $f(x) = x$, and $g(x) = xe^{-x}$ gives

$$\begin{aligned}\mathcal{L}^{-1}\left[\frac{1}{p^2(p+1)^2}\right] &= \int_0^x u(x-u)e^{-x+u}\,du \\ &= (2+x)e^{-x} - 2 + x.\end{aligned}$$

An important function which arises often in applications is the Heaviside step function $H(x-a)$ which is defined by

$$H(x-a) = \begin{cases} 0 & x < a \\ 1 & x \geq a \end{cases} \tag{4.8.5}$$

and is shown in Figure 4.4.

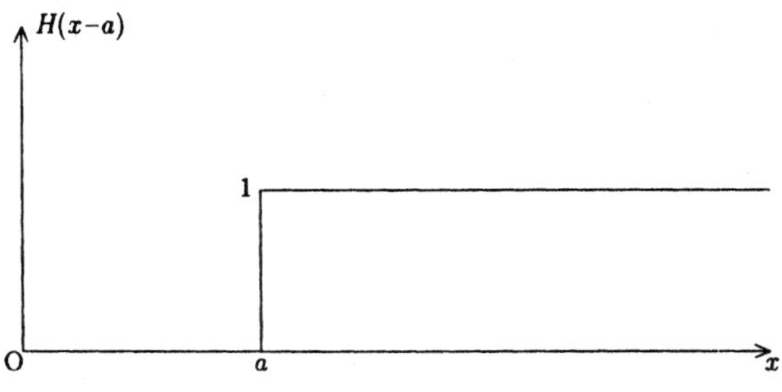

Fig. 4.4.

The Laplace transform of $H(x-a)$ is

$$\begin{aligned} \mathcal{L}(H(x-a)) &= \int_0^\infty e^{-px} H(x-a)\, dx \\ &= -\frac{1}{p} e^{-px} \Big|_a^\infty = \frac{e^{ap}}{p} \qquad (p > 0). \end{aligned} \tag{4.8.6}$$

As an example consider the Laplace transform of

$$f(x) = k[H(x-a) - H(x-b)]$$

which is depicted in Figure 4.5.

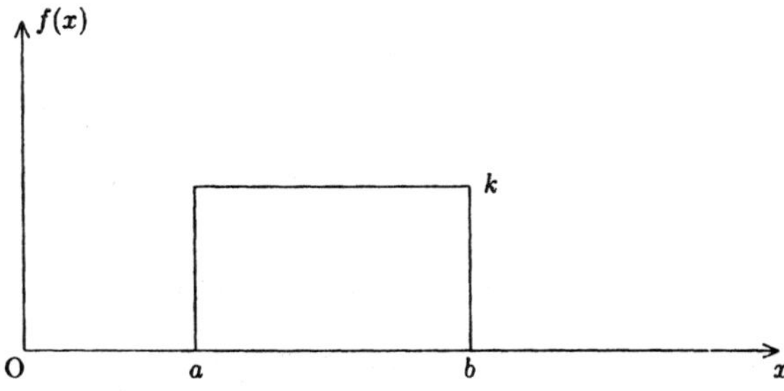

Fig. 4.5.

Then using the result in 4.8.6 gives

$$F(p) = \mathcal{L}(f) = k\left[\frac{e^{-ap}}{p} - \frac{e^{-bp}}{p}\right] = \frac{k}{p}\left[e^{-ap} - \epsilon^{-bp}\right].$$

As a second case, consider finding the Laplace transform of

$$f(x) = k[H(x) - 2H(x-a) + 2H(x-2a) - 2H(x-3a) + 2H(x-4a) - \ldots]$$

which is shown in Figure 4.6.

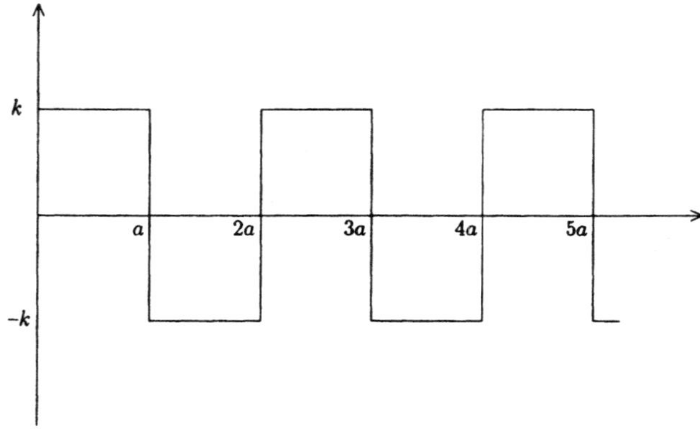

Fig. 4.6.

In this case

$$\begin{aligned}
\mathcal{L}f &= k\left[\frac{1}{p} - \frac{2e^{-pa}}{p} + \frac{2e^{-2pa}}{p} - \frac{2e^{-3pa}}{p} + \ldots\right] \\
&= \frac{k}{p}\left[1 - 2e^{-pa}\left\{1 - e^{-pa} + e^{-2pa} - e^{-3pa}a + \ldots\right\}\right] \\
&= \frac{k}{p}\left[\frac{1-e^{-pa}}{1+e^{-pa}}\right].
\end{aligned}$$

A further device is shifting along the x-axis which yields a further theorem. Suppose the Laplace transform of

$$f(x) = H(x-2)\sin(x-2)$$

is required. These functions are shown in Figure 4.7.

Then simple manipulations yield:

$$\begin{aligned}
\mathcal{L}(H(x-c)f(x-c)) &= \int_0^\infty e^{-px} H(x-c) f(x-c)\, dx \\
&= \int_c^\infty e^{-px} f(x-c)\, dx. \qquad (4.8.7)
\end{aligned}$$

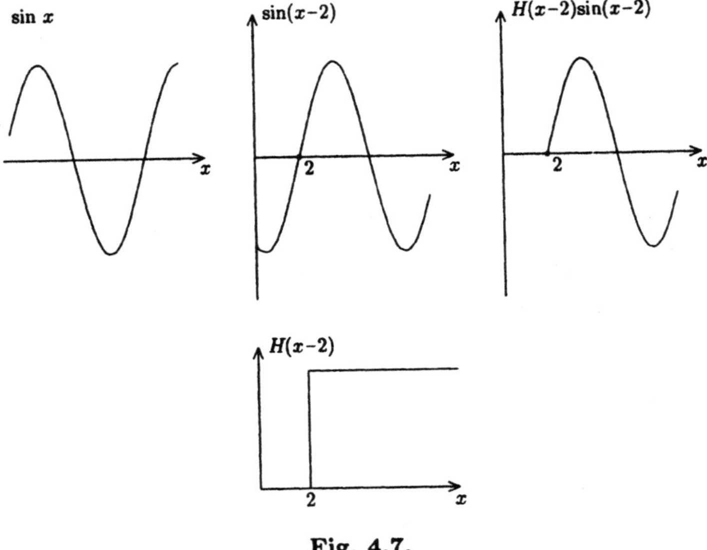

Fig. 4.7.

Further, putting $\tau = x - c$ gives

$$\begin{aligned} \mathcal{L}(H(x-c)f(x-c)) &= \int_0^\infty e^{-p(\tau+c)} f(\tau)\, d\tau \\ &= e^{-pc} \int_0^\infty e^{-p\tau} f(\tau)\, d\tau = e^{-pc}\mathcal{L}[f(x)], \end{aligned}$$

which gives the second shift theorem.

Theorem 4.4 (Second shift theorem)

If $F(p) = \mathcal{L}(f(x))$ then for any positive constant a

$$e^{-ap}F(p) = \mathcal{L}[f(x-a)H(x-a)] \tag{4.8.8}$$

where $H(x-a)$ is the Heaviside unit step.

To illustrate these ideas, consider the following sequence of simple applications

$$\mathcal{L}[H(x-3)(x-3)^2] = e^{-3p}\mathcal{L}(x^2) = \frac{2}{p^3}e^{-3p}$$

(replace x by $x + a$)

$$\begin{aligned}
\mathcal{L}[H(x-2)\sin(x-2)] &= e^{-2p}\mathcal{L}(\sin x) = e^{-2p}\frac{1}{p^2+1} \\
\mathcal{L}[H(x-4)x^2] &= e^{-4p}\mathcal{L}[(x+4)^2] \\
&= e^{-4p}\mathcal{L}(x^2+8x+16) \\
&= e^{-4p}\left(\frac{2!}{p^3}+8\frac{1}{p^2}+\frac{16}{p}\right) \\
\mathcal{L}[H(x-2)e^x] = e^{-2p}\mathcal{L}(e^{x+2}) &= e^{-2p}\mathcal{L}[e^2e^x] = e^{-2p}e^2\frac{1}{p-1}.
\end{aligned}$$

and applying the inverse transform,

$$\mathcal{L}^{-1}\left(\frac{1}{p-5}\right) = e^{5x}$$

yields

$$\mathcal{L}^{-1}\left(\frac{e^{-2p}}{p-5}\right) = e^{5(x-2)}H(x-2).$$

(The e^{-2p} gives the rule of thumb: replace x by $x-2$ in $f(x)$ and multiply by $H(x-2)$.)

As a second example consider

$$\mathcal{L}^{-1}\left(\frac{e^{-p}}{p^2+9}\right).$$

Since

$$\mathcal{L}^{-1}\left(\frac{1}{p^2+9}\right) = \frac{1}{3}\sin 3x,$$

it follows that

$$\mathcal{L}^{-1}\left(\frac{e^{-p}}{p^2+9}\right) = \frac{1}{3}\sin 3(x-1)H(x-1).$$

As a further example, find the inverse Laplace transform of

$$\frac{1}{p(1-e^{-ap})}.$$

Then

$$\begin{aligned}
F(p) &= \frac{1}{p(1-e^{-ap})} = \frac{1}{p}\left(1+e^{-ap}+e^{-2ap}+\ldots\right) \\
&= \frac{1}{p}+\frac{e^{-ap}}{p}+\frac{e^{-2ap}}{p}+\ldots,
\end{aligned}$$

and

$$f(t) = 1 + H(x-a) + H(x-2a) + H(x-3a) + \ldots$$

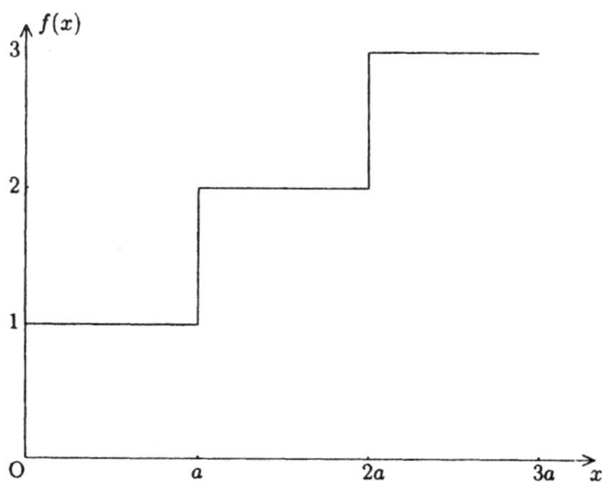

Fig. 4.8.

which is illustrated in Figure 4.8.

By expansion,

$$\frac{c}{(p^2+1)(1+e^{-p\pi})} = \frac{c}{p^2+1}\left[1 - e^{-p\pi} + e^{-2p\pi} - e^{3p\pi} + \ldots\right]$$

which gives

$$\begin{aligned}\mathcal{L}^{-1}\left(\frac{c}{(p^2+1)(1+e^{-p\pi})}\right) &= c[\sin x - \sin(x-\pi)H(x-\pi) \\ &\quad + \sin(x-2\pi)H(x-2\pi) \\ &\quad - \sin(x-3\pi)H(x-3\pi) + \ldots].\end{aligned}$$

If $n\pi < x < (n+1)\pi$ then

$$c[\sin x - \sin(x-\pi) + \sin(x-2\pi) - \sin(x-3\pi) + \ldots + (-1)^n \sin(x-n\pi)]$$

$$= c[\sin x + \sin x + \sin x + \ldots + \sin x] = c(n+1)\sin x.$$

4.9 Use of Laplace Transforms to Solve Partial Differential Equations

As the range of integration for Laplace transforms is infinite, only those independent variables for which the dependent variable of the partial differential equation is defined over an *infinite* range are suitable as variables of transformation. Laplace transforms are Prticularly suited to problems where boundary conditions are given at $t = 0$. Such problems arise in the solution of the heat equation and wave equation, the independent variable t being interpreted there as the time variable. Hence in this section, attention is confined to transforming the time variable t.

Suppose that $u(x,t)$ is an arbitrary function defined for $a \le x \le b$, $t > 0$ where a and b are arbitrary constants, then

$$\begin{aligned}\mathcal{L}\left(\frac{\partial u}{\partial t}\right) &= \int_0^\infty e^{-pt}\frac{\partial u}{\partial t}\,dt = ue^{-pt}\Big|_0^\infty + p\int_0^\infty ue^{-pt}\,dt \\ &= p\,\mathcal{U}(x;p) - u(x,0) \qquad (4.9.1)\end{aligned}$$

where $\mathcal{U}(x;p) = \mathcal{L}(u(x,t))$ is the Laplace transform of $u(x,t)$ with respect to t. Then

$$\begin{aligned}\mathcal{L}\left(\frac{\partial^2 u}{\partial t^2}\right) &= \mathcal{L}\left(\frac{\partial}{\partial t}\left(\frac{\partial u}{\partial t}\right)\right) \\ &= p\mathcal{L}\left(\frac{\partial u}{\partial t}\right) - \frac{\partial u}{\partial t}(x,0) \\ &= p^2\mathcal{U}(x;p) - pu(x,0) - \frac{\partial u}{\partial t}(x,0), \qquad (4.9.2) \\ \mathcal{L}\left(\frac{\partial u}{\partial x}\right) &= \int_0^\infty e^{-pt}\frac{\partial u}{\partial x}\,dt \\ &= \frac{d}{dx}\int_0^\infty e^{-pt}u\,dt \\ &= \frac{d\mathcal{U}}{dx}(x;p), \qquad (4.9.3) \\ x\mathcal{L}\left(\frac{\partial^2 u}{\partial x^2}\right) &= \int_0^\infty e^{-pt}\frac{\partial^2 u}{\partial x^2}\,dt \\ &= \frac{d^2}{dx^2}\int_0^\infty e^{-pt}u\,dt \\ &= \frac{d^2\mathcal{U}}{dx^2}(x;p). \qquad (4.9.4)\end{aligned}$$

As with many ideas in this field, a set of worked examples will explain the techniques in an enlightened manner.

(1) Solve

$$\frac{\partial u}{\partial x} + \frac{\partial u}{\partial t} = x, \qquad x > 0, t > 0,$$

given that $u(x,0) = 0$ for $x > 0$ and $u(0,t) = 0$ for $t > 0$.

Taking the Laplace transform with respect to t gives

$$\mathcal{L}\left(\frac{\partial u}{\partial x}\right) + \mathcal{L}\left(\frac{\partial u}{\partial t}\right) = \mathcal{L}(x)$$

or

$$\frac{d}{dx}\mathcal{U}(x;p) + p\mathcal{U}(x;p) - u(x,0) = \frac{x}{p}$$

which yields

$$\frac{d\mathcal{U}}{dx} + p\mathcal{U} = \frac{x}{p}.$$

This ordinary differential equation has an integrating factor

$$e^{\int p\,dx} = e^{px}$$

and therefore

$$\frac{d}{dx}(\mathcal{U}e^{px}) = \frac{x}{p}e^{px}$$

which integrates to give

$$\begin{aligned}\mathcal{U}e^{px} &= \frac{1}{p}\int xe^{px}\,dx \\ &= \frac{1}{p}\left[\frac{x}{p}e^{px} - \frac{e^{px}}{p^2}\right] + c,\end{aligned}$$

and hence

$$\mathcal{U}(x;p) = \frac{x}{p^2} - \frac{1}{p^3} + ce^{-px}.$$

Now

$$\begin{aligned}\mathcal{U}(0;p) &= \mathcal{L}(u(0,t)) = \int_0^\infty u(0,t)e^{-pt}\,dt = 0, \\ \mathcal{U}(0;p) &= 0 \Rightarrow 0 = 0 - \frac{1}{p^3} + c, \qquad c = \frac{1}{p^3}\end{aligned}$$

and hence the solution of the transformed problem is

$$\mathcal{U}(x;p) = \frac{x}{p^2} - \frac{1}{p^3} + \frac{1}{p^3}e^{-px}.$$

Taking the inverse Laplace transform gives

$$u(x,t) = xt - \frac{t^2}{2} + \frac{(t-x)^2}{2}H(t-x)$$

and hence if $t > x$,

$$\begin{aligned} u(x,t) &= xt - \frac{t^2}{2} + \frac{(t-x)^2}{2} = xt - \frac{t^2}{2} + \frac{t^2}{2} - xt + \frac{x^2}{2} \\ &= \frac{x^2}{2} \end{aligned}$$

and if $x > t$, then

$$u(x,t) = xt - \frac{t^2}{2}.$$

(2) Solve

$$\frac{\partial^2 u}{\partial x^2} = \frac{\partial u}{\partial t} \qquad \text{for} \quad 0 \le x \le 2,\ t > 0,$$

given that

$$u(0,t) = 0, \qquad u(2,t) = 0, \qquad u(x,0) = 3\sin 2\pi x.$$

Taking the Laplace transform with respect to t gives

$$\mathcal{L}\left(\frac{\partial^2 u}{\partial x^2}\right) = \mathcal{L}\left(\frac{\partial u}{\partial t}\right)$$

which gives the ordinary differential equation

$$\frac{d^2\mathcal{U}}{dx^2} = p\mathcal{U} - u(x,0)$$

or

$$\frac{d^2\mathcal{U}}{dx^2} - p\mathcal{U} = -u(x,0) = -3\sin 2\pi x.$$

The complementary function satisfies

$$\frac{d^2\mathcal{U}}{dx^2} - p\mathcal{U} = 0$$

with solution

$$\mathcal{U}(x;p) = Ae^{\sqrt{p}x} + Be^{-\sqrt{p}x}$$

and the particular integral can be found by trying

$$\mathcal{U}(x;p) = C\sin 2\pi x + D\cos 2\pi x$$

which gives

$$-4\pi^2 C\sin 2\pi x - 4\pi^2 D\cos 2\pi x - pC\sin 2\pi x - pD\cos 2\pi x = -3\sin 2\pi x.$$

Equating coefficients of $\sin 2\pi x$ yields

$$-4\pi^2 C - pC = -3 \qquad \text{or} \qquad C = \frac{3}{p + 4\pi^2}$$

and that of $\cos 2\pi x$

$$-4\pi^2 D - pD = 0 \qquad \text{or} \qquad D = 0.$$

Hence the particular integral is

$$\frac{3}{p+4\pi^2}\sin 2\pi x$$

and

$$\mathcal{U}(x;p) = Ae^{\sqrt{p}x} + Be^{-\sqrt{p}x} + \frac{3}{p+4\pi^2}\sin 2\pi x.$$

Now apply the boundary conditions

$$u(0,t) = 0 \qquad \text{to yield} \quad \mathcal{U}(0;p) = \mathcal{L}(u(0,t)) = \mathcal{L}(0) = 0$$

and

$$u(2,t) = 0 \qquad \text{to yield} \quad \mathcal{U}(2;p) = \mathcal{L}(u(2,t)) = \mathcal{L}(0) = 0$$

which give

$$\begin{aligned} \mathcal{U}(0;p) &= 0 \Rightarrow 0 = A + B \\ \mathcal{U}(2;p) &= 0 \Rightarrow 0 = Ae^{2\sqrt{p}} + Be^{-2\sqrt{p}} \\ & \Rightarrow A = B = 0 \end{aligned}$$

and hence

$$\mathcal{U}(x;p) = \frac{3}{p+4\pi^2}\sin 2\pi x.$$

Taking the inverse transform

$$u(x,t) = 3\sin 2\pi x e^{-4\pi^2 t}.$$

Some further exercises are presented on the last sections which include practice using the Laplace transform on specific partial differential equations.

EXERCISES

4.29 Solve

$$x\frac{\partial u}{\partial t} + \frac{\partial u}{\partial x} = x \qquad x > 0, t > 0$$

given that $u(x,0) = 0$ for $x > 0$ $u(0,t) = 0$ for $t > 0$.

4.30 Solve

$$\frac{\partial^2 u}{\partial x^2} = \frac{1}{c^2}\frac{\partial^2 u}{\partial t^2} - k\sin \pi x$$

for $0 < x < 1$, $t > 0$ given that

$$\begin{aligned} u(x,0) &= 0, \qquad \left.\frac{\partial u}{\partial t}\right|_{t=0} = 0, \\ u(0,t) &= 0, \qquad u(1,t) = 0. \end{aligned}$$

4.31 Solve

$$\frac{\partial u}{\partial t} + x\frac{\partial u}{\partial x} = x, \qquad x > 0,\ t > 0$$

given that

$$u(x,0) = 0,\ x > 0, \qquad u(0,t) = 0,\ t > 0.$$

4.32 Solve

$$x\frac{\partial \phi}{\partial t} + \frac{\partial \phi}{\partial x} = x^3, \qquad t > 0, x > 0.$$

given that

$$\phi(x,0) = 0, \qquad \phi(0,t) = 0$$

4.33 Solve

$$x\frac{\partial \phi}{\partial t} + \frac{\partial \phi}{\partial x} = x, \qquad t > 0, x > 0,$$

given that

$$\phi = \phi_0(x) \qquad \text{when} \quad t = 0$$

and

$$\phi = 0 \qquad \text{when} \quad x = 0.$$

(Assume $\phi'(x)$ exists for all x.)

4.34 Solve

$$\frac{\partial^2 u}{\partial x^2} = 6\frac{\partial^2 u}{\partial t^2}, \quad 0 < x < 2, \quad t > 0,$$

given that

$$\begin{aligned} u(x,0) &= 0, \qquad \frac{\partial u}{\partial t}(x,0) = \sin \pi x \\ u(0,t) &= 0, \qquad u(2,t) = 0. \end{aligned}$$

4.35 Solve

$$\frac{\partial y}{\partial t} + x\frac{\partial y}{\partial x} + y = x, \qquad x > 0, \qquad t > 0,$$

subject to

$$y(0,t) = y(x,0) = 0.$$

5
Green's Functions

5.1 Introduction

Green's functions are named after the mathematician and physicist George Green who was born in Nottingham in 1793 and "invented" the Green's function in 1828. This invention was developed in an essay written by Green entitled "Mathematical Analysis to the Theories of Electricity and Magnetism" originally published in Nottingham in 1828 and reprinted by the George Green Memorial Committee to mark the bicentenary of the birth of George Green in 1993. In this essay, Green's function solutions to the Laplace and Poisson equation are formulated (but not in the manner considered in this chapter, in which the Green's function is defined using the delta function).

The Green's function is a powerful mathematical tool rather than a physical concept and was successfully applied to classical electromagnetism and acoustics in the late 19th century. More recently, the Green's function has been the working tool of calculations in particle physics, condensed matter and solid state physics, quantum mechanics and many other topics of advanced applied mathematics and mathematical physics. Just as the Green's function revolutionised classical field theory in the 19th century (hydrodynamics, electrostatics and magnetism) so it revolutionised quantum field theory in the mid 20th century through the introduction of quantum Green's functions. This provided the essential link between the theories of quantum electrodynamics in the 1940s and 1950s and has played a major role in theoretical physics ever since. It is interesting to note that the pioneering work of Richard Feynman in the 1950s and 1960s which led to the development of the Feynman diagram was based on the Green's function; in fact, the Feynman diagram can be considered to be a pictorial representation of a Green's function (a Green's function associated with wave operators) - what Feynman referred to as a "propagator".

The Green's function is possibly one of the most powerful tools for solving partial differential equations - a tool that is all the more enigmatic in that the work of George Green was neglected for nearly 30 years after his death in 1841 and to this day no one knows what he looked like or how and why he developed his revolutionary ideas.

Green's functions are used mainly to solve certain types of linear inhomogeneous partial differential equations (although homogeneous partial differential equations can also be solved using this approach). In principle, the Green's function technique can be applied to any linear constant coefficient inhomogeneous partial differential equation (either scalar or vector) in any number of independent variables although in practice difficulties can arise in computing the Green's function analytically. In fact, Green's functions provide more than just a solution; it could be said that they transform a partial differential equation representation of a physical problem into a integral equation representation - the kernel of the integral equation being composed (completely or partly) of the Green's function associated with the partial differential equation. This is why Green's function solutions are considered to be one of the most powerful analytical tools we have for solving partial differential equations - equations that arise in areas such as electromagnetism (Maxwell's equations), wave mechanics (elastic wave equation), optics (Helmholtz equation), quantum mechanics (Schrödinger and Dirac equations), fluid dynamics (viscous fluid equation), relativistic particle dynamics (Klein–Gordon equation), general relativity (Einstein equations) to name but a few.

In this chapter, we provide an introduction to the method and consider the wave equation, the diffusion equation and finally the Poisson and Laplace equations. Most of this material is related to Green's function solutions to the wave equation which is considered first. This reflects the rôle that the Green's function has played in quantum field theory and wave mechanics in general over the past 30–40 years. It is an area in which Green's functions have, and continue to play a central part. In particular, we consider Green's function solutions to the Schrödinger equation, the inhomogeneous Helmholtz equation and provide a short introduction to volume scattering theory. We also consider in some detail the Green's function solution to the homogeneous Helmholtz equation as applied to physical optics (surface scattering theory).

The large majority of the material contained in this chapter is concerned with the use of "Free Space" Green's functions which provide a general solution in the infinite domain or over a finite domain to which boundary conditions are applied.

By way of a short introduction to help the reader understand the principle of using Green's functions we now consider two short examples. The first example is based on considering point sources to generate a solution to an ordinary differential equation and is based on a "qualitative analysis". The second example makes specific use of the delta function and its properties to develop a solution which is based on a more systematic analysis – as used throughout this chapter.

Example 5.1

Consider the following inhomogeneous ordinary differential equation

$$\hat{L}u(x) = f(x) \tag{5.1.1}$$

where $\hat{L}$ is a linear differential operator and $f(x)$ is a given function (the source term), the solution being required on the interval $0 \leq x \leq a$ where a is some constant.

Instead of considering $f(x)$ as a continuous source function, let us approximate it by a set of discrete source functions $f(\xi_1), f(\xi_2), \ldots, f(\xi_n)$ acting at the points $x = \xi_1, x = \xi_2, \ldots, x = \xi_n$, all for $x \in [0, a]$. Now define the function $g(x; \xi_i)$ to be the solution to equation 5.1.1 due to a point source acting at ξ_i. The solution due to the single effect of this point source is $g(x; \xi_i)f(\xi_i)$. The solution for $u(x)$ is then obtained by summing the results for all the n-point source terms acting over the interval $0 \leq x \leq a$, and takes the form

$$u(x) = \sum_{i=1}^{n} g(x; \xi_i)f(\xi_i).$$

As n becomes larger, so that the number of point source functions $f(\xi_i)$ increases, and a better and better approximation to $f(x)$ is obtained. In the limit as $n \to \infty$, so $|\xi_i - \xi_{i+1}| \to 0 \ \forall i$ the summation in the equation above may be replaced by an integral to give the required solution to equation 5.1.1 in the form

$$u(x) = \int_0^a g(x; \xi)f(\xi)d\xi.$$

The function $g(x; \xi)$ is called the Green's function of the problem.

The notation used to write a Green's function changes considerable from author to author. They are usually written in the form $g(x; \xi)$ (as in this example) or $g(x \mid \xi)$ (as used throughout this Chapter) both forms actually being equivalent to $g(\mid x - \xi \mid)$ as we shall see later.

Similar results to the one given above may be obtained for linear partial differential equations: for example, the solution of the Poisson equation in two dimensions.

$$\nabla^2 u(x, y) = f(x, y); \qquad x \in [0, a], \qquad y \in [0, b]$$

may be written as

$$u(x,y) = \int_0^a \int_0^b g(x,y;\xi,\eta) f(\xi,\eta) d\xi d\eta$$

where $g(x,y;\xi,\eta)$ is the Green's function of the problem.

The approach for developing a Green's function solution discussed in this example is based on considering point sources to provide a set of elementary results and then summing up the results to give the required solution. In optics and acoustics, this principle is often referred to as Huygens' principle which allows the optical or acoustic field generated by a given source to be computed by considering the field generated from a single point on the source and then summing up the field generated from a large collection of such points. In this sense, the principle behind a Green's function solution is effectively the same as Huygens' principle, i.e. find the solution to the problem for a single point and then integrate over all such points. The relationship between a point source which is described mathematically by the delta function and the Green's function is an important one which is why the delta function together with some other generalised functions have been discussed in Chapter 1.

By way of a short introduction to the use of the delta function for solving partial differential equations using Green's functions, we consider the following example which in comparison with the first example, provides a more complete form of analysis to develop a Green's function solution for the one-dimensional inhomogeneous wave equation

Example 5.2

Consider the equation

$$\left(\frac{\partial^2}{\partial x^2} + k^2\right) u(x,k) = f(x) \tag{5.1.2}$$

where k (the wavenumber) is a constant and $f(x)$ is the source term, the solution being required over all space $x \in (-\infty, \infty)$ subject to the conditions that u and $\partial u/\partial x$ are zero at $\pm\infty$. This equation describes the behaviour of "steady waves" (constant wavelength $\lambda = 2\pi/k$) due to a source $f(x)$.

Define the Green's function as being the solution to the equation obtained by replacing the source term with a delta function which represents a point source at x_0 say, giving the equation

$$\left(\frac{\partial^2}{\partial x^2} + k^2\right) g(x \mid x_0, k) = \delta(x - x_0) \tag{5.1.3}$$

where δ has the following fundamental property

$$\int_{-\infty}^{\infty} u(x)\delta(x - x_0)dx = u(x_0).$$

Multiplying equation 5.1.2 by g gives

$$g\left(\frac{\partial^2}{\partial x^2} + k^2\right)u = gf$$

and multiplying equation 5.1.3 by u gives

$$u\left(\frac{\partial^2}{\partial x^2} + k^2\right)g = u\delta(x - x_0).$$

Now subtract the two results and integrate to obtain

$$\int_{-\infty}^{\infty}\left(g\frac{\partial^2 u}{\partial x^2} - u\frac{\partial^2 g}{\partial x^2}\right)dx = \int_{-\infty}^{\infty} fgdx - \int_{-\infty}^{\infty} u\delta(x - x_0)dx.$$

Using the generalized sampling property of the delta function given above and rearranging the result, we obtain

$$u(x_0, k) = \int_{-\infty}^{\infty} fgdx - \int_{-\infty}^{\infty}\left(g\frac{\partial^2 u}{\partial x^2} - u\frac{\partial^2 g}{\partial x^2}\right)dx.$$

Evaluating the second integral on the right-hand side,

$$\int_{-\infty}^{\infty}\left(g\frac{\partial^2 u}{\partial x^2} - u\frac{\partial^2 g}{\partial x^2}\right)dx$$

$$= \int_{-\infty}^{\infty}\left[\frac{\partial}{\partial x}\left(g\frac{\partial u}{\partial x}\right) - \frac{\partial g}{\partial x}\frac{\partial u}{\partial x} - \frac{\partial}{\partial x}\left(u\frac{\partial g}{\partial x}\right) + \frac{\partial u}{\partial x}\frac{\partial g}{\partial x}\right]dx$$

$$= \int_{-\infty}^{\infty}\frac{\partial}{\partial x}\left(g\frac{\partial u}{\partial x}\right)dx - \int_{-\infty}^{\infty}\frac{\partial}{\partial x}\left(u\frac{\partial g}{\partial x}\right)dx$$

$$= \left[g\frac{\partial u}{\partial x}\right]_{-\infty}^{\infty} - \left[u\frac{\partial g}{\partial x}\right]_{-\infty}^{\infty}$$

$$= 0,$$

provided u and $\partial u/\partial x$ are zero at $x = \pm\infty$. With these conditions, we obtain the Green's function solution to equation 5.1.2 in the form

$$u(x_0, k) = \int_{-\infty}^{\infty} f(x)g(x \mid x_0, k)dx.$$

Note, that this result is a "type" of convolution – the convolution of the source function f with the Green's function g. Physically, Green's functions associated with wave type problems as in this example, represents the way in which a wave propagates from one point in space to another. For this reason, they are sometimes referred to as propagators. In this case, the Green's function is a function of the "path length" between x and x_0 irrespective of whether $x > x_0$ or $x < x_0$. The path length is given by $\mid x - x_0 \mid$ and the Green's function is a function of this path length which is why, using the notation $x \mid x_0 \equiv \mid x - x_0 \mid$, we write $g(x \mid x_0)$.

The solution given above is of little value unless the Green's function can be computed. Before the computational techniques associated with Green's functions can be studied, it is necessary to be familiar with a class of functions known as generalised functions and in particular, one of the most commonly used generalised function, the delta function. These functions were covered in Chapter 1, Section 1.8 which should now be studied by those readers who are not familiar with the delta function and its properties.

5.2 Green's Functions for the Time-independent Wave Equation

In this section, we shall concentrate on the computation of Green's functions for the time-independent wave equation in one, two and three dimensions. The solution is over all space and the Green's function is not constrained to any particular boundary conditions (except those at $\pm\infty$). It is therefore referred to as a free space Green's function.

The One-dimensional Green's Function

We start by reconsidering Example 2 given in Section 5.1 which, through the application of the sampling property of the delta function together with some relatively simple analysis, demonstrated that the solution to the inhomogeneous wave equation

$$\left(\frac{\partial^2}{\partial x^2} + k^2\right) u(x,k) = f(x)$$

for constant k and $x \in (-\infty, \infty)$ subject to the boundary conditions

$$u(x,k) \mid_{\pm\infty} = 0 \qquad \text{and} \qquad \left[\frac{\partial}{\partial x} u(x,k)\right]_{\pm\infty} = 0$$

is given by

$$u(x_0,k)=\int\limits_{-\infty}^{\infty} f(x)g(x\mid x_0,k)dx$$

where g is the Green's function. This solution is of course worthless without an expression for the Green's function which is given by the solution to the equation

$$\left(\frac{\partial^2}{\partial x^2}+k^2\right)g(x\mid x_0,k)=-\delta(x-x_0) \qquad (5.2.1)$$

subject to $g(x|x_0,k)|_{\pm\infty}=0$ and $[\partial g(x|x_0,k)/\partial x]_{\pm\infty}=0$. We shall therefore start by looking at the evaluation of the Green's function for this case. Note that in this case, the Green's function is defined for $-\delta$ on the right-hand side instead of $+\delta$. This is for convenience only in the computations which follow; it does not effect the analysis and is merely a convention which ultimately reduces the number of negative signs associated with the calculation. For this reason, many authors define the Green's function with $-\delta$ – a definition which is used throughout the rest of this chapter.

The solution to this equation is based on employing the properties of the Fourier transform. Writing $X=\mid x-x_0\mid$, we express g and δ as Fourier transforms, that is

$$g(X,k)=\frac{1}{2\pi}\int\limits_{-\infty}^{\infty} G(u,k)\exp(iuX)du \qquad (5.2.2)$$

and

$$\delta(X)=\frac{1}{2\pi}\int\limits_{-\infty}^{\infty}\exp(iuX)du.$$

Substituting these expressions into equation 5.2.1 and differentiating gives

$$\frac{1}{2\pi}\int\limits_{-\infty}^{\infty}(-u^2+k^2)G(u,k)\exp(iuX)du=-\frac{1}{2\pi}\int\limits_{-\infty}^{\infty}\exp(iuX)du$$

from which it follows that

$$G(u,k)=\frac{1}{u^2-k^2}.$$

Substituting this result back into equation 5.2.2, we obtain

$$\begin{aligned} g(X,k) &= \frac{1}{2\pi}\int\limits_{-\infty}^{\infty}\frac{\exp(iuX)}{u^2-k^2}du \\ &= \frac{1}{2\pi}\int\limits_{-\infty}^{\infty}\frac{\exp(iuX)}{(u-k)(u+k)}du. \end{aligned}$$

The problem is therefore reduced to that of evaluating the above integral. This can be done using Cauchy's integral formula,

$$\oint_C f(z)dz = 2\pi i \times (\text{sum of the residues enclosed by } C)$$

where C is the contour defining the path of integration. In order to evaluate the integral explicitly using this formula, we must consider the singular nature or poles of the integrand at $z = -k$ and $z = k$. For now let us consider a contour which encloses both poles. The residue at $z = k$ is given by $\exp(ikX)/(2k)$ and at $z = -k$ by $-\exp(-ikX)/(2k)$. Hence the Green's function is given by

$$g(X,k) = 2\pi i\left(\frac{\exp(ikX)}{4\pi k} - \frac{\exp(-ikX)}{4\pi k}\right) = -\frac{1}{k}\sin(kX).$$

This Green's function represents the propagation of waves travelling away from the point disturbance at $x = x_0$ or "outgoing waves" and also waves travelling towards the point disturbance or "incoming waves". Since x and x_0 are points along a line, we can consider the result to be the sum of waves travelling to the left of $\delta(x - x_0)$ in which $x < x_0$ and to the right of $\delta(x - x_0)$ where $x > x_0$. In most applications it is convenient to consider the Green's function for outgoing or (more rarely) incoming waves but not both. Here, the Green's function for incoming waves is given by

$$g(x \mid x_0, k) = -\frac{i}{2k}\exp(-ik \mid x - x_0 \mid)$$

and for outgoing waves is

$$g(x \mid x_0, k) = \frac{i}{2k}\exp(ik \mid x - x_0 \mid).$$

The Two-dimensional Green's Function

In two dimensions, the same method can be used to obtain the (free space) Green's function, that is to solve the equation

$$(\nabla^2 + k^2)g(\mathbf{r} \mid \mathbf{r}_0, k) = -\delta^2(\mathbf{r} - \mathbf{r}_0)$$

subject to some boundary conditions at $\mid \mathbf{r} \mid = \infty$, where

$$\mathbf{r} = \hat{\mathbf{x}}x + \hat{\mathbf{y}}y; \qquad \mathbf{r}_0 = \hat{\mathbf{x}}x_0 + \hat{\mathbf{y}}y_0$$

and

$$\nabla^2 = \frac{\partial^2}{\partial x^2} + \frac{\partial^2}{\partial y^2}.$$

Note that

$$\delta^2(\mathbf{r}-\mathbf{r}_0) \equiv \delta(x-x_0)\delta(y-y_0).$$

Also note that g is a function of the path length $\mid \mathbf{r}-\mathbf{r}_0 \mid$. Writing $\mathbf{R}=\mathbf{r}-\mathbf{r}_0$ and using the same technique as before, namely the one used to derive an integral representation of the one-dimensional Green's function, we obtain

$$g(R,k)=\frac{1}{(2\pi)^2}\int_{-\infty}^{\infty}\frac{\exp(i\mathbf{u}\cdot\mathbf{R})}{u^2-k^2}d^2\mathbf{u}.$$

In polar coordinates this result becomes

$$g(R,k)=\frac{1}{(2\pi)^2}\int_{0}^{\pi}\int_{-\infty}^{\infty}\frac{\exp(iuR\cos\theta)}{u^2-k^2}udud\theta.$$

Integrating over u first and using Cauchy's residue theorem, we have

$$\oint_C \frac{z\exp(izR\cos\theta)}{(z+k)(z-k)}dz = i\pi\exp(ikR\cos\theta)$$

where the contour of integration C has been chosen to enclose just one of the poles at $z=k$. This provides an expression for the "outgoing" Green's function in which the wave propagates away from the point disturbance at $\mathbf{r}_0$. A solution for the pole at $z=-k$ would provide a solution which represents a wavefield converging on $\mathbf{r}_0$. The "outgoing" Green's function is usually the most physically significant result (accept for an implosion for example). Thus, the (outgoing) Green's function can be written in the form

$$g(R,k)=\frac{i}{4\pi}\int_{0}^{\pi}\exp(ikR\cos\theta)d\theta.$$

Writing the Green's function in this form allows us to employ the result

$$H_0^{(1)}(kR)=\frac{1}{\pi}\int_{0}^{\pi}\exp(ikR\cos\theta)d\theta$$

where $H_0^{(1)}$ is the Hankel function (of the first kind and of order zero). This is the integral representation for the Hankel transform and it can be used to write the two-dimensional Green's function as

$$g(\mathbf{r}\mid\mathbf{r}_0,k)=\frac{i}{4}H_0^{(1)}(k\mid\mathbf{r}-\mathbf{r}_0\mid).$$

A useful form of this function can be obtained by employing the asymptotic approximation

$$H_0^{(1)}(kR)\simeq\sqrt{\frac{2}{\pi}}\exp(-i\pi/4)\frac{\exp(-ikR)}{\sqrt{kR}}$$

which is valid when

$$kR \gg 1.$$

This condition means that the wavelength of the wave originating from $\mathbf{r}_0$ is very small compared with the distance between $\mathbf{r}_0$ and $\mathbf{r}$ which is physically reasonable in many cases and so a two-dimensional Green's function of the following form can be used:

$$g(\mathbf{r} \mid \mathbf{r}_0, k) = \frac{1}{\sqrt{8\pi}} \exp(i\pi/4) \frac{\exp(ik \mid \mathbf{r} - \mathbf{r}_0 \mid)}{\sqrt{k \mid \mathbf{r} - \mathbf{r}_0 \mid}}.$$

The Three-dimensional Green's Function

In three dimensions, the free space Green's function is given by the solution to the equation

$$(\nabla^2 + k^2) g(\mathbf{r} \mid \mathbf{r}_0, k) = -\delta^3(\mathbf{r} - \mathbf{r}_0)$$

with boundary conditions at $\mid \mathbf{r} \mid = \infty$, where

$$\mathbf{r} = \hat{\mathbf{x}}x + \hat{\mathbf{y}}y + \hat{\mathbf{z}}z; \qquad \mathbf{r}_0 = \hat{\mathbf{x}}x_0 + \hat{\mathbf{y}}y_0 + \hat{\mathbf{z}}z_0;$$
$$\delta^3(\mathbf{r} - \mathbf{r}_0) \equiv \delta(x - x_0)\delta(y - y_0)\delta(z - z_0)$$

and

$$\nabla^2 = \frac{\partial^2}{\partial x^2} + \frac{\partial^2}{\partial y^2} + \frac{\partial^2}{\partial z^2}.$$

In this case

$$g(R, k) = \frac{1}{(2\pi)^3} \int_{-\infty}^{\infty} \frac{\exp(i\mathbf{u} \cdot \mathbf{R})}{u^2 - k^2} d^3\mathbf{u}.$$

It proves convenient to evaluate this integral using spherical polar coordinates which gives

$$g(R, k) = \frac{1}{(2\pi)^3} \int_0^{2\pi} d\phi \int_{-1}^{1} d(\cos\theta) \int_0^{\infty} \frac{\exp(iuR\cos\theta) u^2}{u^2 - k^2} du.$$

Integrating over ϕ and θ we then obtain

$$g(R, k) = \frac{1}{2\pi^2 R} \int_0^{\infty} \frac{u \sin(uR)}{u^2 - k^2} du.$$

Since the integrand is an even function we may extend the integration to include the interval $-\infty$ to 0 by writing

$$g(R, k) = \frac{1}{4\pi^2 R} \int_{-\infty}^{\infty} \frac{u \sin(uR)}{u^2 - k^2} du.$$

This is done in anticipation of using Cauchy's residue theorem to evaluate the contour integral

$$\oint_C \frac{z\exp(izR)}{(z-k)(z+k)}dz$$

which has simple poles at $z = \pm k$. Choosing the contour C to enclose the pole at $z = k$, the residue is $\exp(ikR)/2$ and thus, the outgoing Green's becomes

$$g(\mathbf{r} \mid \mathbf{r}_0, k) = \frac{1}{4\pi \mid \mathbf{r} - \mathbf{r}_0 \mid} \exp(ik \mid \mathbf{r} - \mathbf{r}_0 \mid).$$

Observe, that in one, two and three dimensions the Green's function is singular. The precise nature of the singularity changes from one dimension to the next. In three dimensions, the Green's function is spatially singular when $\mathbf{r} = \mathbf{r}_0$ whereas in one dimension the singularity is temporal, that is the singularity occurs when $k = 0$. In two dimensions, the Green's function is a Hankel function whose argument is $k \mid \mathbf{r} - \mathbf{r}_0 \mid$ and has both a temporal and spatial singularity which occurs when either $k = 0$ or $\mathbf{r} = \mathbf{r}_0$. A good example of this two-dimensional Green's function is observed when a small stone falls vertically into a large pool of water. The symmetrical expanding wavefront represents the result of applying a short impulse to the surface of the water. What is observed is a good approximation to a Hankel function!

Asymptotic Forms

Although the Green's functions for the inhomogeneous wave equation can be computed in the manner given above, their algebraic form is not always easy or useful to work with. For this reason, it is worth considering their asymptotic forms which relate to the case when the field generated by a point source is a long distance away from that source. Thus, asymptotic approximations for these Green's functions are based on considering the case where the source at $\mathbf{r}_0$ is moved further and further away from the observer at $\mathbf{r}$. There are two approximations which are important in this respect which are often referred to as the Fraunhofer and Fresnel approximations. These approximations are usually associated with the applications of Green's functions in optics (in which both Fraunhofer and Fresnel undertook their original work) but are in fact of general applicability and will be used later in this chapter. Joseph Fraunhofer started out as a lens grinder and established an optics company in the early 19th century. He developed the theory of Fraunhofer diffraction in 1823.

In one dimension, we do not have an asymptotic approximation as such. We note however, that

$$|x - x_0| = \begin{cases} x_0 - x, & x_0 > x; \\ x - x_0, & x > x_0 \end{cases}$$

so that the Green's function for a left-travelling wave can be written as

$$g(x \mid x_0, k) = \frac{i}{2k} \exp(ikx_0) \exp(-ikx)$$

and for a right-travelling wave,

$$g(x \mid x_0, k) = \frac{i}{2k} \exp(-ikx_0) \exp(ikx).$$

In two and three dimensions, we expand the path length between the source and observer in terms of their respective coordinates. First, let us look at the result in two dimensions. In this case,

$$\begin{aligned} \mid \mathbf{r} - \mathbf{r}_0 \mid &= \sqrt{r_0^2 + r^2 - 2\mathbf{r} \cdot \mathbf{r}_0} \\ &= r_0 \left(1 - \frac{2\mathbf{r} \cdot \mathbf{r}_0}{r_0^2} + \frac{r^2}{r_0^2}\right)^{\frac{1}{2}} \end{aligned}$$

where $\mathbf{r} = \hat{\mathbf{x}}x + \hat{\mathbf{y}}y$, $r = \mid \mathbf{r} \mid$ and $r_0 = \mid \mathbf{r}_0 \mid$. A binomial expansion of this result gives

$$\mid \mathbf{r} - \mathbf{r}_0 \mid = r_0 \left(1 - \frac{\mathbf{r} \cdot \mathbf{r}_0}{r_0^2} + \frac{r^2}{2r_0^2} + \cdots\right) \tag{5.2.3}$$

which under the condition

$$\frac{r}{r_0} \ll 1$$

reduces to

$$\mid \mathbf{r} - \mathbf{r}_0 \mid \simeq r_0 - \hat{\mathbf{n}}_0 \cdot \mathbf{r}$$

where

$$\hat{\mathbf{n}}_0 = \frac{\mathbf{r}_0}{r_0}.$$

It is sufficient to let

$$\frac{1}{\mid \mathbf{r} - \mathbf{r}_0 \mid} \simeq \frac{1}{r_0}, \qquad r_0 \gg r$$

because small changes in $\hat{\mathbf{n}} \cdot \mathbf{r}$ compared with r_0 are not significant in an expression of this type. However, with the exponential function

$$\exp[ik(r_0 - \hat{\mathbf{n}}_0 \cdot \mathbf{r})]$$

a relatively small change in the value of $r_0 - \hat{\mathbf{n}}_0 \cdot \mathbf{r}$ compared with r_0 will still cause this term to oscillate rapidly, particularly if the value of k is large. We therefore write

$$\exp(ik \mid \mathbf{r} - \mathbf{r}_0 \mid) = \exp(ikr_0) \exp(-ik\hat{\mathbf{n}}_0 \cdot \mathbf{r}).$$

The asymptotic form of the two-dimensional Green's function is then given by

$$g(\mathbf{r} \mid \mathbf{r}_0, k) = \frac{\exp(i\pi/4)}{\sqrt{8\pi}} \frac{1}{\sqrt{kr_0}} \exp(ikr_0) \exp(-ik\hat{\mathbf{n}}_0 \cdot \mathbf{r}), \qquad kr_0 \gg 1.$$

In three dimensions, the result is (using exactly the same arguments as in the two-dimensional case)

$$g(\mathbf{r} \mid \mathbf{r}_0, k) = \frac{1}{4\pi r_0} \exp(ikr_0) \exp(-ik\hat{\mathbf{n}}_0 \cdot \mathbf{r})$$

where

$$\mathbf{r} = \hat{\mathbf{x}}x + \hat{\mathbf{y}}y + \hat{\mathbf{z}}z.$$

Notice that when we observe the field described by a Green's function at large distances (that is the field generated by a point source a long distance away), it behaves like a plane wave $\exp(-ik\hat{\mathbf{n}}_0 \cdot \mathbf{r})$. Approximating the Green's function in this way provides a description for the wave in what is commonly referred to as the far field or Fraunhofer zone (or plane). This approximation is often referred to as the Fraunhofer approximation in physical optics. In this zone, the wavefront which reaches the observer is a plane wavefront because the divergence of the field is so small. Observations of a field in this zone are said to be in the Fourier plane because, as we shall see later in this chapter, they lead to expressions for the wavefield which involve a Fourier transform. This is the basis for Fraunhofer diffraction theory which is important in applications such as x-ray crystallography, electromagnetic and acoustic imaging and of course modern optics.

When the source is brought closer to the observer, the wavefront ceases to be a plane wavefront. In this case, the Fraunhofer approximation is inadequate and another approximation for the Green's function must be used. This is known as the Fresnel approximation and is based on incorporating the next term in the binomial expansion of $\mid \mathbf{r} - \mathbf{r}_0 \mid$, namely the quadratic term $r^2/2r_0^2$ in equation 5.2.3. In this case, it is assumed that $r^2/r_0^2 \ll 1$ rather than $r/r_0 \ll 1$ so that all the terms in the binomial expansion of $\mid \mathbf{r} - \mathbf{r}_0 \mid$ that occur after the quadratic term can be neglected. As before, $\mid \mathbf{r} - \mathbf{r}_0 \mid^{-1}$ is approximated by $1/r_0$ but the exponential term now possesses an additional feature, namely a quadratic phase factor. In this case, the two- and three-dimensional Green's functions are given by

$$g(\mathbf{r} \mid \mathbf{r}_0, k) = \frac{\exp(i\pi/4)}{\sqrt{8\pi}} \frac{\exp(ikr_0)}{\sqrt{kr_0}} \exp(-ik\hat{\mathbf{n}}_0 \cdot \mathbf{r}) \exp(ir^2/2r_0), \qquad kr_0 \gg 1$$

and

$$g(\mathbf{r} \mid \mathbf{r}_0, k) = \frac{\exp(ikr_0)}{4\pi r_0} \exp(-ik\hat{\mathbf{n}}_0 \cdot \mathbf{r}) \exp(ir^2/2r_0)$$

respectively. This type of approximation is used in the study of systems (optical systems for example) in which the divergence of the field is a measurable quantity. If the source is moved even closer to the observer then neither the Fraunhofer nor the Fresnel approximations will apply. In such cases, it is usually easier to retain the Green's function in full rather than consider another term in the binomial expansion of the path length. Analysis of a wavefield that is produced when a non-asymptotic form of the Green's function is used is referred to as near-field analysis. Thus, the Green's function solution to two- and three-dimensional wave-type partial differential equations usually falls into one of the following three categories:

(i) near-field analysis;

(ii) intermediate field (Fresnel zone) analysis;

(iii) far-field (Fraunhofer zone of Fourier plane) analysis.

In practice, the far-field approximation is much easier to use. This is because it leads to solutions that can be written in terms of a Fourier transform which is a relatively easy transform to work with and invert. Using the Fresnel approximation leads to solutions which involve a class of integral known as the Fresnel integral. The nonlinear behaviour of this integral, because of the quadratic phase factor, makes it more difficult to evaluate compared with the Fourier integral. There are relatively few applications in wavefield theory which require a full near-field analysis. This is fortunate because near-field analysis presents some formidable computational problems.

5.3 Green's Function Solution to the Three-dimensional Inhomogeneous Wave Equation

In the previous section, the free space Green's functions for the inhomogeneous time-independent wave equation were considered in one, two and three dimensions. In this section, we turn our attention to the more general problem of developing a solution for the wavefield $u(\mathbf{r}, k)$ generated by an arbitrary and time-independent source function $f(\mathbf{r})$. Working in three dimensions, our aim is to solve

$$(\nabla^2 + k^2)u(\mathbf{r}, k) = -f(\mathbf{r}), \qquad \mathbf{r} \in V$$

for u where V is the volume of the source function which is of compact support. Note that we define the source term as $-f$ rather than $+f$. This is done so that there is consistency with the definition of the Green's function which is defined in terms of $-\delta$ by convention. We start by writing the equation for a Green's function, i.e.

$$(\nabla^2 + k^2)g(\mathbf{r} \mid \mathbf{r}_0, k) = -\delta^3(\mathbf{r} - \mathbf{r}_0).$$

If we now multiply both sides of the first equation by g and both sides of the second equation by u, then by subtracting the two results we obtain

$$g\nabla^2 u - u\nabla^2 g = -gf + u\delta^3.$$

We assume that the source is confined to a finite region of space with a finite volume V. Outside this region, it is assumed that the source function is zero. By integrating the last equation over V, we can exploit the result

$$\int u(\mathbf{r},k)\delta^3(\mathbf{r}-\mathbf{r}_0)d^3\mathbf{r} = u(\mathbf{r}_0,k)$$

and therefore write

$$\begin{aligned} u(\mathbf{r}_0,k) &= \int_V f(\mathbf{r})g(\mathbf{r}\mid\mathbf{r}_0,k)d^3\mathbf{r} \\ &+ \int_V [g(\mathbf{r}\mid\mathbf{r}_0,k)\nabla^2 u(\mathbf{r},k) - u(\mathbf{r},k)\nabla^2 g(\mathbf{r}\mid\mathbf{r}_0,k)]d^3\mathbf{r}. \end{aligned}$$

Observe, that this expression is not a proper solution for u because this function occurs in both the left- and right-hand sides. We require a solution for u in terms of known quantities on the right-hand side of the above equation. To this end, we can simplify the second term by using Green's theorem:

$$\int_V (g\nabla^2 u - u\nabla^2 g)d^3\mathbf{r} = \oint_S (g\nabla u - u\nabla g)\cdot\hat{\mathbf{n}}d^2\mathbf{r}.$$

Here, S defines the surface enclosing the volume V and $d^2\mathbf{r}$ is an element of this surface. The unit vector $\hat{\mathbf{n}}$ points out of the surface and is perpendicular to the surface element $d^2\mathbf{r}$. Green's theorem is a special but important consequence of Gauss' divergence theorem as shown below.

Green's Theorem

Let u and g be any two piecewise continuous functions of position and S be a closed surface surrounding a volume V. If u, g and their first and second partial derivatives are single-valued and continuous within and on S, then

$$\int_V (g\nabla^2 u - u\nabla^2 g)d^3\mathbf{r} = \oint_S \left(g\frac{\partial u}{\partial\hat{\mathbf{n}}} - u\frac{\partial g}{\partial\hat{\mathbf{n}}}\right)d^2\mathbf{r}$$

where $\partial/\partial\hat{\mathbf{n}}$ is a partial derivative in the outward normal direction on S.

The proof of this result stems from noting that since

$$\nabla\cdot(g\nabla u) = \nabla g\cdot\nabla u + g\nabla^2 u$$

and

$$\nabla\cdot(u\nabla g) = \nabla u\cdot\nabla g + u\nabla^2 g$$

then

$$\int_V \nabla \cdot (g\nabla u - u\nabla g)d^3\mathbf{r} = \int_V (g\nabla^2 u - u\nabla^2 g)d^3\mathbf{r}.$$

However from Gauss' theorem

$$\int_V \nabla \cdot \mathbf{F}d^3\mathbf{r} = \oint_S \mathbf{F} \cdot \hat{\mathbf{n}}d^2\mathbf{r}$$

for any vector $\mathbf{F}$. Hence,

$$\int_V \nabla \cdot (g\nabla^2 u - u\nabla^2 g)d^3\mathbf{r} = \oint_S (g\nabla u - u\nabla g) \cdot \hat{\mathbf{n}}d^2\mathbf{r}$$

which provides the basic result, a result can be written in an alternative (and arguably more elegant way) by defining

$$\nabla u \cdot \hat{\mathbf{n}} \equiv \frac{\partial u}{\partial \hat{\mathbf{n}}}$$

and

$$\nabla g \cdot \hat{\mathbf{n}} \equiv \frac{\partial g}{\partial \hat{\mathbf{n}}}$$

so that we can write

$$\int_V (g\nabla^2 u - u\nabla^2 g)d^3\mathbf{r} = \oint_S \left(g\frac{\partial u}{\partial \hat{\mathbf{n}}} - u\frac{\partial g}{\partial \hat{\mathbf{n}}} \right) d^2\mathbf{r}.$$

This theorem provides a solution for the wavefield u at $\mathbf{r}_0$ of the form

$$u(\mathbf{r}_0, k) = \int_V fgd^3\mathbf{r} + \oint_S (g\nabla u - u\nabla g) \cdot \hat{\mathbf{n}}d^2\mathbf{r}. \tag{5.3.1}$$

The Dirichlet and Neumann Boundary Conditions

Although Green's theorem allows us to simplify the solution for u (in the sense that we now have a two-dimensional instead of a three-dimensional integral), we still do not have a proper solution for u since this field variable is present on both the left- and right-hand sides of equation 5.3.1. However, as a result of applying Green's theorem we now only need to specify u and ∇u on the surface S. Therefore, if we know, *a priori*, the behaviour of u and ∇u on S we can compute u at any other observation point $\mathbf{r}_0$ from equation 5.3.1. Clearly, some sort of statement about the behaviour of u and ∇u on S is required, that is the boundary conditions need to be specified.

In general, the type of conditions that may be applied depends on the applications that are involved. In practice, two types of boundary conditions are commonly considered. The first one is known as the homogeneous Dirichlet boundary condition which states that u is zero on S and the second one is known as the homogeneous Neumann condition which states that ∇u is zero on S. Taken together, these boundary conditions are known as the "homogeneous conditions" and are referred to as such throughout the rest of this chapter. When u satisfies these homogeneous boundary conditions, the solution for u is given by

$$u(\mathbf{r}_0, k) = \int_V f(\mathbf{r})g(\mathbf{r} \mid \mathbf{r}_0, k)d^3\mathbf{r} \tag{5.3.2}$$

because

$$\oint_S (g\nabla u - u\nabla g) \cdot \hat{\mathbf{n}} d^2\mathbf{r} = 0.$$

If the field generated by a source is measured a long distance away from the location of the source, then by using the far field approximation for the Green's function discussed in Section 5.2, we have (in three dimensions)

$$u(\hat{\mathbf{n}}_0, k) = \frac{1}{4\pi r_0}\exp(ikr_0)\int_V f(\mathbf{r})\exp(-ik\hat{\mathbf{n}}_0 \cdot \mathbf{r})d^3\mathbf{r}.$$

In this case, the field generated by the source is given by the three-dimensional Fourier transform of the source function f. By measuring the radiation pattern produced by a source in the near field, the structure or spatial distribution of the source may be recovered by solving equation 5.3.2 for $f(\mathbf{r})$. In the far field, the source function can be recovered by taking the inverse Fourier transform of the observed field. This is an example of a solution to a class of problem known as an inverse source problem.

Reciprocity Theorem

We shall close this section with an important theorem which applies to all Green's functions associated with any linear partial differential equation. Here, the reciprocity theorem will be proved for the three-dimentsional Green's function corresponding to the time-independent wave equation. The theorem states that if $\mathbf{r}_1$ and $\mathbf{r}_2$ are two points in space then

$$g(\mathbf{r}_1 \mid \mathbf{r}_2, k) = g(\mathbf{r}_2 \mid \mathbf{r}_1, k).$$

The proof of this result can be obtained by considering the equations

$$(\nabla^2 + k^2)g(\mathbf{r} \mid \mathbf{r}_1, k) = -\delta^3(\mathbf{r} - \mathbf{r}_1)$$

and

$$(\nabla^2 + k^2)g(\mathbf{r} \mid \mathbf{r}_2, k) = -\delta^3(\mathbf{r} - \mathbf{r}_2).$$

Then

$$g(\mathbf{r} \mid \mathbf{r}_2, k)\nabla^2 g(\mathbf{r} \mid \mathbf{r}_1, k) - g(\mathbf{r} \mid \mathbf{r}_1, k)\nabla^2 g(\mathbf{r} \mid \mathbf{r}_2, k)$$
$$= g(\mathbf{r} \mid \mathbf{r}_1, k)\delta^3(\mathbf{r} - \mathbf{r}_2) - g(\mathbf{r} \mid \mathbf{r}_2, k)\delta^3(\mathbf{r} - \mathbf{r}_1).$$

Integrating over V, using Green's theorem and assuming homogeneous boundary conditions (i.e. Dirichlet and Neumann boundary conditions) on the surface of V we have

$$\int_V g(\mathbf{r} \mid \mathbf{r}_1, k)\delta^3(\mathbf{r} - \mathbf{r}_1)d^3\mathbf{r} - \int_V g(\mathbf{r} \mid \mathbf{r}_2, k)\delta^3(\mathbf{r} - \mathbf{r}_2)d^3\mathbf{r} = 0$$

or

$$g(\mathbf{r}_2 \mid \mathbf{r}_1, k) = g(\mathbf{r}_1 \mid \mathbf{r}_2, k).$$

Thus, the propagation of a wave from a point at $\mathbf{r}_1$ to $\mathbf{r}_2$ is the same as the propagation of a wave from a point at $\mathbf{r}_2$ to $\mathbf{r}_1$.

5.4 Green's Function Solutions to the Inhomogeneous Helmholtz and Schrödinger Equations: An Introduction to Scattering Theory

The inhomogeneous Helmholtz and Schrödinger equations have been studied for many years and have a wide range of applications in solid state physics, quantum mechanics, electromagnetic and acoustic wave propagation and scattering to name but a few.

In quantum mechanics, elementary particles of matter appear (to the human observer) to behave like waves called deBroglie waves. The mechanics becomes that of wave motion and wavefunctions are used to describe the behaviour of quantum systems. The square modulus of the wavefunction is taken to represent the probability of a particle existing at a given point in space. For this reason, the wavefunctions are sometimes called probability waves which can be scattered by an atomic or nuclear potential $V(\mathbf{r})$. If the potential is an elastic scatterer and the deBroglie waves describe non-relativistic particles then the (time-independent) partial differential equation that best describes this (scattering) effect is

$$(\nabla^2 + k^2)u(\mathbf{r}, k) = V(\mathbf{r})u(\mathbf{r}, k)$$

where k is the wavenumber and u is the scattered field and $|u|^2$ is its intensity. This equation is known as the Schrödinger equation after the Austrian physicist Erwin Schrödinger who postulated it in the 1920s. Comparing this equation with the partial differential equation studied in the previous section, it is clear that the equation for a deBroglie wavefield is produced by replacing the source function f with $-Vu$.

Another fundamental wave equation of importance in electromagnetics and acoustics is the inhomogeneous Helmholtz equation which is given by

$$(\nabla^2 + k^2)u(\mathbf{r}, k) = -k^2\gamma(\mathbf{r})u(\mathbf{r}, k)$$

where γ is an inhomogeneity which is responsible for scattering the wavefield u and is therefore sometimes referred to as a scatterer – usually considered to be of compact support.

In this section we shall consider the solution to these types of wave equations which for k constant are essentially the same. We shall start by investigating the Green's function solution to the inhomogeneous Helmholtz equation.

Basic Solution to the Inhomogeneous Helmholtz Equation

The same Green's function method that has already been presented in Section 5.3 can be used to solve the inhomogeneous Helmholtz equation. The basic solution is (under the assumption that γ is of compact support $\mathbf{r} \in V$)

$$u(\mathbf{r}_0, k) = k^2 \int\limits_V g\gamma u d^3\mathbf{r} + \oint\limits_S (g\nabla u - u\nabla g) \cdot \hat{\mathbf{n}} d^2\mathbf{r}.$$

Once again, to compute the surface integral, a condition for the behaviour of u on the surface S of γ must be chosen. Consider the case where the incident wavefield u_i is a simple plane wave of unit amplitude

$$\exp(i\mathbf{k} \cdot \mathbf{r})$$

satisfying the homogeneous wave equation

$$(\nabla^2 + k^2)u_i(\mathbf{r}, k) = 0.$$

By choosing the condition $u(\mathbf{r}, k) = u_i(\mathbf{r}, k)$ on the surface of γ, we obtain the result

$$u(\mathbf{r}_0, k) = k^2 \int\limits_V g\gamma u d^3\mathbf{r} + \oint\limits_S (g\nabla u_i - u_i\nabla g) \cdot \hat{\mathbf{n}} d^2\mathbf{r}.$$

Now, using Green's theorem to convert the surface integral back into a volume integral, we have

$$\oint\limits_S (g\nabla u_i - u_i\nabla g) \cdot \hat{\mathbf{n}} d^2\mathbf{r} = \int\limits_V (g\nabla^2 u_i - u_i\nabla^2 g) d^3\mathbf{r}.$$

Noting that

$$\nabla^2 u_i = -k^2 u_i$$

and that

$$\nabla^2 g = -\delta^3 - k^2 g$$

we obtain

$$\int_V (g\nabla^2 u_i - u_i \nabla^2 g) d^3\mathbf{r} = \int \delta^3 u_i d^3\mathbf{r} = u_i.$$

Hence, by choosing the field u to be equal to the incident wavefield u_i on the surface of γ, we obtain a solution of the form

$$u = u_i + u_s$$

where

$$u_s = k^2 \int_V g\gamma u d^3\mathbf{r}.$$

The wavefield u_s is often referred to as the scattered wavefield.

The Born Approximation

From the last result it is clear that in order to compute the scattered field u_s, we must define u inside the volume integral. Unlike the surface integral, a boundary condition will not help here because it is not sufficient to specify the behaviour of u at a boundary. In this case, the behaviour of u throughout V needs to be known. In general, it is not possible to do this (i.e. to compute the scattered wavefield exactly – at least to date) and we are forced to choose a model for u inside V that is compatible with a particular physical problem in the same way that an appropriate set of boundary conditions are required to evaluate the surface integral. The simplest model for the internal field is based on assuming that u behaves like u_i for $\mathbf{r} \in V$. The scattered field is then given by

$$u_s(\mathbf{r}_0, k) = k^2 \int_V g(\mathbf{r} \mid \mathbf{r}_0, k)\gamma u_i(\mathbf{r}, k) d^3\mathbf{r}.$$

This assumption provides an approximate solution for the scattered field and is known as the Born approximation after Max Born who first introduced the approximation in the study of quantum mechanics in the 1920s.

There is another way of deriving this result that is instructive and helps us to obtain criteria for the validity of this approximation which will be considered shortly. We start with the inhomogeneous Helmholtz equation

$$(\nabla^2 + k^2)u = -k^2 \gamma u$$

and consider a solution for u in terms of a sum of the incident and scattered fields, i.e.

$$u = u_i + u_s.$$

The wave equation then becomes

$$(\nabla^2 + k^2)u_s + (\nabla^2 + k^2)u_i = -k^2\gamma(u_i + u_s).$$

If the incident field satisfies

$$(\nabla^2 + k^2)u_i = 0,$$

then

$$(\nabla^2 + k^2)u_s = -k^2\gamma(u_i + u_s).$$

Assuming that

$$u_i + u_s \simeq u_i, \qquad \mathbf{r} \in V$$

we obtain

$$(\nabla^2 + k^2)u_s \simeq -k^2\gamma u_i.$$

Solving for u_s and using the homogeneous boundary conditions (i.e. $u_s = 0$ on S and $\nabla u_s = 0$ on S) we obtain

$$u_s = \oint_S (g\nabla u_s - u_s\nabla g) \cdot \hat{\mathbf{n}} d^2\mathbf{r} + k^2 \int_V g\gamma u_i d^3\mathbf{r}$$

$$= k^2 \int_V g\gamma u_i d^3\mathbf{r}.$$

This is a first-order approximation to the problem. Higher-order approximations are discussed in Section 5.7 on the Born series solution.

Conditions for the Validity of the Born Approximation

In general, the Born approximation requires that u_s is "small" compared with u_i. The question is: what do we really mean by the term "small" and how can we quantify it? One way to answer this question is to compute an appropriate measure for both the incident and scattered fields and compare the two results. Consider the case where we compute the root mean square modulus (i.e. the ℓ_2 norm) of each field. We then require

$$\left(\int_V | u_s(\mathbf{r}_0, k) |^2 d^3\mathbf{r}_0\right)^{\frac{1}{2}} \ll \left(\int_V | u_i(\mathbf{r}_0, k) |^2 d^3\mathbf{r}_0\right)^{\frac{1}{2}}$$

or

$$\frac{\|u_s\|}{\|u_i\|} \ll 1. \tag{5.4.1}$$

Essentially, this condition means that the average intensity u_s in V is small compared with the average intensity of u_i in V.

Let us now look in more detail at the nature of this condition. Ideally, what we want is a version of this condition in terms of a set of physical parameters (such as the wavelength and the physical extent of γ for example). The Born scattered field at $\mathbf{r}_0$ is given by

$$u_s(\mathbf{r}_0, k) = k^2 \int_V g(\mathbf{r} \mid \mathbf{r}_0, k)\gamma(\mathbf{r})u_i(\mathbf{r}, k)d^3\mathbf{r}.$$

By taking the ℓ_2 norm of this equation we can write

$$\begin{aligned} \|u_s(\mathbf{r}_0, k)\| &= \left\| k^2 \int_V g(\mathbf{r} \mid \mathbf{r}_0, k)\gamma(\mathbf{r})u_i(\mathbf{r}, k)d^3\mathbf{r} \right\| \\ &\leq k^2 \|u_i(\mathbf{r}_0, k)\| \times \left\| \int_V g(\mathbf{r} \mid \mathbf{r}_0, k)\gamma(\mathbf{r})d^3\mathbf{r} \right\|. \end{aligned}$$

Using this result, the condition required for the Born approximation to hold (i.e. condition 5.4.1) can be written as

$$k^2 \left\| \int_V g(\mathbf{r} \mid \mathbf{r}_0, k)\gamma(\mathbf{r})d^3\mathbf{r} \right\| \ll 1, \qquad \mathbf{r}_0 \in V. \tag{5.4.2}$$

Here, the norm involves integration over the spatial variable $\mathbf{r}_0$ in the scattering volume V. To emphasize this we write $\mathbf{r}_0 \in V$. Although condition 5.4.2 provides an credible condition for the Born approximation, it is not, in itself, a particularly useful result.

To achieve a more useful result in terms of an appropriate set of variables, we resort to comparing just the moduli of u_i and u_s. In this case, we require that

$$\frac{| u_s |}{| u_i |} \ll 1, \qquad \forall \ \mathbf{r}_0 \in V.$$

Since

$$\begin{aligned} | u_s(\mathbf{r}_0, k) | &= \left| k^2 \int_V g(\mathbf{r} \mid \mathbf{r}_0, k)\gamma(\mathbf{r})u_i(\mathbf{r}, k)d^3\mathbf{r} \right| \\ &\leq k^2 \mid u_i(\mathbf{r}_0, k) \mid \times \left| \int_V g(\mathbf{r} \mid \mathbf{r}_0, k)\gamma(\mathbf{r})d^3\mathbf{r} \right| \end{aligned}$$

we can write this condition in the form

$$I(\mathbf{r}_0) \ll 1$$

where

$$\begin{aligned} I(\mathbf{r}_0) &= k^2 \left| \int_V g(\mathbf{r} \mid \mathbf{r}_0, k)\gamma(\mathbf{r}) d^3\mathbf{r} \right| \\ &\leq k^2 \left(\int_V \mid g(\mathbf{r} \mid \mathbf{r}_0, k) \mid^2 d^3\mathbf{r} \int_V \mid \gamma(\mathbf{r}) \mid^2 d^3\mathbf{r} \right)^{\frac{1}{2}} . \end{aligned}$$

Substituting the expression for the three-dimensional Green's function into the above expression, we have

$$I(\mathbf{r}_0) \leq k^2 \left(\frac{1}{16\pi^2} \int_V \frac{1}{\mid \mathbf{r} - \mathbf{r}_0 \mid^2} d^3\mathbf{r} \int_V \mid \gamma(\mathbf{r}) \mid^2 d^3\mathbf{r} \right)^{\frac{1}{2}} .$$

A relatively simple calculation can now be performed if we consider γ to be a sphere of volume V and radius R. However, even in this case, it is still difficult to evaluate the first integral over $\mathbf{r}$ for all values of $\mathbf{r}_0$. We therefore resort to calculating its least upper bound which occurs when $\mathbf{r}_0 = 0$. Changing to spherical polar coordinates, we have

$$\begin{aligned} \sup \int_V \frac{1}{\mid \mathbf{r} - \mathbf{r}_0 \mid^2} d^3\mathbf{r} &= \int_V \frac{1}{r^2} d^3\mathbf{r} = \int_0^{2\pi} \int_{-1}^{1} \int_0^{R} dr d(\cos\theta) du \\ &= 4\pi R \end{aligned}$$

where sup denotes the **supremum** over all $\mathbf{r}_0 \in V$. Using this result, we can write

$$\sup I(\mathbf{r}_0) \leq k^2 \left(\frac{R}{4\pi} \int_V \mid \gamma(\mathbf{r}) \mid^2 d^3\mathbf{r} \right)^{\frac{1}{2}}$$

and noting that

$$V = \int_V d^3\mathbf{r} = \frac{4}{3}\pi R^3$$

we obtain

$$\sup I(\mathbf{r}_0) \leq \frac{1}{\sqrt{3}} k^2 R^2 \bar{\gamma}$$

where

$$\bar{\gamma} = \sqrt{\frac{\int \mid \gamma \mid^2 d^3\mathbf{r}}{\int d^3\mathbf{r}}}.$$

Hence, the condition for the Born approximation to apply becomes (ignoring $\sqrt{3}$)

$$k^2R^2\bar{\gamma} \ll 1$$

or

$$\bar{\gamma} \ll \frac{1}{k^2R^2}.$$

This condition demonstrates that in principle, large values of γ can occur so long as its root mean square value over the volume V is small compared with $1/k^2R^2$. In scattering theory, γ is said to be a "weak scatterer". Note that when k or R approach zero, this condition is easy to satisfy and that Born scattering is more likely to occur in situations where

$$\frac{\lambda}{R} \gg 1$$

where λ is the wavelength (note that $k = 2\pi/\lambda$). If

$$\frac{\lambda}{R} \sim 1$$

then the value of $\bar{\gamma}$ must be small for Born scattering to occur.

By repeating the method given above, it is easy to show that in two dimensions the condition required for the Born approximation to apply is given by

$$\bar{\gamma} \ll \frac{1}{(kR)^{3/2}}$$

where R is the radius of a disc of area A and $\bar{\gamma}$ is now defined as the root mean square over A.

In one dimension, the result is

$$\bar{\gamma} \ll \frac{1}{kL}$$

where L is the length of the scatterer and $\bar{\gamma}$ is defined as the root mean square over L. In both cases we use the same Green's function solution to solve the two-dimensional and one-dimensional inhomogeneous Helmholtz equations, respectively. In each case, we assume that the scattered field is, on average, weak compared with the incident field. We may consider the term "weak", to imply that the total energy associated with u_s inside the inhomogeneity γ is small compared with u_i outside the scatterer.

Asymptotic Born Scattering

By measuring u_s, we can attempt to invert the relevant integral equation and hence recover or reconstruct γ. This type of problem is known as the inverse scattering problem, and solutions to this problem are called inverse scattering solutions. This subject is one of the most fundamental and difficult problems to solve in mathematical physics and is the subject of continuing research in the area of inverse problems in general. The simplest type of inverse scattering problem occurs when a Born scattered wavefield is measured in the far field or Fraunhofer zone (i.e. when the Green's functions takes on its asymptotic form discussed in Section 5.2).

From previous results, working in three dimensions, we know that when the incident field is a (unit) plane wave

$$u_i = \exp(ik\hat{\mathbf{n}}_i \cdot \mathbf{r})$$

where $\hat{\mathbf{n}}_i$ points in the direction of the incident field, the Born scattered field observed at $\mathbf{r}_s$ becomes

$$u_s(\hat{\mathbf{n}}_s, \hat{\mathbf{n}}_i, k) = \frac{k^2}{4\pi r_s} \exp(ikr_s) \int\limits_V \exp[-ik(\hat{\mathbf{n}}_s - \hat{\mathbf{n}}_i) \cdot \mathbf{r}]\gamma(\mathbf{r})d^3\mathbf{r}, \qquad \mathbf{r} \in V$$

where $\hat{\mathbf{n}}_s(= \mathbf{r}_s/r_s)$ denotes the direction in which u_s propagates. From this result, it is clear, that the function γ can be recovered from u_s by three-dimensional Fourier inversion. The scattered field produced by a two-dimensional Born scatterer in the far field is given by

$$u_s(\hat{\mathbf{n}}_i, \hat{\mathbf{n}}_s, k) = \frac{\exp(i\pi/4)}{\sqrt{8\pi}} \frac{k^2}{\sqrt{kr_s}} \exp(ikr_s) \int\limits_A \exp[-ik(\hat{\mathbf{n}}_s - \hat{\mathbf{n}}_i) \cdot \mathbf{r}]\gamma(\mathbf{r})d^2\mathbf{r},$$
$$\mathbf{r} \in A.$$

In one dimension, the equivalent result is (for a right-travelling wave)

$$u_s(x_s, k) = \frac{ik}{2} \exp(ikx_s) \int\limits_L \gamma(x)dx, \qquad x \in L.$$

Notice that when $\hat{\mathbf{n}}_s = \hat{\mathbf{n}}_i$,

$$u_s = \frac{k^2}{4\pi r_0} \exp(ikr_0) \int\limits_V \gamma(\mathbf{r})d^3\mathbf{r}.$$

This is called the forward-scattered field. In terms of Fourier analysis, it represents the zero frequency or DC level of the spectrum of γ. Another special case arises when $\hat{\mathbf{n}}_s = -\hat{\mathbf{n}}_i$. The scattered field that is produced in this case is called the back-scattered field and in three dimensions is given by

$$u_s(\hat{\mathbf{n}}_s, k) = \frac{k^2}{4\pi r_s} \int\limits_V \exp(-2ik\hat{\mathbf{n}}_s \cdot \mathbf{r})\gamma(\mathbf{r})d^3\mathbf{r}.$$

In one dimension, the result is (for a left-travelling wave)

$$u_s(k) = \frac{ik}{2}\exp(ikx_s)\int_L \exp(-2ikx)\gamma(x)dx.$$

Observe, that in one dimension, the scattering function can only be recovered (via Fourier inversion) by measuring the back-scattered spectrum whereas in two and three dimensions, the scattering function can be recovered by either keeping k fixed or varying k. The choice available in two and three dimensions leads to a range of applications in non-destructive evaluation, for example.

Some Examples of Born Scattering: Rutherford and Rayleigh Scattering

By way of a short introduction to the applications and uses of the Born approximation, some well known examples are now presented in which it is used to derive expressions for the scattered intensity associated with two physically different but "Green's function related" scattering phenomena – Rayleigh scattering and Rutherford scattering. We shall consider Rutherford scattering first.

Rutherford Scattering

Rutherford scattering ranks as one of the most important experiments of the 20th century, because it was the basis for developing the basic "visual model" for the atom we except today – a positively charged nucleus with negatively charged orbiting electrons.

In Rutherford's famous experiment (which dates from 1910), α-particles (or helium nuclei) were scattered by gold leaf. The differential cross section denoted by $d\sigma/d\Omega$ (i.e. the number of particles scattered into a solid angle $d\Omega$ per unit time divided by the number of particles incident per unit area per unit time) was then measured at different scattering angles θ. By treating the α-particles as classical Newtonian particles, Rutherford showed that if the scattering potential (i.e. the nucleus of the atoms in the gold leaf) is a repulsive Coulomb potential, then

$$\frac{d\sigma}{d\Omega} \propto \frac{1}{\sin^4(\theta/2)}.$$

This was before the development of quantum mechanics and the emergence of Schrödinger's equation as a governing partial differential equation in quantum mechanics. In this section, we shall derive Rutherford's result by solving Schrödinger's equation using a Green's function.

In terms of quantum mechanics we can consider Rutherford's scattering experiment to consist of a source of plane waves (i.e. the deBroglie or probability waves associated with the α-particles), a scattering function (the potential associated with the nucleus of the atoms which make up the gold leaf) and a measuring device which allows us to record the intensity of the scattered radiation at different angles to the incident beam. From Schrödinger's equation, the Born scattered wave in the far field due to a scattering potential V which is influential over all space is given by

$$u_s(\hat{\mathbf{n}}_s, \hat{\mathbf{n}}_i, k) = -\frac{\exp(ikr_s)}{4\pi r_s} \int_{-\infty}^{\infty} \exp[-ik(\hat{\mathbf{n}}_s - \hat{\mathbf{n}}_i) \cdot \mathbf{r}] V(\mathbf{r}) d^3\mathbf{r}.$$

For fixed k and r_s (the distance at which the scattered wavefield is measured from the scattering event), the measured intensity I of the scattered wavefield is given by

$$I = u_s u_s^* = \frac{1}{16\pi^2 r_s^2} \mid A \mid^2$$

where A is the scattering amplitude,

$$A(\hat{\mathbf{n}}_s, \hat{\mathbf{n}}_i, k) = \int_{-\infty}^{\infty} \exp[-ik(\hat{\mathbf{n}}_s - \hat{\mathbf{n}}_i) \cdot \mathbf{r}] V(\mathbf{r}) d^3\mathbf{r}.$$

The differential cross section measures the flux of particles through a given area in specific period of time. It is essentially a measure of the wavefield intensity so thus,

$$\frac{d\sigma}{d\Omega} = I.$$

Hence, using quantum mechanics (i.e. Schrödinger's equation), the differential cross-section for Rutherford's scattering experiment can be obtained by evaluating the Fourier transform of the potential V. For a radially symmetric potential $V(r)$, the scattering amplitude becomes (switching to spherical polar coordinates r, ψ, ϕ)

$$A(\hat{\mathbf{n}}_s, \hat{\mathbf{n}}_i) = \int_0^{2\pi} d\phi \int_{-1}^{1} d(\cos\psi) \int_0^{\infty} dr \quad r^2 \exp(-ik \mid \hat{\mathbf{n}}_s - \hat{\mathbf{n}}_i \mid r\cos\psi) V(r)$$

The modulus of $\hat{\mathbf{n}}_s - \hat{\mathbf{n}}_i$ is given by

$$\mid \hat{\mathbf{n}}_s - \hat{\mathbf{n}}_i \mid = \sqrt{(\hat{\mathbf{n}}_s - \hat{\mathbf{n}}_i) \cdot (\hat{\mathbf{n}}_s - \hat{\mathbf{n}}_i)} = \sqrt{2(1 - \cos\theta)}$$

where

$$\cos\theta = \hat{\mathbf{n}}_s \cdot \hat{\mathbf{n}}_i$$

and θ is the scattering angle (the angle between the incident and scattered fields). Using the half angle formula,

$$1 - \cos\theta = 2\sin^2(\theta/2)$$

we can write

$$|\hat{\mathbf{n}}_s - \hat{\mathbf{n}}_i| = 2\sin(\theta/2)$$

and integrating over ϕ and ψ the scattering amplitude as a function θ can be written as

$$A(\theta) = \frac{2\pi}{k\sin(\theta/2)} \int_0^\infty \sin[2kr\sin(\theta/2)]V(r)rdr.$$

All we need to do now is compute the remaining integral over r. If we use a simple Coulomb potential where $V(r) \propto 1/r$, then we run in to a problem because the integrand does not converge as $r \to \infty$. For this reason, another radially symmetric potential is introduced which is given by

$$V(r) = \frac{\exp(-ar)}{r}$$

where $a > 0$ is a constant. This type of potential is known as a screened Coulomb potential, the parameter a determining the range over which the potential is influential. It allows us to evaluate the scattering amplitude analytically. We can then observe the behaviour of $|A|^2$ for a Coulomb potential by letting a approach zero. The scattering amplitude becomes

$$A(\theta) = \frac{2\pi}{k\sin(\theta/2)} \int_0^\infty \sin[2kr\sin(\theta/2)]\exp(-ar)dr.$$

This integral is given by

$$\frac{2k\sin(\theta/2)}{a^2 + [2k\sin(\theta/2)]^2}$$

and we can write

$$A(\theta) = \frac{\pi}{k^2\sin^2(\theta/2)}\left(1 + \frac{a^2}{[2k\sin(\theta/2)]^2}\right)^{-1}.$$

Hence, as a approaches zero, we obtain

$$A(\theta) \simeq \frac{\pi}{k^2\sin^2(\theta/2)}$$

and the intensity of the scattered field is

$$I = |A(\theta)|^2 \propto \frac{1}{\sin^4(\theta/2)}.$$

One may think of Rutherford's scattering experiment as an inverse scattering problem in the sense that he deduced the potential of the nucleus by recording the way in which it scattered α-particles. However, he did not actually solve the inverse problem directly because he assumed that the scattering potential acted like a repulsive Coulomb potential *a priori* and justified this hypothesis later by showing that the theoretical and experimental results were compatible. One final and interesting point to note is that in order to undertake the experiment, Rutherford required a very thin foil which was only relatively a few atoms thick. Gold leaf was the best possible technical solution to this problem at the time. The reason for this was that the α-particles needed (on average) to scatterer from one nucleus at a time in order to investigate the repulsive Coulomb potential theory. If a thicker foil had been used, the α-particles may have scattered from a number of atoms as they passed through it. Multiple scattering would have led to an indeterminacy in the results. It is interesting to note that the Born approximation used here to verify Rutherford's results using a Green's function solution to Schrödinger's equation is consistent with the concept of single or weak scattering.

Rayleigh Scattering

Rayleigh scattering is the scattering of electromagnetic radiation by small dielectric scatterers. It is named after the English scientist Lord Rayleigh who was one of the 19th century's most prolific physical scientists and made contributions in many areas of mathematics, physics and chemistry including some of the earliest studies on the scattering of light following the development of Clerk Maxwell's theory of electromagnetism.

If we consider a scalar electromagnetic wave theory, then we can take an equation of the form

$$(\nabla^2 + k^2)u(\mathbf{r}, k) = -k^2\gamma(\mathbf{r})u(\mathbf{r}, k), \qquad \gamma = \epsilon_r - 1; \qquad \mathbf{r} \in V$$

to describe the behaviour of the electric field u where ϵ_r is the relative permittivity of a dielectric of compact support V. This is of course a highly idealised case, but it helps to provide another demonstration of Born scattering in a form that is pertinent to use of Green's functions for solving physically significant problems.

In the context of electromagnetic scattering problems, the Born approximation is sometimes referred to as the Rayleigh–Gan approximation – just a different name for an identical mathematical technique. Using this approximation, the asymptotic form of the the scattered electric field is given by

$$u_s(\hat{\mathbf{n}}_s, \hat{\mathbf{n}}_i, k) = \frac{k^2}{4\pi r_s}\exp(ikr_s)\int\limits_V \exp[-ik(\hat{\mathbf{n}}_s - \hat{\mathbf{n}}_i)\cdot\mathbf{r}]\gamma(\mathbf{r})d^3\mathbf{r}.$$

There are two important differences between this equation and its counterpart in quantum mechanics (i.e. the Schrödinger equation). First, the coefficient in front of the integral possesses a factor k^2. Second, the integral itself is over a finite volume of space V which is determined by the spatial extent of γ.

In quantum mechanics, the influence of the potential is felt over all space so that the integral is over $\pm\infty$. This is an important distinction between scattering problems in quantum mechanics which involve asymptotic potentials (potentials which go to zero at infinity) and classical scattering problems of the type considered here. In the latter case, the scatterer γ has a finite spatial extent (it is of compact support).

Let us consider a model where a plane electromagnetic wave is incident on a homogeneous spherical dielectric object of radius R and relative permittivity ϵ_r. The theory which describes this type of scattering (scattering of light from uniform spheres) is called Mie theory. In this case, the Born scattered amplitude is given by (following the same methods as those used earlier)

$$A(\theta) = \frac{2\pi k\gamma}{\sin(\theta/2)} \int_0^R \sin[2kr\sin(\theta/2)]rdr. \tag{5.4.3}$$

If the dimensions of the scatterer are small compared with the wavelength, then

$$kR \ll 1$$

and

$$\sin[2kr\sin(\theta/2)] \simeq 2kr\sin(\theta/2), \qquad 0 \leq r \leq R.$$

The scattering amplitude is then given by

$$A(\theta) \simeq 4\pi k^2\gamma \int_0^R r^2 dr = k^2\gamma V$$

where $V = 4\pi R^3/3$ is the volume of the scatterer. In this case, the scattering is entirely isotropic (i.e. the scattering amplitude is independent of the scattering angle). The intensity is proportional to k^4 or

$$\mid A(\theta) \mid^2 \propto \frac{1}{\lambda^4}.$$

Note that there is a large inverse dependence on the wavelength. This result is characteristic of Rayleigh scattering and of the spectra produced by light scattering from small sub-wavelength structures. In the visible part of the spectrum, the intensity is greatest for blue light (the colour associated with the smallest wavelength of the visible spectrum). This is why the sky is blue – sunlight is being scattered by small sub-wavelength particles in the upper atmosphere.

When $kR \sim 1$, the scattering amplitude is obtained by evaluating the integral in equation 5.4.3. This is easy to do, the scattering amplitude being given by

$$A(\theta) = 3V\gamma k^2 \frac{J_1[2kR\sin(\theta/2)]}{2kR\sin(\theta/2)}$$

where J_1 is the spherical Bessel function

$$J_1(x) = \frac{\sin(x)}{x^2} - \frac{\cos(x)}{x}.$$

In this case, the scattering is not isotropic but strongly dependent on the scattering angle.

Discussion

This section has been concerned with the use of Green's functions for solving two fundamental inhomogeneous partial differential equations and has been designed to provide an introduction to the role that Green's functions play in an important aspect of mathematical physics – scattering theory. It has been concerned with the use of Green's functions for solving time-independent inhomogeneous wave equations with homogeneous boundary conditions. In the following section we consider the time-dependent case.

EXERCISES

5.1 By means of Laplace transforms, find the general Green's function solution to the equation

$$\left(\frac{\partial^2}{\partial x^2} + k^2\right) u(x,k) = -f(x)$$

where k is a constant, the solution being required in the positive half space $x \in [0, \infty)$.

5.2 Find the Green's function solution to the equation

$$\left(\frac{\partial^2}{\partial x^2} + k^2\right) u(x,k) = 0, \qquad x \in [0, L],$$

subject to the boundary conditions $u(0,k) = 0$ and $u(L,k) = 0$ by first finding a Green's function solution for the infinite domain and then adding a solution of the homogeneous equation to fit the boundary conditions.

5.3 Show, by taking the Laplace transform with respect to x of the equation

$$\left(\frac{\partial^2}{\partial x^2} + k^2\right) g(x \mid x_0, k) = -\delta(x - x_0)$$

that the Green's function g on the interval $[0, 1]$ with the boundary conditions

$$g(0 \mid x_0, k) = 0 \qquad \text{and} \qquad \left[\frac{\partial g(x \mid x_0, k)}{\partial x}\right]_{x=0} = g(1 \mid x_0, k)$$

is

$$g(x \mid x_0, k) = \frac{\sin(kx)\sin[k(1 - x_0)]}{k(\sin k - k)} - \frac{\sin[k(x - x_0)]}{k} H(x - x_0).$$

5.4 Use Fourier transforms to show that the solution of

$$\nabla^2 g(\mathbf{r} \mid \mathbf{r}_0, k) - \lambda g(\mathbf{r} \mid \mathbf{r}_0, k) = -\delta^3(\mathbf{r} - \mathbf{r}_0)$$

is

$$g(\mathbf{r} \mid \mathbf{r}_0, k) = \frac{\exp(-\sqrt{\lambda} \mid \mathbf{r} - \mathbf{r}_0 \mid)}{4\pi \mid \mathbf{r} - \mathbf{r}_0 \mid}.$$

5.5 Show that if $u(\mathbf{r}, k)$ satisfies the partial differential equation

$$[\nabla^2 + k^2 + V(\mathbf{r})]u(\mathbf{r}, k) = 0$$

then u also satisfies the integral equation

$$u(\mathbf{r}_0, k) = f(\mathbf{r}_0, k) + \int g(\mathbf{r} \mid \mathbf{r}_0, k) u(\mathbf{r}, k) V(\mathbf{r}) d^3\mathbf{r}$$

where $f(\mathbf{r})$ is a solution to $(\nabla^2 + k^2)u(\mathbf{r}, k) = 0$ and $g(\mathbf{r} \mid \mathbf{r}_0, k)$ is a Green's function for the same equation.

5.5 Green's Function Solution to Maxwell's Equations and Time-dependent Problems

This section introduces the Green's function as a tool for predicting the behaviour of electromagnetic fields. We start by considering the basic equations of electromagnetism (Maxwell's equation) and show how these can be manipulated (under certain conditions) to form inhomogeneous wave equations for related electromagnetic field potentials. The Green's function is then used to solve these equations which leads directly to a description of the properties of an electromagnetic field.

In addition to providing a short introduction to a set of partial differential equations which are of particular significance in science and engineering, this chapter necessarily considers the rôle of Green's functions for solving time-dependent inhomogeneous wave equations; all previous discussions in this chapter having been related to the time-independent case.

The motion of electrons (and other charged particles) gives rise to electric **e** and magnetic **b** fields. These fields are related by the (microscopic) Maxwell's equations which are as follows.

– Coulomb's law

$$\nabla \cdot \mathbf{e} = 4\pi\rho. \tag{5.5.1}$$

where ρ is the charge density.

– Faraday's law

$$\nabla \times \mathbf{e} = -\frac{1}{c}\frac{\partial \mathbf{b}}{\partial t} \tag{5.5.2}$$

where c is the speech of light (in a vacuum).

– No free magnetic poles law

$$\nabla \cdot \mathbf{b} = 0 \tag{5.5.3}$$

– Modified (by Maxwell) Ampere's law

$$\nabla \times \mathbf{b} = \frac{1}{c}\frac{\partial \mathbf{e}}{\partial t} + \frac{4\pi}{c}\mathbf{j} \tag{5.5.4}$$

where **j** is the current density. These microscopic Maxwell's equations are used to predict the pointwise electric and magnetic fields given the charge and current densities (ρ and **j** respectively). They are linear equations since if

$$\rho_1, \;\; \mathbf{j}_1 \rightarrow \mathbf{e}_1, \;\; \mathbf{b}_1$$

and

$$\rho_2, \;\; \mathbf{j}_2 \rightarrow \mathbf{e}_2, \;\; \mathbf{b}_2$$

then

$$\rho_1 + \rho_2, \;\; \mathbf{j}_1 + \mathbf{j}_2 \rightarrow \mathbf{e}_1 + \mathbf{e}_2, \;\; \mathbf{b}_1 + \mathbf{b}_2$$

because $\nabla\cdot$, $\nabla\times$ and the time derivatives are all linear operators.

Wave Equation Solution of Maxwell's Equations

If we take the curl of equation 5.5.2

$$\nabla \times \nabla \times \mathbf{e} = -\frac{1}{c}\nabla \times \frac{\partial \mathbf{b}}{\partial t}$$

and use the identity

$$\nabla \times \nabla \times = \nabla\nabla \cdot - \nabla^2$$

then from equations 5.5.1 and 5.5.4, we obtain

$$\nabla(4\pi\rho) - \nabla^2\mathbf{e} = -\frac{1}{c}\frac{\partial}{\partial t}\left(\frac{1}{c}\frac{\partial \mathbf{e}}{\partial t} + \frac{4\pi \mathbf{j}}{c}\right)$$

or after rearranging,

$$\nabla^2\mathbf{e} - \frac{1}{c^2}\frac{\partial^2\mathbf{e}}{\partial t^2} = 4\pi\nabla\rho + \frac{4\pi}{c^2}\frac{\partial\mathbf{j}}{\partial t}. \tag{5.5.5}$$

Taking the curl of equation 5.5.4, using the same identity as above, substituting for equations 5.5.2 and 5.5.3 and rearranging the result gives

$$\nabla^2\mathbf{b} - \frac{1}{c^2}\frac{\partial^2\mathbf{b}}{\partial t^2} = -\frac{4\pi}{c}\nabla\times\mathbf{j}. \tag{5.5.6}$$

Equations 5.5.5 and 5.5.6 are the inhomogeneous wave equations for **e** and **b**. They are related or coupled with the vector field **j** (which is related to **b**).

If we define a region of free space where $\rho = 0$ and $\mathbf{j} = 0$, then both **e** and **b** satisfy

$$\nabla^2\mathbf{f} - \frac{1}{c^2}\frac{\partial^2\mathbf{f}}{\partial t^2} = 0$$

which is the homogeneous wave equation. One possible solution of this equation (in Cartesian coordinates) is

$$f_x = F(z - ct); \qquad f_y = 0, \qquad f_z = 0$$

which describes a wave or distribution F moving along z at velocity c (d'Alembert solution to the time-dependent wave equation).

General Solution to Maxwell's Equations

The basic method of solving Maxwell's equations (i.e. finding **e** and **b** given ρ and **j**) involves the following.

- Expressing **e** and **b** in terms of two other fields U and $\mathbf{A}$.
- Obtaining two separate equations for U and $\mathbf{A}$.
- Solving these equations for U and $\mathbf{A}$ from which **e** and **b** can then be computed.

For any vector field $\mathbf{A}$

$$\nabla\cdot\nabla\times\mathbf{A} = 0.$$

Hence, if we write

$$\mathbf{b} = \nabla\times\mathbf{A} \tag{5.5.7}$$

equation 5.5.3 remains unchanged, and equation 5.5.2 can then be written as

$$\nabla\times\mathbf{e} = -\frac{1}{c}\frac{\partial}{\partial t}\nabla\times\mathbf{A}$$

or

$$\nabla \times \left(\mathbf{e} + \frac{1}{c} \frac{\partial \mathbf{A}}{\partial t} \right) = 0.$$

The field $\mathbf{A}$ is called the magnetic vector potential. Similarly, for any scalar field U

$$\nabla \times \nabla U = 0$$

and thus equation 5.5.2 is satisfied if we write

$$\pm \nabla U = \mathbf{e} + \frac{1}{c} \frac{\partial \mathbf{A}}{\partial t}$$

or

$$\mathbf{e} = -\nabla U - \frac{1}{c} \frac{\partial \mathbf{A}}{\partial t} \tag{5.5.8}$$

where the minus sign is taken by convention. The field U is called the electric scalar potential. Substituting equation 5.5.8 into Maxwell's equation 5.5.1 gives

$$\nabla \cdot \left(\nabla U + \frac{1}{c} \frac{\partial \mathbf{A}}{\partial t} \right) = -4\pi\rho$$

or

$$\nabla^2 U + \frac{1}{c} \frac{\partial}{\partial t} \nabla \cdot \mathbf{A} = -4\pi\rho. \tag{5.5.9}$$

Substituting equations 5.5.7 and 5.5.8 into Maxwell's equation 5.5.4 gives

$$\nabla \times \nabla \times \mathbf{A} + \frac{1}{c} \frac{\partial}{\partial t} \left(\nabla U + \frac{1}{c} \frac{\partial \mathbf{A}}{\partial t} \right) = \frac{4\pi}{c} \mathbf{j}.$$

Using the identity

$$\nabla \times \nabla \times \mathbf{A} = \nabla \nabla \cdot \mathbf{A} - \nabla^2 \mathbf{A}$$

this becomes

$$\nabla^2 \mathbf{A} - \frac{1}{c^2} \frac{\partial^2 \mathbf{A}}{\partial t^2} - \nabla \left(\nabla \cdot \mathbf{A} + \frac{1}{c} \frac{\partial U}{\partial t} \right) = -\frac{4\pi}{c} \mathbf{j}. \tag{5.5.10}$$

If we could solve equations 5.5.9 and 5.5.10 for U and $\mathbf{A}$ then $\mathbf{e}$ and $\mathbf{b}$ could be computed. However, these equations are coupled. They can be decoupled if we introduce a condition known as the Lorentz condition:

$$\nabla \cdot \mathbf{A} + \frac{1}{c} \frac{\partial U}{\partial t} = 0. \tag{5.5.11}$$

Substituting equation 5.5.11 into equations 5.5.9 and 5.5.10 gives

$$\begin{aligned} \nabla^2 U - \frac{1}{c^2} \frac{\partial^2 U}{\partial t^2} &= -4\pi\rho \\ \nabla^2 \mathbf{A} - \frac{1}{c^2} \frac{\partial^2 \mathbf{A}}{\partial t^2} &= -\frac{4\pi \mathbf{j}}{c} \end{aligned}$$

respectively. These equations are uncoupled inhomogeneous wave equations. Unlike the wave equations that we have considered so far in this chapter, these equations are time dependent and it is therefore pertinent at this point to consider the Green's function for a time-dependent wave equation.

Green's Functions for Time-dependent Inhomogeneous Wave Equations

We shall consider the three-dimensional problem first but stress that the methods of solution discussed here apply directly to problems in one and two dimensions. Thus, consider the case in which a time varying source function $f(\mathbf{r},t)$ produces a wavefield U which is taken to be the solution to the equation.

$$\left(\nabla^2 - \frac{1}{c^2}\frac{\partial^2}{\partial t^2}\right)U(\mathbf{r},t) = -f(\mathbf{r},t). \tag{5.5.12}$$

As with the time-independent problem, the Green's function for this equation is defined as the solution to the equation obtained by replacing $f(\mathbf{r},t)$ with $\delta^3(\mathbf{r}-\mathbf{r}_0)\delta(t-t_0)$, that is the solution to the equation

$$\left(\nabla^2 + \frac{1}{c^2}\frac{\partial^2}{\partial t^2}\right)G(\mathbf{r}\mid\mathbf{r}_0,t\mid t_0) = -\delta^3(\mathbf{r}-\mathbf{r}_0)\delta(t-t_0) \tag{5.5.13}$$

where G is used to denote the time-dependent Green's function, $\mathbf{r}_0$ is the position of the source and $t\mid t_0 \equiv t-t_0$. To obtain the equation for the time-independent Green's function, we write G and $\delta(t-t_0)$ as Fourier transforms,

$$G(\mathbf{r}\mid\mathbf{r}_0,t\mid t_0) = \frac{1}{2\pi}\int\limits_{-\infty}^{\infty} g(\mathbf{r}\mid\mathbf{r}_0,\omega)\exp[i\omega(t-t_0)]d\omega$$

and

$$\delta(t-t_0) = \frac{1}{2\pi}\int\limits_{-\infty}^{\infty}\exp[i\omega(t-t_0)]d\omega$$

where ω is the angular frequency. Substituting these equations into equation 5.5.13 we then obtain

$$(\nabla^2+k^2)g(\mathbf{r}\mid\mathbf{r}_0,k) = -\delta^3(\mathbf{r}-\mathbf{r}_0)$$

which is the same equation as that used previously to define the time-independent Green's function. Thus, once g has been obtained, the time-dependent Green's function can be derived by computing the Fourier integral given above. Using the expression for g derived earlier,

$$\begin{aligned} G(\mathbf{r}\mid\mathbf{r}_0,t\mid t_0) &= \frac{1}{2\pi}\int\limits_{-\infty}^{\infty}\frac{1}{4\pi\mid\mathbf{r}-\mathbf{r}_0\mid}\exp(ik\mid\mathbf{r}-\mathbf{r}_0\mid)\exp[i\omega(t-t_0)]d\omega \\ &= \frac{1}{4\pi\mid\mathbf{r}-\mathbf{r}_0\mid}\delta(t-t_0+\mid\mathbf{r}-\mathbf{r}_0\mid/c). \end{aligned}$$

In two dimensions, the point source (which depends on x and y), can be treated as a line source, that is a uniform source extending from $z_0=-\infty$ to $z_0=\infty$

along a line parallel to the z-axis and passing through the point (x_0, y_0). Thus, a simple way of computing the two-dimensional Green's function is to integrate the three-dimensional Green's function from $z_0 = -\infty$ to $z_0 = \infty$, i.e.

$$G(\mathbf{s} \mid \mathbf{s}_0, t \mid t_0) = \int_{-\infty}^{\infty} \frac{\delta(t - t_0 + \mid \mathbf{r} - \mathbf{r}_0 \mid /c)}{4\pi \mid \mathbf{r} - \mathbf{r}_0 \mid} dz_0$$

where

$$\mathbf{s} = \hat{\mathbf{x}}x + \hat{\mathbf{y}}y$$

and

$$\mathbf{s}_0 = \hat{\mathbf{x}}x_0 + \hat{\mathbf{y}}y_0.$$

Writing $\tau = (t - t_0)c$, $\xi = z_0 - z$, $S = \mid \mathbf{s} - \mathbf{s}_0 \mid$ and $R = \mid \mathbf{r} - \mathbf{r}_0 \mid$ we have

$$R^2 = \xi^2 + S^2$$

and

$$\frac{dR}{dz_0} = \frac{\xi}{R}$$

and so the Green's function can be written in the form

$$\begin{aligned} G(S, \tau) &= \frac{1}{4\pi} \int_{-\infty}^{\infty} \frac{\delta(\tau + R)}{\sqrt{R^2 - S^2}} dR \\ &= \begin{cases} \dfrac{1}{4\pi} \dfrac{1}{\sqrt{\tau^2 - S^2}}, & \tau > S; \\ 0, & \tau < S. \end{cases} \end{aligned}$$

Similarly, in one dimension, the time-dependent Green's function can be calculated by integrating the three-dimensional Green's function over z_0 and y_0. Alternatively, we can use the expression for $g(x \mid x_0, k)$ giving

$$G(x \mid x_0, t \mid t_0) = \frac{1}{2\pi} \int_{-\infty}^{\infty} \frac{i}{2k} \exp(ik \mid x - x_0 \mid) \exp[i\omega(t - t_0)] d\omega.$$

This equation is the inverse Fourier transform of the product of two functions (noting that $k = \omega/c$), namely $i/2k$ and $\exp(ik \mid x - x_0 \mid)$. Thus, using the convolution theorem and noting that

$$\frac{1}{2\pi} \int_{-\infty}^{\infty} \frac{i}{2k} \exp[i\omega(t - t_0)] d\omega = \frac{c}{4} \operatorname{sgn}(t - t_0)$$

and

$$\frac{1}{2\pi} \int_{-\infty}^{\infty} \exp(ik \mid x - x_0 \mid) \exp[i\omega(t - t_0)] d\omega = \delta(t - t_0 + \mid x - x_0 \mid /c),$$

we obtain

$$\begin{aligned} G(x \mid x_0, t \mid t_0) &= \frac{c}{4}\text{sgn}(t-t_0) \otimes \delta(t-t_0+\mid x-x_0 \mid /c) \\ &= \frac{c}{4}\text{sgn}[t-t_0+\mid x-x_0 \mid /c] \end{aligned}$$

where $\otimes$ denotes the convolution integral and sgn is defined by

$$\text{sgn}(x) = \begin{cases} 1, & x > 0; \\ -1, & x < 0. \end{cases}$$

Discussion

There is a striking difference between the time-dependent Green's functions derived above. In three dimensions, the effect of an impulse after a time $t - t_0$ is found concentrated on a sphere of radius $c(t - t_0)$ whose centre is the source point. The effect of the impulse can therefore only be experienced by an observer at one location over an infinitely short period of time. After the pulse has passed by an observer, the disturbance ceases. In two dimensions, the disturbance is spread over the entire plane $\mid \mathbf{s} - \mathbf{s}_0 \mid$. At $\mid \mathbf{s} - \mathbf{s}_0 \mid = c(t - t_0)$, there is a singularity which defines the position of the two-dimensional wavefront as it propagates outwards from the source point at $\mathbf{s}_0$. For $\mid \mathbf{s} - \mathbf{s}_0 \mid < c(t - t_0)$, the Green's function is still finite and therefore, unlike the three-dimensional case, the disturbance is still felt after the wavefront has passed by the observer. In one dimension, the disturbance is uniformly distributed over all points of observation through which the wavefront has passed, since for all values of $\mid x - x_0 \mid$ and $c(t - t_0)$, the Green's function is either $c/4$ or $-c/4$.

Compared with the Green's function in one and two dimensions, the three-dimensional Green's function possesses the strongest singularity. Compared with the delta function, the singularity of the two-dimensional Green's function at $\mid \mathbf{s} - \mathbf{s}_0 \mid = c(t - t_0)$ is very weak. In one dimension, the time-dependent Green's function is not singular but discontinuous when $\mid x - x_0 \mid = c(t - t_0)$.

Green's Function Solution to Maxwell's Equation

Having briefly discussed the time-dependent Green's functions for the wave equation, we can now investigate the general solution to Maxwell's equation under the Lorentz condition. In particular, we shall now consider the solution for the electric scalar potential U – equation 5.5.12 with $f = 4\pi\rho$. The form of analysis is the same as that used before throughout this chapter. Solving for U, using Green's theorem (with homogeneous boundary conditions) and the conditions that u and $\partial u/\partial t$ are zero at $t = \pm\infty$ gives

$$\begin{aligned} U(\mathbf{r}_0, t_0) &= \int_{-\infty}^{\infty} \int \rho(\mathbf{r}, t) G(\mathbf{r} \mid \mathbf{r}_0, t \mid t_0) d^3\mathbf{r} dt \\ &= \int d^3\mathbf{r} \int_{-\infty}^{\infty} dt \frac{\rho(\mathbf{r}, t)}{R} \delta\left[\frac{R}{c} + t - t_0\right] \\ &= \int d^3\mathbf{r} \frac{\rho\left(\mathbf{r}, t_0 - \frac{R}{c}\right)}{R} \end{aligned}$$

where $R = \mid \mathbf{r} - \mathbf{r}_0 \mid$ or

$$U(\mathbf{r}_0, t_0) = \int \frac{\rho(\mathbf{r}, \tau)}{R} d^3\mathbf{r}$$

where

$$\tau = t_0 - \frac{R}{c}.$$

The solution for the magnetic vector potential $\mathbf{A}$ can be found by solving for the components A_x, A_y and A_z separately. These are all scalar equations of exactly the same type and therefore have identical solutions. The wavefields U and $\mathbf{A}$ are called the retarded potentials. The current value of U at $(\mathbf{r}_0, t_0)$ depends on ρ at earlier times $\tau = t_0 - R/c$. A change in ρ or $\mathbf{j}$ affects U and $\mathbf{A}$ (and hence $\mathbf{e}$ and $\mathbf{b}$) R/c seconds later – the change propagates outward at velocity c. This is the principle of electromagnetic wave propagation.

EXERCISES

5.6 The electric field potential U satisfies the equation

$$\nabla^2 U(\mathbf{r}, t) - \frac{1}{c^2} \frac{\partial^2}{\partial t^2} U(\mathbf{r}, t) = -4\pi\rho(\mathbf{r}) \exp(i\omega t)$$

where ρ is the charge density, ω is the angular frequency and c is the speed of electromagnetic waves in a vacuum. Use a Green's function to compute the amplitude of the electric field potential produced by a thin antenna radiating 10 m wavelength electromagnetic radiation at a distance of 1000 m from the antenna when $\rho(\mathbf{r}) = 1/r^2$.

(Hint: compute the Green's function solution to this equation in the far field and then use spherical polar coordinates (r, θ, ϕ) noting that $d^3\mathbf{r} = r^2 dr d(\cos\theta) d\phi$ in spherical polars and that $\int_0^\infty \frac{\sin x}{x} dx = \frac{\pi}{2}$.)

5.7 Compute the three-dimensional Green's functions for the following time-dependent wavefield operators (where σ is a constant).

(i) The Klein–Gordon operator:

$$\nabla^2 - \frac{1}{c^2}\frac{\partial^2}{\partial t^2} - \sigma^2.$$

(Hint: the Laplace transform of the function

$$f(x) = \begin{cases} J_0\left(a\sqrt{x^2 - b^2}\right), & x > b; \\ 0, & x < b \end{cases}$$

is

$$F(p) = \frac{\exp\left(-b\sqrt{p^2 + a^2}\right)}{\sqrt{p^2 + a^2}}$$

where J_0 is the Bessel function (of order 0) and a and b are positive constants.)

(ii) Electromagnetic wave propagation in a conducting medium:

$$\nabla^2 - \frac{1}{c^2}\frac{\partial^2}{\partial t^2} - \sigma\frac{\partial}{\partial t}.$$

(Hint: the Laplace transform of the function

$$f(x) = \begin{cases} I_0\left(a\sqrt{x^2 - b^2}\right), & x > b; \\ 0, & x < b \end{cases}$$

is

$$F(p) = \frac{\exp\left(-b\sqrt{p^2 - a^2}\right)}{\sqrt{p^2 - a^2}}$$

where I_0 is the modified Bessel function (of order zero) and a and b are positive constants.)

5.6 Green's Functions and Optics: Kirchhoff Diffraction Theory

The phenomenon of diffraction is common to one-dimensional transverse (and hence scalar) waves such as water waves, true scalar waves such as in acoustics and vector waves as in optics. All three cases appear to exhibit the same magnitude of diffraction phenomena. In order to approximately describe optical diffraction, it is reasonable to first adopt a scalar model for light. This type of model is concerned with a monochromatic scalar "disturbance"

$$U(\mathbf{r}, t) = u(\mathbf{r})\exp(-i\omega t)$$

where u is the scalar complex amplitude. In free space U satisfies the homogeneous wave equation

$$\nabla^2 U - \frac{1}{c^2}\frac{\partial^2 U}{\partial t^2} = 0$$

where c is the (constant) velocity. In this case u satisfies the homogeneous Helmholtz equation

$$\nabla^2 u + k^2 u = 0$$

where

$$k = \frac{\omega}{c}.$$

Scalar diffraction theory should be regarded as a first approximation to optical diffraction. The observed intensity I (the observed quantity at optical frequencies) can be taken to be given by

$$I = | u |^2 .$$

Except in free space, u is not (in general) a Cartesian component of the vector electric or magnetic field. Scalar diffraction theory is accurate if:

(i) the diffracting aperture is large compared with the wavelength.

(ii) the diffracted fields are observed at a reasonable distance from the screen.

This section is devoted to solving the homogeneous Helmholtz equation using a Green's function by implementing the Kirchhoff theory of diffraction – a theory which is fundamental to modern optics.

In Section 5.3, the surface integral obtained using a Green's functions solution and Green's theorem was discarded under the assumption of homogeneous boundary conditions or that $u = u_i$ (the incident field) of the surface S. We were then left with a volume integral (i.e. volume scattering). In this section, we make explicit use of this surface integral to develop a solution to the homogeneous Helmholtz equation. This leads to the theory of surface scattering of which Kirchhoff diffraction theory is a special (but very important) case. Kirchhoff developed a rigourous theory of diffraction in 1887 and demonstrated that previous results and ideas in optics could be obtained from the wave equation. This theory is interesting in that it provided a well formulated theory of optics (i.e. one based on the solution to a partial differential equation) but it is also one of the first theories to make explicit use of Green's theorem and the Green's function combined.

Kirchhoff Diffraction Theory

Consider a scalar wavefield u described by the homogeneous Helmholtz equation

$$(\nabla^2 + k^2)u = 0.$$

Let u_i be the field incident on a surface S and consider the following (Kirchhoff) boundary conditions

$$u = u_i, \qquad \frac{\partial u}{\partial \hat{\mathbf{n}}} = \frac{\partial u_i}{\partial \hat{\mathbf{n}}} \qquad \text{on S.}$$

The results which derive from a solution to the Helmholtz equation based on these boundary conditions is called Kirchhoff diffraction theory. The theory can be applied to surfaces of different topology but is commonly associated with plane surfaces such as those representing a plane aperture in a screen which is the case we will considered in this section.

The Green's Function Solution

Consider the Green's function g which is the solution to

$$(\nabla^2 + k^2)g = -\delta^3(\mathbf{r} - \mathbf{r}_0)$$

given by

$$g(\mathbf{r} \mid \mathbf{r}_0, k) = \frac{1}{4\pi \mid \mathbf{r} - \mathbf{r}_0 \mid} \exp(ik \mid \mathbf{r} - \mathbf{r}_0 \mid).$$

We can construct two equations:

$$g\nabla^2 u + k^2 gu = 0$$

and

$$u\nabla^2 g + k^2 ug = -u\delta^3.$$

Subtracting these equations and integrating over a volume V we obtain a solution for the field u at $\mathbf{r}_0$,

$$u(\mathbf{r}_0, k) = \oint_S \left(g\frac{\partial u}{\partial \hat{\mathbf{n}}} - u\frac{\partial g}{\partial \hat{\mathbf{n}}} \right) d^2\mathbf{r}$$

where we have used Green's theorem to write the solution in terms of a (closed) surface integral. We must consider the surface integration carefully to obtain a sensible result for an aperture in a screen. Consider the case where the surface of integration is made up of the following three surface patches:

(i) the surface covering the aperture S_1;

(ii) a plane surface adjacent to the screen S_2 (not covering the aperture);

(iii) a semispherical surface S_3 connected to S_2;

where in each case, the surfaces are considered to exist in the diffraction domain – the side of the screen which the incident field does not illuminate. Thus $S = S_1 + S_2 + S_3$.

On S_2 (the screen itself) u and $\partial u/\partial\hat{\mathbf{n}}$ are identically zero. In the aperture (over the surface S_1) the values of u and $\partial u/\partial\hat{\mathbf{n}}$ will have the values they would have if the screen were not there (u_i and $\partial u_i/\partial\hat{\mathbf{n}}$). Evaluating of the behaviour of the field over S_3 requires some attention which is compounded in the computation of

$$\frac{\partial g}{\partial\hat{\mathbf{n}}} = \hat{\mathbf{n}}\cdot\nabla g.$$

Evaluating ∇g we obtain

$$\begin{aligned}
\nabla g &= \hat{\mathbf{x}}\frac{\partial}{\partial x}\frac{\exp\left(ik\sqrt{(x-x_0)^2+\ldots}\right)}{4\pi\sqrt{(x-x_0)^2+\ldots}}+\ldots\\
&= -\hat{\mathbf{x}}\frac{1}{4\pi}\exp\left(ik\sqrt{(x-x_0)^2+\ldots}\right)[(x-x_0)^2+\cdots]^{-\frac{3}{2}}(x-x_0)\\
&\quad+\hat{\mathbf{x}}\frac{ik}{4\pi}\frac{(x-x_0)}{(x-x_0)^2+\cdots}\exp\left(ik\sqrt{(x-x_0)^2+\ldots}\right)+\ldots\\
&= \hat{\mathbf{x}}\frac{\exp\left(ik\sqrt{(x-x_0)^2+\ldots}\right)}{4\pi\sqrt{(x-x_0)^2+\ldots}}\frac{(x-x_0)}{\sqrt{(x-x_0)^2+\ldots}}\\
&\quad\times\left(ik-\frac{1}{\sqrt{(x-x_0)^2+\ldots}}\right)+\ldots\\
&= \hat{\mathbf{m}}\left(ik-\frac{1}{|\mathbf{r}-\mathbf{r}_0|}\right)g
\end{aligned}$$

where

$$\hat{\mathbf{m}} = \frac{\mathbf{r}-\mathbf{r}_0}{|\mathbf{r}-\mathbf{r}_0|}.$$

Therefore,

$$\frac{\partial g}{\partial\hat{\mathbf{n}}} = \hat{\mathbf{n}}\cdot\hat{\mathbf{m}}\left(ik-\frac{1}{|\mathbf{r}-\mathbf{r}_0|}\right)g.$$

In most practical circumstances the diffracted field is observed at distances $|\mathbf{r}-\mathbf{r}_0|$ where

$$|\mathbf{r}-\mathbf{r}_0|\gg\lambda.$$

This condition allows us to introduce the simplification

$$\nabla g\simeq ik\hat{\mathbf{m}}g$$

so that

$$\frac{\partial g}{\partial\hat{\mathbf{n}}}\simeq ik\hat{\mathbf{n}}\cdot\hat{\mathbf{m}}g.$$

The surface integral over S_3 can therefore be written as

$$\int\limits_{S_3} g\left(\frac{\partial u}{\partial\hat{\mathbf{n}}}-ik\hat{\mathbf{n}}\cdot\hat{\mathbf{m}}u\right)d^2\mathbf{r}.$$

For simplicity, if we consider S_3 to be a hemisphere of radius $R=|\mathbf{r}-\mathbf{r}_0|$ with origin at O say, then we may write this integral in the form

$$\int_{\Omega} \frac{\exp(ikR)}{4\pi R}\left(\frac{\partial u}{\partial \hat{\mathbf{n}}} - ik\hat{\mathbf{n}}\cdot\hat{\mathbf{m}}u\right) R^2 d\Omega$$

where Ω is the solid angle subtended by S_3 at O. If we now assume that

$$\lim_{R\to\infty} R\left(\frac{\partial u}{\partial \hat{\mathbf{n}}} + ik\hat{\mathbf{n}}\cdot\hat{\mathbf{m}}u\right) = 0$$

uniformly with angle, then the surface integral over S_3 can be neglected. This limiting condition is called the Sommerfeld radiation condition and is satisfied if $u \to 0$ as fast as $|\mathbf{r}-\mathbf{r}_0|^{-1}\to 0$. With this requirement met, the only contribution to the surface integral will be in the plane of the aperture and using the Kirchhoff boundary conditions we have

$$u(\mathbf{r}_0, k) = \int_S \left(g\frac{\partial u_i}{\partial \hat{\mathbf{n}}} - u_i\frac{\partial g}{\partial \hat{\mathbf{n}}}\right) d^2\mathbf{r}$$

where S is taken to be S_1. This equation is referred to as the Kirchhoff integral. Note that in deriving this result, we have failed to take into account the finite width of the aperture and therefore the effect of the edges of the aperture on the field within the aperture. Thus, the model can only apply to apertures much larger than the wavelength of the field and for apertures which are "thin".

To compute the diffracted field using the Kirchhoff integral, an expression for u_i must be introduced and the derivatives $\partial/\partial\hat{\mathbf{n}}$ with respect to u_i and g computed. Let us consider the case where the incident field is a plane wavefield of unit amplitude (with wavenumber $k\equiv|\mathbf{k}|$, $\hat{\mathbf{k}}=\mathbf{k}/k$). Then

$$\begin{aligned} u_i &= \exp(i\mathbf{k}\cdot\mathbf{r}), \\ \frac{\partial u_i}{\partial \hat{\mathbf{n}}} = \hat{\mathbf{n}}\cdot\nabla\exp(i\mathbf{k}\cdot\mathbf{r}) &= i\mathbf{k}\cdot\hat{\mathbf{n}}\exp(i\mathbf{k}\cdot\mathbf{r}) = ik\hat{\mathbf{n}}\cdot\hat{\mathbf{k}}\exp(i\mathbf{k}\cdot\mathbf{r}) \end{aligned}$$

and the Kirchhoff diffraction formula reduces to the form

$$u(\mathbf{r}_0, k) = ik\int_S \exp(i\mathbf{k}\cdot\mathbf{r})(\hat{\mathbf{n}}\cdot\hat{\mathbf{k}} - \hat{\mathbf{n}}\cdot\hat{\mathbf{m}})g(\mathbf{r}\mid\mathbf{r}_0, k)d^2\mathbf{r}.$$

Fraunhofer Diffraction

Fraunhofer diffraction assumes that the diffracted wavefield is observed a large distance away from the screen and as in previous sections is based on the asymptotic form of the Green's function. For this reason, Fraunhofer diffraction is sometimes called diffraction in the "far field". The basic idea is to exploit the simplifications that can be made to the Kirchhoff diffraction integral by considering the case when

$$r_0 \gg r$$

where $r \equiv\mid \mathbf{r} \mid$ and $r_0 \equiv\mid \mathbf{r}_0 \mid$. In this case,

$$\frac{1}{\mid \mathbf{r}-\mathbf{r}_0 \mid} \simeq \frac{1}{r_0}$$

and

$$\hat{\mathbf{n}}\cdot\hat{\mathbf{k}}-\hat{\mathbf{n}}\cdot\hat{\mathbf{m}} \simeq \hat{\mathbf{n}}\cdot\hat{\mathbf{k}}+\hat{\mathbf{n}}\cdot\hat{\mathbf{r}}_0$$

where

$$\hat{\mathbf{r}}_0=\frac{\mathbf{r}_0}{r_0}.$$

With regard to the term $\exp(ik \mid \mathbf{r}-\mathbf{r}_0 \mid)$,

$$\mid \mathbf{r}-\mathbf{r}_0 \mid = r_0\left(1-2\frac{\mathbf{r}\cdot\mathbf{r}_0}{r_0^2}+\frac{r^2}{r_0^2}\right)^{\frac{1}{2}} \simeq r_0-\mathbf{r}\cdot\hat{\mathbf{r}}_0.$$

Thus, the Kirchhoff diffraction integral reduces to

$$u(\mathbf{r}_0,k) \simeq \frac{ik\alpha}{4\pi r_0}\exp(ikr_0)\int\limits_S \exp(i\mathbf{k}\cdot\mathbf{r})\exp(-ik\hat{\mathbf{r}}_0\cdot\mathbf{r})d^2\mathbf{r}$$

where $\alpha=\hat{\mathbf{n}}\cdot\hat{\mathbf{k}}+\hat{\mathbf{n}}\cdot\hat{\mathbf{r}}_0$. This is the Fraunhofer diffraction integral.

Fresnel Diffraction

Fresnel diffraction is based on considering the binomial expansion of $\mid \mathbf{r}-\mathbf{r}_0 \mid$ in the function $\exp(ik \mid \mathbf{r}-\mathbf{r}_0 \mid)$ to second order and retaining the term $r^2/2r_0$;

$$\begin{aligned}\mid \mathbf{r}-\mathbf{r}_0 \mid &= r_0-\mathbf{r}\cdot\hat{\mathbf{r}}_0+\frac{r^2}{2r_0}+\cdots\\ &\simeq r_0-\mathbf{r}\cdot\hat{\mathbf{r}}_0+\frac{r^2}{2r_0}.\end{aligned}$$

This approximation is necessary when the diffraction pattern is observed in what is called the intermediate field or Fresnel zone in which

$$u(\mathbf{r}_0,k) \simeq \frac{ik\alpha}{4\pi r_0}\exp(ikr_0)\int\limits_S \exp(i\mathbf{k}\cdot\mathbf{r})\exp(-ik\mathbf{r}_0\cdot\mathbf{r})\exp\left(ik\frac{r^2}{2r_0}\right)d^2\mathbf{r}.$$

This is the Fresnel diffraction formula.

EXERCISES

5.8 Using the Fraunhofer diffraction integral, show that the diffraction pattern observed on a plane screen generated by a plane aperture described by a function $f(x,y)$ is determined by the two-dimensional Fourier transform of this function under the following conditions.

(i) The aperture is illuminated by a plane wave at normal incidence.

(ii) The diffraction pattern is observed at small angles only.

(iii) The aperture is "infinitely thin".

5.9 Using the Fresnel diffraction formula, and the same conditions given in question 5.8 above, show that the diffraction pattern generated in the Fresnel zone by an aperture $f(x,y)$ is determined by the two-dimensional convolution of $f(x,y)$ with the function

$$\exp\left[\frac{i\pi}{\lambda z_0}(x^2+y^2)\right].$$

5.7 Approximation Methods and the Born Series

The solutions considered so far have been based on the application of the Born approximation (Born scattering theory) or Kirchhoff approximation (Kirchhoff diffraction theory) to Green's function solutions of wave equations, taking as examples the Schrödinger equation and the inhomogeneous and homogeneous Helmholtz equations.

In this section, we consider the Wenzel–Kramers–Brillouin (WKB) and the Rytov approximations for solving inhomogeneous wave equations, taking the inhomogeneous Helholtz equation by way of an example (albeit an important one). The WKB method is based on the idea that if the wavelength of the wavefield u is very small compared with variations in γ then a suitable approximation can be introduced which provides an appropriate solution. A similar idea is used for the Rytov approximation. In both cases, the result is based on the use of an exponential type or "eikonal" transformation where a solution of the type $A(\mathbf{r},k)\exp[is(\mathbf{r},k)]$ or just $A(\mathbf{r},k)\exp[s(\mathbf{r},k)]$ is considered which is analogous (in the latter case) to a plane wave solution of the type $A\exp(i\mathbf{k}\cdot\mathbf{r})$. In this transform, the scalar field s is known as the eikonal from the Greek meaning image.

The WKB and Rytov approximations are based on a similar idea – an idea which has a long history dating back to Huygens. In his book "Treatise on Light", Huygens suggested that the reflection and refraction properties of light can be explained on the basis of a sequence of wavefronts spreading out from a source much as ripples spread out from a stone thrown into water and lets each point on such a wavefront act as a new disturbance source. Although Huygens does not seem to specify exactly what is meant by a wavefront he emphasised that the spacing between successive wavefronts need not be uniform, which is one way of considering the physical interpretation of the WKB approximation. Another example of the WKB approximation being used earlier was in a paper by George Green on "The motion of waves in a variable canal of small depth and width" (published in the Transactions of the Cambridge Philosophical Society in 1837) who developed a solution for waves along a narrow (to make the problem one dimensional) but variable channel. His solution involves an approach which is similar to the WKB method used in quantum mechanics. It is therefore arguable that the approximation should be called Green's approximation!

The WKB Approximation

To illustrate the idea behind the WKB approximation, let us consider a general solution to

$$\left(\frac{\partial^2}{\partial x^2} + k^2\right) u(x,k) = -k^2\gamma(x)u(x,k). \tag{5.7.1}$$

The Green's function solution to this equation is given by

$$u = u_i + u_s$$

where u_i is the incident wavefield (typically a unit amplitude plane wave) and u_s is given by

$$u_s(x_0,k) = k^2 \int \gamma(x)g(x \mid x_0,k)u(x,k)dx.$$

Instead of considering the solution to be the sum of two wavefields u_i and u_s, suppose we introduce the eikonal transform

$$u(x,k) = u_i(x,k)\exp[s(x,k)].$$

Substituting this result into equation 5.7.1 and differentiating, we obtain

$$\frac{\partial^2 u_i}{\partial x^2} + 2\frac{\partial s}{\partial x}\frac{\partial u_i}{\partial x} + u_i\left(\frac{\partial s}{\partial x}\right)^2 + u_i\frac{\partial^2 s}{\partial x^2} + k^2 u_i = -k^2\gamma u_i.$$

Now, if we consider u_i to be a solution to $\partial^2 u_i/\partial x^2 + k^2 u_i = 0$ (i.e. $u_i = \exp(ikx)$) then after differentiating u_i and rearranging we have

$$2ik\frac{\partial s}{\partial x} + \left(\frac{\partial s}{\partial x}\right)^2 + \frac{\partial^2 s}{\partial x^2} = -k^2\gamma. \tag{5.7.2}$$

This is a nonlinear Riccatian equation for s and at first sight appears to be more complicated than the original (as a result of introducing a nonlinear transformation). However, if we introduce the condition that the wavelength $\lambda = 2\pi/k$ (assumed to be a constant) is significantly smaller than the spatial extent over which s varies, then the nonlinear term and the second derivative can be ignored and we can write

$$2ik\frac{ds}{dx} = -k^2\gamma$$

whose general solution is (ignoring the constant of integration)

$$s(x) = \frac{ik}{2}\int^{x}\gamma(x)dx.$$

The solution for u is therefore given by

$$u(x,k) = u_i \exp\left(\frac{ik}{2}\int^{x}\gamma(x)dx\right) = \exp\left[ik\left(x + \frac{1}{2}\int^{x}\gamma(x)dx\right)\right].$$

This is an example of the WKB approximation and is based on the idea that if k is large compared with the magnitudes of the terms $(\partial s/\partial x)^2$ and $\partial^2 s/\partial x^2$ then the only terms in equation 5.7.2 that matter are $2ik(\partial s/\partial x)$ and $-k^2\gamma$. In other words, if L is the characteristic scale length over which s varies then

$$\frac{\lambda}{L} \ll 1.$$

The solution describes a plane wavefield whose phase kx is modified by $k\int \gamma dx/2$. A similar approach can be used in higher dimensions which leads to an interpretation of the solutions in terms of the characteristics or rays and the geometric properties associated with them.

The WKB approximation, as illustrated here, does not in itself make use of a Green's function. We shall now consider the Rytov approximation which is based on a similar idea to the WKB approximation and makes explicit use of the Green's function.

The Rytov Approximation

Consider the three-dimensional inhomogeneous Helmholtz equation

$$(\nabla^2 + k^2)u(\mathbf{r},k) = -k^2\gamma(\mathbf{r})u(\mathbf{r},k), \qquad \mathbf{r} \in V.$$

If we substitute $u = u_i \exp(s)$ into this equation and differentiate, we obtain the nonlinear Riccatian equation

$$\nabla^2 s + 2\frac{\nabla u_i}{u_i}\cdot\nabla s + \nabla s\cdot\nabla s = -k^2\gamma$$

where u_i is taken to satisfy the equation

$$\nabla^2 u_i + k^2 u_i = 0, \qquad \text{i.e.} \;\; u_i = \exp(i\mathbf{k}\cdot\mathbf{r}).$$

Suppose we assume that s varies sufficiently slowly for the nonlinear term $\nabla s \cdot \nabla s$ to be neglected compared with the other terms, then we can write (approximately)

$$u_i \nabla^2 s + 2\nabla u_i \cdot \nabla s = -k^2 \gamma u_i. \tag{5.7.3}$$

This is the Rytov approximation. To facilitate a Green's function solution, we substitute $s = w/u_i$ into equation 5.7.3. Differentiating, we have

$$\begin{aligned} u_i \nabla^2 s + 2\nabla u \cdot \nabla s &= \nabla^2 w + 2u_i \nabla w \cdot \nabla \left(\frac{1}{u_i}\right) + u_i w \nabla^2 \left(\frac{1}{u_i}\right) \\ &\quad + 2\frac{\nabla u_i}{u_i} \cdot \nabla w + 2w\nabla u_i \cdot \nabla \left(\frac{1}{u_i}\right) \\ &= \nabla^2 w + k^2 w \end{aligned}$$

and thus, equation 5.7.3 reduces to

$$\nabla^2 w + k^2 w = -k^2 \gamma u_i.$$

The Green's function solution to this equation (subject to homogeneous boundary conditions) is

$$w(\mathbf{r}_0, k) = k^2 \int\limits_V u_i(\mathbf{r}, k)\gamma(\mathbf{r}) g(\mathbf{r} \mid \mathbf{r}_0, k) d^3\mathbf{r}$$

and thus we arrive at the solution

$$u(\mathbf{r}_0, k) = u_i(\mathbf{r}_0, k) \exp\left[\frac{k^2}{u_i(\mathbf{r}_0, k)} \int\limits_V u_i(\mathbf{r}, k)\gamma(r) g(\mathbf{r} \mid \mathbf{r}_0, k) d^3\mathbf{r}\right].$$

Note that we can write this result as

$$\begin{aligned} u &= u_i \left(1 + \frac{k^2}{u_i} \int\limits_V u_i \gamma g d^3\mathbf{r} + \cdots\right) \\ &\simeq u_i + k^2 \int\limits_V u_i \gamma g d^3\mathbf{r} \end{aligned}$$

which is the solution under the Born approximation.

Condition for the Rytov Approximation

The condition required for the validity of the Rytov approximation can be investigated by considering a Green's function solution with the nonlinear term $\nabla s \cdot \nabla s$ included. In this case equation 5.7.3 becomes

$$u_i \nabla^2 s + 2\nabla u_i \cdot \nabla s = -k^2 \gamma u_i - u_i \nabla s \cdot \nabla s.$$

Substituting $s = w/u_i$ into this equation (accept for the second term on the right-hand side) we have

$$\nabla^2 w + k^2 w = -k^2 \gamma u_i - u_i \nabla s \cdot \nabla s$$

whose Green's function solution is

$$w = k^2 \int_V u_i \gamma g d^3\mathbf{r} + \int_V u_i (\nabla s \cdot \nabla s) g d^3\mathbf{r}$$

so that we can write

$$s = \frac{k^2}{u_i} \int_V u_i \gamma g d^3\mathbf{r} + \frac{k^2}{u_i} \int_V u_i \gamma g \left(\frac{\nabla s \cdot \nabla s}{k^2 \gamma} \right) d^3\mathbf{r}.$$

In order for the second term on the right-hand side to be neglected, we must introduce the condition

$$\frac{\nabla s \cdot \nabla s}{k^2 \gamma} \ll 1$$

or

$$\| k^2 \gamma \| \gg \| \nabla s \cdot \nabla s \|.$$

The interpretation of this condition is not trivial. Clearly, the larger the value of k (i.e. the smaller the value of the wavelength) for a given magnitude of γ and ∇s, the more appropriate the condition becomes. Thus, the condition would appear to be valid if the wavelength of the field is small compared with γ. Since s can be taken to be the phase of the wavefield solution u, another physical interpretation of the condition is that the characteristic scale length over which a change in phase occurs ∇s is small compared with the wavelength for a given γ.

Born Series Solution

In Section 5.4, the Born approximation was introduced and used to solve some elementary problems. We shall now consider a natural extension to the Born approximation which is based on generating a series solution to the problem, known generally as the Neumann series solution.

Consider the three-dimensional Green's function solution to the equation

$$(\nabla^2 + k^2)u(\mathbf{r}, k) = -k^2\gamma u(\mathbf{r}, k)$$

which is given by (as derived in Section 5.4)

$$u(\mathbf{r}_0, k) = u_i(\mathbf{r}_0, k) + u_s(\mathbf{r}_0, k)$$

where

$$u_s(\mathbf{r}_0, k) = k^2 \int_V g(\mathbf{r} \mid \mathbf{r}_0, k)\gamma(\mathbf{r})u(\mathbf{r}, k)d^3\mathbf{r},$$

u_i is the incident field satisfying the equation

$$(\nabla^2 + k^2)u_i(\mathbf{r}, k) = 0$$

and g is the outgoing Green's function given by

$$g(\mathbf{r} \mid \mathbf{r}_0, k) = \frac{\exp(ik \mid \mathbf{r} - \mathbf{r}_0 \mid)}{4\pi \mid \mathbf{r} - \mathbf{r}_0 \mid}.$$

We have seen that the Born approximation to this equation is given by considering $u \sim u_i$, $\mathbf{r} \in V$ which is valid provided $|u_s| \ll |u_i|$. We then obtain an approximate solution u_1 say of the form

$$u_1(\mathbf{r}_0, k) = u_i(\mathbf{r}_0, k) + k^2 \int_V g(\mathbf{r} \mid \mathbf{r}_0, k)\gamma(\mathbf{r})u_i(\mathbf{r}, k)d^3\mathbf{r}.$$

This result can be considered to be the first approximation to a series solution, in which the second approximation, u_2 say, is given by

$$u_2(\mathbf{r}_0, k) = u_i(\mathbf{r}_0, k) + k^2 \int_V g(\mathbf{r} \mid \mathbf{r}_0, k)\gamma(\mathbf{r})u_1(\mathbf{r}, k)d^3\mathbf{r}$$

and the third approximation u_3 is given by

$$u_3(\mathbf{r}_0, k) = u_i(\mathbf{r}_0, k) + k^2 \int_V g(\mathbf{r} \mid \mathbf{r}_0, k)\gamma(\mathbf{r})u_2(\mathbf{r}, k)d^3\mathbf{r}$$

and so on. In general, we can consider the iteration

$$u_{j+1}(\mathbf{r}_0, k) = u_i(\mathbf{r}_0, k) + k^2 \int_V g(\mathbf{r} \mid \mathbf{r}_0, k)\gamma(\mathbf{r})u_j(\mathbf{r}, k)d^3\mathbf{r}, \qquad j = 0, 1, 2, 3, \ldots.$$

where $u_0 = u_i$. In principle, if this series converges, then it must converge to the solution. To investigate its convergence, it is convenient to use operator notation and write

$$u_{j+1} = u_i + \hat{I}u_j$$

where $\hat{I}$ is the integral operator

$$\hat{I} = k^2 \int_V d^3\mathbf{r} g\gamma.$$

At each iteration j we can consider the solution to be given by

$$u_j = u + \epsilon_j$$

where ϵ_j is the error associated with the solution at iteration j and u is the exact solution. A necessary condition for convergence is that $\epsilon_j \to 0$ as $j \to \infty$. Now

$$u + \epsilon_{j+1} = u_i + \hat{I}(u + \epsilon_j) = u_i + \hat{I}u + \hat{I}\epsilon_j$$

and therefore we can write

$$\epsilon_{j+1} = \hat{I}\epsilon_j$$

since $u = u_i + \hat{I}u$. Thus

$$\epsilon_1 = \hat{I}\epsilon_0; \qquad \epsilon_2 = \hat{I}\epsilon_1 = \hat{I}(\hat{I}\epsilon_0); \qquad \epsilon_3 = \hat{I}\epsilon_2 = \hat{I}[\hat{I}(\hat{I}\epsilon_0)]; \ldots$$

or

$$\epsilon_j = \hat{I}^j \epsilon_0$$

from which it follows that

$$\|\epsilon_j\| = \|\hat{I}^j \epsilon_0\| \leq \|\hat{I}^j\| \ \ \|\epsilon^0\| \leq \|\hat{I}\|^j \|\epsilon^0\|.$$

The condition for convergence therefore becomes

$$\lim_{j \to \infty} \|\hat{I}\|^j = 0.$$

This is only possible if

$$\|\hat{I}\| < 1$$

or

$$k^2 \left\| \int_V g(\mathbf{r} \mid \mathbf{r}_0, k)\gamma(\mathbf{r}) d^3\mathbf{r} \right\| < 1.$$

Comparing this result with condition 5.4.2 and the analysis associated with it given in Section 5.4, it is clear that

$$\bar{\gamma} < \frac{1}{k^2 R^2}$$

must be satisfied for the series to converge where R is the radius of a sphere of volume V.

This series solution, which can be written as

$$
\begin{aligned}
u(\mathbf{r}_0,k) &= u_i(\mathbf{r}_0,k) + k^2 \int\limits_V g(\mathbf{r} \mid \mathbf{r}_0,k)\gamma(\mathbf{r})u_i(\mathbf{r},k)d^3\mathbf{r} \\
&\quad + k^2 \int\limits_V g(\mathbf{r} \mid \mathbf{r}_0,k)\gamma(\mathbf{r}) \left[k^2 \int\limits_V g(\mathbf{r}_1 \mid \mathbf{r},k)\gamma(\mathbf{r}_1)u_i(\mathbf{r}_1,k)d^3\mathbf{r}_1 \right] d^3\mathbf{r} \\
&\quad + k^2 \int\limits_V g(\mathbf{r} \mid \mathbf{r}_0)\gamma(\mathbf{r}) \left[k^2 \int\limits_V g(\mathbf{r}_1 \mid \mathbf{r},k)\gamma(\mathbf{r}_1) \right. \\
&\quad \times \left. \left(k^2 \int\limits_V g(\mathbf{r}_2 \mid \mathbf{r}_1,k)\gamma(\mathbf{r}_2)u_i(\mathbf{r}_2,k)d^3\mathbf{r}_2 \right) d^3\mathbf{r}_1 \right] d^3\mathbf{r} + \cdots \\
&= u_i(\mathbf{r}_0,k) + k^2 \int\limits_V d^3\mathbf{r} g(\mathbf{r} \mid \mathbf{r}_0,k)\gamma(\mathbf{r})u_i(\mathbf{r},k) \\
&\quad + k^4 \int\limits_V \int\limits_V d^3\mathbf{r} d^3\mathbf{r}_1 g(\mathbf{r} \mid \mathbf{r}_0,k)\gamma(\mathbf{r})g(\mathbf{r}_1 \mid \mathbf{r},k)\gamma(\mathbf{r}_1)u_i(\mathbf{r}_1,k) \\
&\quad + k^6 \int\limits_V \int\limits_V \int\limits_V d^3\mathbf{r} d^3\mathbf{r}_1 d^3\mathbf{r}_2 g(\mathbf{r} \mid \mathbf{r}_0,k)\gamma(\mathbf{r})g(\mathbf{r}_1 \mid \mathbf{r},k)\gamma(\mathbf{r}_1) \\
&\quad g(\mathbf{r}_2 \mid \mathbf{r}_1,k)\gamma(\mathbf{r}_2)u_i(\mathbf{r}_2,k) \\
&\quad + \cdots
\end{aligned}
$$

is an example of a Neumann series solution to a Fredholm integral equation and is known as the Born series.

Another approach to deriving this result can be taken by considering the inverse operator. Writing

$$u = u_i + \hat{I}u$$

we have

$$(1 - \hat{I})u = u_i$$

or

$$
\begin{aligned}
u &= (1 - \hat{I})^{-1}u_i \\
&= (1 + \hat{I} + \hat{I}^2 + \hat{I}^3 + \cdots)u_i.
\end{aligned}
$$

EXERCISES

5.10 The propagation of a scalar wavefield u through a one-dimensional inhomogeneous continuum with time-independent velocity profile

$c(x)$ can be described by the equation

$$\left(\frac{\partial^2}{\partial x^2}+\frac{\omega^2}{c^2}\right)u(x,\omega)=0.$$

(i) Given that $c=c_0+v$ where $v/c_0 \ll 1$ and c_0 is a constant and by writing u in the form $u=\exp(-ikx)+w(x,k)$ where $\mid w \mid\ll 1$, show that this equation can be reduced to the approximate form

$$\left(\frac{\partial^2}{\partial x^2}+k^2\right)w(x,k)=\frac{2k^2v}{c_0}\exp(-ikx)$$

where $k=\omega/c_0$.

(ii) Solve the equation above using the appropriate Green's function and asymptotic formulation. Hence show that

$$w(a,\tau)=-\frac{1}{2c_0}\frac{dv}{d\tau}*\delta(\tau+a), \qquad a\to\infty$$

where $\tau=2t$ is the two-way travel time and $*$ denotes the convolution integral. Explain the physical significance of the conditions $v/c_0\ll 1$ and $\mid w \mid\ll 1$ used to derive this result.

(iii) Let $u(x,k)=\exp[iks(x)]$. Show that if the continuum is characterised by a velocity profile given by $c=\alpha c_0/x$ where α is a constant, then as $\omega\to\infty$, the solution for u defined by the original equation is given by

$$u(x,k)=\exp\left[ik\left(\beta+\frac{x^2}{2\alpha}\right)\right]$$

where β is an arbitrary constant. Explain the physical significance of the limiting condition (on the frequency) used to derive this result.

5.11 Find the Green's function solution to the equation

$$(\nabla^2+k^2)u(\mathbf{r}\mid\mathbf{r}_0,k)=-k^2\gamma(\mathbf{r})u(\mathbf{r}\mid\mathbf{r}_0,k)-\delta^3(\mathbf{r}-\mathbf{r}_0)$$

using the eikonal transform

$$u(\mathbf{r}\mid\mathbf{r}_0,k)=g(\mathbf{r}\mid\mathbf{r}_0,k)\exp[s(\mathbf{r}\mid\mathbf{r}_0,k)]$$

and the Rytov approximation (with homogeneous boundary conditions) where g is the solution of

$$(\nabla^2+k^2)g(\mathbf{r}\mid\mathbf{r}_0,k)=-\delta^3(\mathbf{r}-\mathbf{r}_0).$$

Compute the back-scattered field in this case.

5.12 Solve the following integral equation using (i) Laplace transforms; (ii) the Neumann series

$$u(x) = x - \int_0^x (t - x)u(t)dt.$$

5.8 Green's Function Solution to the Diffusion Equation

The homogeneous diffusion equation

$$\nabla^2 u(\mathbf{r}, t) = \sigma \frac{\partial}{\partial t} u(\mathbf{r}, t)$$

where σ is a constant differs in many aspects from the scalar wave equation and the Green's functions exhibit these differences. The most important single feature is the asymmetry of the diffusion equation with respect to time. For the wave equation, if $u(\mathbf{r}, t)$ is a solution, so is $u(\mathbf{r}, -t)$. However, if $u(\mathbf{r}, t)$ is a solution of

$$\nabla^2 u = \sigma \frac{\partial u}{\partial t}$$

the function $u(\mathbf{r}, -t)$ is not; it is a solution of the quite different equation,

$$\nabla^2 u(\mathbf{r}, -t) = -\sigma \frac{\partial}{\partial t} u(\mathbf{r}, -t).$$

Thus, unlike the wave equation, the diffusion equation differentiates between past and future. This is because the diffusing field u represents the behaviour of some average property of an ensemble of many particles which cannot in general return to their original state. Thus causality must be considered in the solution to the diffusion equation. This in turn leads to the use of the Laplace transform for solving the equation with respect to t (compared with the Fourier transform used to solve the wave equation with respect to t).

Evaluation of the Green's Function

As in the case of the scalar wave equation, it is possible to evaluate the Green's function for the diffusion equation which satisfies homogeneous boundary conditions and the causality condition

$$G(\mathbf{r} \mid \mathbf{r}_0, t \mid t_0) = 0 \qquad \text{if} \quad t < t_0.$$

This can be accomplished for one, two and three dimension simultaneously. Thus with $R=\mid \mathbf{r}-\mathbf{r}_0 \mid$ and $\tau = t-t_0$ we require the solution of the equation

$$\left(\nabla^2-\sigma\frac{\partial}{\partial\tau}\right)G(R,\tau)=-\delta^n(R)\delta(\tau), \qquad \tau>0 \qquad (5.8.1)$$

where n is 1, 2 or 3 depending on the number of dimensions. One way of solving this equation is to first take the Laplace transform with respect to τ, then solve for G (in Laplace space) and inverse Laplace transform. This requires an initial condition to be specified, in particular the value of G at $\tau=0$. Another way to solve this equation is to take its Fourier transform with respect to R, solve for G (in Fourier space) and then inverse Fourier transform. Here, we adopt the latter approach. Let

$$G(R,\tau)=\frac{1}{(2\pi)^n}\int\limits_{-\infty}^{\infty}\widetilde{G}(\mathbf{k},\tau)\exp(i\mathbf{k}\cdot\mathbf{R})d^n\mathbf{k}$$

and

$$\delta^n(R)=\frac{1}{(2\pi)^n}\int\limits_{-\infty}^{\infty}\exp(i\mathbf{k}\cdot\mathbf{R})d^n\mathbf{k}$$

then equation 5.8.1 reduces to

$$\sigma\frac{\partial\widetilde{G}}{\partial\tau}+k^2\widetilde{G}=\delta(\tau)$$

which has the solution

$$\widetilde{G}=\frac{1}{\sigma}\exp(-k^2\tau/\sigma)H(\tau)$$

where $H(\tau)$ is the step function

$$H(\tau)=\begin{cases}1, & \tau>0\\ 0, & \tau<0.\end{cases}$$

Hence, the Green's functions are given by

$$\begin{aligned}G(R,\tau) &= \frac{1}{\sigma(2\pi)^n}H(\tau)\int\limits_{-\infty}^{\infty}\exp(i\mathbf{k}\cdot\mathbf{R})\exp(-k^2\tau/\sigma)d^n\mathbf{k}\\ &= \frac{1}{\sigma(2\pi)^n}H(\tau)\left(\int\limits_{-\infty}^{\infty}\exp(ik_xR_x)\exp(-k_x^2\tau/\sigma)dk_x\right)\cdots.\end{aligned}$$

By rearranging the exponent in the integral, it becomes possible to evaluate each integral exactly. Thus,

$$ik_xR_x-k_x^2\frac{\tau}{\sigma}=-\left(k_x\sqrt{\frac{\tau}{\sigma}}-i\frac{R_x}{2}\sqrt{\frac{\sigma}{\tau}}\right)^2-\left(\frac{\sigma R_x^2}{4\tau}\right)=-\frac{\tau}{\sigma}\xi^2-\left(\frac{\sigma R_x^2}{4\tau}\right)$$

where

$$\xi = k_x - i\frac{\sigma R_x}{2\tau}.$$

Hence the integral over k_x becomes

$$\int_{-\infty}^{\infty} \exp\left[-\left(\frac{\tau}{\sigma}\xi^2\right) - \left(\frac{\sigma R_x}{4\tau}\right)\right] d\xi = e^{-(\sigma R_x^2/4\tau)} \int_{-\infty}^{\infty} e^{-(\tau\xi^2/\sigma)} d\xi$$

$$= \sqrt{\frac{\pi\sigma}{\tau}} \exp\left[-\left(\frac{\sigma R_x^2}{4\tau}\right)\right].$$

Introducing this result into the expression for G, we obtain

$$G(R,\tau) = \frac{1}{\sigma}\left(\frac{\sigma}{4\pi\tau}\right)^{\frac{n}{2}} \exp\left[-\left(\frac{\sigma R^2}{4\tau}\right)\right] H(\tau), \qquad \tau > 0.$$

The function G satisfies an important integral property which is valid for all n:

$$\int_{-\infty}^{\infty} g(R,\tau) d^n\mathbf{r} = \frac{1}{\sigma}; \qquad \tau > 0.$$

This is the expression for the conservation of the Green's function associated with the diffusion equation. For example, if we consider the diffusion of heat, then if at a time t_0 and at a point in space $\mathbf{r}_0$ a source of heat is introduced, then the heat diffuses out through the medium characterised by σ in such a way that the total heat energy is unchanged.

General Solution to the Inhomogeneous Diffusion Equation

Working in three dimensions, let us consider the general solution to the equation

$$\left(\nabla^2 - \sigma\frac{\partial}{\partial t}\right) u(\mathbf{r},t) = -f(\mathbf{r},t)$$

where f is a source of compact support ($\mathbf{r} \in V$) and define the Green's function as the solution to the equation

$$\left(\nabla^2 - \sigma\frac{\partial}{\partial t}\right) G(\mathbf{r} \mid \mathbf{r}_0, t \mid t_0) = -\delta^3(\mathbf{r} - \mathbf{r}_0)\delta(t - t_0).$$

It is convenient to first take the Laplace transform of these equations with respect to $\tau = t - t_0$ to obtain

$$\nabla^2\bar{u} - \sigma[-u_0 + p\bar{u}] = -\bar{f} \qquad 5.8.2$$

and

$$\nabla^2\bar{G} + \sigma[-G_0 + p\bar{G}] = -\delta^3 \qquad 5.8.3$$

where

$$\bar{u}(\mathbf{r},p)=\int_0^\infty u(\mathbf{r}\mid\mathbf{r}_0,\tau)\exp(-p\tau)d\tau$$

$$\bar{G}(\mathbf{r}\mid\mathbf{r}_0,p)=\int_0^\infty G(\mathbf{r}\mid\mathbf{r}_0,\tau)\exp(-p\tau)d\tau$$

$$\bar{f}(\mathbf{r},p)=\int_0^\infty f(\mathbf{r},\tau)\exp(-p\tau)d\tau$$

$$u_0\equiv u(\mathbf{r},\tau=0)\qquad\text{and}\qquad G_0\equiv G(\mathbf{r}\mid\mathbf{r}_0,\tau=0)=0.$$

Multiplying equation 5.8.2 by $\bar{G}$ and equation 5.8.3 by $\bar{u}$, subtracting the two results and integrating over V we obtain

$$\int_V(\bar{G}\nabla^2\bar{u}-\bar{u}\nabla^2\bar{G})d^3\mathbf{r}+\sigma\int_V u_0\bar{G}d^3\mathbf{r}=-\int_V\bar{f}\bar{G}d^3\mathbf{r}+\bar{u}(\mathbf{r}_0,p).$$

Using Green's theorem and rearranging the result gives

$$\bar{u}(\mathbf{r}_0,\tau)=\int_V\bar{f}(\mathbf{r},p)\bar{G}(\mathbf{r}\mid\mathbf{r}_0,p)d^3\mathbf{r}+\sigma\int_V u_0(\mathbf{r})\bar{G}(\mathbf{r}\mid\mathbf{r},p)d^3\mathbf{r}+\oint_S(\bar{g}\nabla\bar{u}-\bar{u}\nabla\bar{g})\cdot\hat{\mathbf{n}}d^2\mathbf{r}.$$

Finally, taking the inverse Laplace transform and using the Convolution Theorem for Laplace transforms, we can write

$$\begin{aligned}u(\mathbf{r}_0,\tau) &= \int_0^\tau\int_V f(\mathbf{r},\tau')G(\mathbf{r}\mid\mathbf{r}_0,\tau-\tau')d^3\mathbf{r}d\tau'+\sigma\int_V u_0(\mathbf{r})G(\mathbf{r}\mid\mathbf{r}_0,\tau)d^3\mathbf{r}\\ &+\int_0^\tau\oint_S[G(\mathbf{r}\mid\mathbf{r}_0,\tau')\nabla u(\mathbf{r},\tau-\tau')\\ &-u(\mathbf{r},\tau')\nabla G(\mathbf{r}\mid\mathbf{r}_0,\tau-\tau')]\cdot\hat{\mathbf{n}}d^2\mathbf{r}d\tau'.\end{aligned}$$

Example: A Solution to the One-dimensional Homogeneous Diffusion Equation

Consider the case when the source term is zero and the volume of interest is the infinite domain, so that the surface integral is zero. Then we have (replacing τ by t)

$$u(\mathbf{r}_0,t)=\sigma\int_V u_0(\mathbf{r})G(\mathbf{r}\mid\mathbf{r}_0,t)d^3\mathbf{r}.$$

In one dimension, this reduces to

$$u(x_0,t) = \sqrt{\frac{\sigma}{4\pi t}} \int_{-\infty}^{\infty} \exp\left[-\frac{\sigma(x_0-x)^2}{4t}\right] u_0(x)\,dx, \qquad t>0.$$

Thus we see that the field u at a time $t>0$ is given by the convolution of the field at time $t=0$ with the Gaussian function

$$\sqrt{\frac{\sigma}{4\pi t}} \exp\left(-\frac{\sigma x^2}{4t}\right).$$

EXERCISES

5.13 Find the Green's function defined by the equation

$$\left(\nabla^2 + \sigma\frac{\partial}{\partial t}\right) G(\mathbf{r}\mid\mathbf{r}_0, t\mid t_0) = -\delta^3(\mathbf{r}-\mathbf{r}_0)\delta(t-t_0)$$

by first taking the Laplace transform with respect to t and using the initial condition

$$G\mid_{t=t_0} = 0$$

and the result

$$\int_0^{\infty} t^{-3/2} e^{a/t} dt = \sqrt{\frac{\pi}{a}} e^{-2\sqrt{ap}} dt.$$

5.14 Investigate the reciprocity theorem for the diffusion equation and in particular, show that

$$G(\mathbf{r}\mid\mathbf{r}_0, t\mid t_0) = G(\mathbf{r}_0\mid\mathbf{r}, -t_0\mid -t).$$

5.9 Green's Function Solution to the Laplace and Poisson Equations

The two- and three-dimensional Laplace and Poisson equations are given by

$$\nabla^2 u = 0$$

and

$$\nabla^2 u = -f \qquad (5.9.1)$$

respectively. We consider the Poisson equation first. The general approach is identical to that used to derive a solution to the inhomogeneous Helmholtz equation. Thus, working in three dimensions and defining the Green's function to be the solution of

$$\nabla^2 g(\mathbf{r} \mid \mathbf{r}_0) = -\delta^3(\mathbf{r} - \mathbf{r}_0) \tag{5.9.2}$$

from equation 5.9.1 we obtain the following result

$$u = \oint_S (g\nabla u - u\nabla g) \cdot \hat{\mathbf{n}} d^2\mathbf{r} + \int_V g f d^3\mathbf{r}$$

where we have used Green's theorem to obtain the surface integral on the right-hand side. The problem now is to find the Green's function for this problem. Clearly, since the solution to the equation

$$(\nabla^2 + k^2)g = -\delta^3(\mathbf{r} - \mathbf{r}_0)$$

is

$$g(\mathbf{r} \mid \mathbf{r}_0, k) = \frac{1}{4\pi \mid \mathbf{r} - \mathbf{r}_0 \mid} \exp(ik \mid \mathbf{r} - \mathbf{r}_0 \mid),$$

we should expect the Green's function for the three-dimensional Poisson equation (and the Laplace equation) to be of the form

$$g(\mathbf{r} \mid \mathbf{r}_0) = \frac{1}{4\pi \mid \mathbf{r} - \mathbf{r}_0 \mid}. \tag{5.9.3}$$

This can be shown by taking the Fourier transform of equation 5.9.2 which gives

$$k^2 G(k) = 1$$

where

$$G(k) = \int g(R) \exp(i\mathbf{k} \cdot \mathbf{R}) d^3\mathbf{R}, \qquad R = \mid \mathbf{r} - \mathbf{r}_0 \mid.$$

Therefore

$$\begin{aligned} g(R) &= \frac{1}{(2\pi)^3} \int \frac{\exp(i\mathbf{k} \cdot \mathbf{R})}{k^2} d^3\mathbf{k} \\ &= \frac{1}{(2\pi)^3} \int_0^{2\pi} d\phi \int_{-1}^{1} d(\cos\theta) \int_0^{\infty} dk \exp(ikR\cos\theta) \\ &= \frac{1}{2\pi^2 R} \int_0^{\infty} \frac{\sin(kR)}{k} dk \\ &= \frac{1}{4\pi R} \end{aligned}$$

using spherical polar coordinates and the result

$$\int_0^\infty \frac{\sin x}{x} dx = \frac{\pi}{2}.$$

Thus, we obtain the following fundamental result:

$$\nabla^2 \left(\frac{1}{4\pi R} \right) = -\delta^3(R).$$

With homogeneous boundary conditions, the solution to the Poisson equation is

$$u(\mathbf{r}_0) = \frac{1}{4\pi} \int_V \frac{f(\mathbf{r})}{\mid \mathbf{r} - \mathbf{r}_0 \mid} d^3\mathbf{r}.$$

In two dimensions the solution is of the same form, but with a Green's function given by

$$g(\mathbf{r} \mid \mathbf{r}_0) = \frac{1}{2\pi} \ln \mid \mathbf{r} - \mathbf{r}_0 \mid .$$

The general solution to Laplace's equation is

$$u = \oint_S (g\nabla u - u\nabla g) \cdot \hat{\mathbf{n}} d^2\mathbf{r}$$

with g given by equation 5.9.3. These solutions to the Laplace and Poisson equations are analogous to those for the homogeneous and inhomogeneous Helmholtz equations.

5.10 Discussion

This chapter has provided a brief introduction to the use of Green's functions for solving partial differential equations in different dimensions and for time-dependent and time-independent problems. A more detailed consideration of the role of Green's functions for solving partial differential equations is beyond the scope of this book. For further reading, the reader may wish to look at Morse and Feshbach (1953) or Roach (1970).

The material discussed in this chapter has been based almost exclusively on the use of free space Green's functions in which a solution is developed over the infinite domain to which boundary conditions can be applied. There are a number of techniques for computing the Green's function G for finite or bounded domain problems provided the geometry is simple enough. For example, we can consider

$$G(\mathbf{r} \mid \mathbf{r}_0) = g(\mathbf{r} \mid \mathbf{r}_0) + F(\mathbf{r} \mid \mathbf{r}_0)$$

where g is the free space Green's function and F represents the boundary effects. F cannot have a singularity within the bounded domain and so as $r \to r_0, G(\mathbf{r} \mid \mathbf{r}_0) \to g(\mathbf{r} \mid \mathbf{r}_0)$. In the "Imaging Method" F is determined by considering a mirror image of the source point $\mathbf{r}_0$ in the opposite side of the boundary $\mathbf{r}_1$ say. F is then given by $-g(\mathbf{r} \mid \mathbf{r}_1)$ satisfying Dirichlet boundary conditions or $+g(\mathbf{r} \mid \mathbf{r}_1)$ satisfying Neumann boundary conditions. Application of this method to two or more boundaries leads naturally to expressions for G which involve infinite series as the effect of one boundary on another is taken into account in order to generate a complete solution on a bounded domain.

Finally, it is worth mentioning that Green's functions are used in certain numerical methods of solution to partial differential equations by utilising Green's theorem and Green's functions. These are the boundary element methods in which a numerical solution is devised by discretising the surface integral considered in this chapter into surface patches and computing the surface integral numerically.

A
Solutions of Exercises

Chapter 1

1.1 In equation 1.1.15 there will be an extra term $-\rho\, dx\, r\frac{\partial u}{\partial t}$ which expresses the resistance as proportional to the string velocity. Hence after reduction the term $-r\frac{\partial u}{\partial t}$ remains.

1.2 For a transverse elastic force, the force is now proportional to the displacement u and the extra term in 1.1.15 is $\rho\, ds\, ku$ which then reduces to the final extra term $-ku$.

1.3 For a general external force $f(x,t)$ apply the extra term as in the previous exercises.

1.4 See 1.1.10 in the text.

1.5 When a fluid passes over a surface the force is tangential to the surface and proportional to the normal velocity gradient. Hence for the tip of a string in a viscous fluid the balance yields

$$\frac{\partial u}{\partial n} + b\frac{\partial u}{\partial t} = 0.$$

1.6 Eliminate i by the following lines:

$$\begin{aligned} v_x + Ri + Li_t &= 0 \\ i_x + Cv_t + Gv &= 0. \end{aligned}$$

Hence

$$v_{xx} + Ri_x + Li_{xt} = 0$$

$$i_{xt} + Cv_{tt} + Gv_t \quad = \quad 0$$

by differentiating, and then elimination gives

$$v_{xx} + R(-Cv_t - Gv) + L(-Cv_{tt} - Gv_t) = 0$$

as required.

1.7 From the solution:

$$\begin{aligned} \frac{\partial u}{\partial t} &= (1/n)(-n^2k)e^{-n^2kt}\sin nx \\ \frac{\partial u}{\partial x} &= e^{-n^2kt}\cos nt \\ \frac{\partial^2 u}{\partial x^2} &= -ne^{-n^2kt}\sin nt, \end{aligned}$$

from which it is obvious that the solution given satisfies $u_t = ku_{xx}$.

1.8 (i) The characteristic equation satisfies:

$$\frac{dy}{dx} = \frac{x \pm \sqrt{x^2 + 4y^2}}{-2}$$

and hence whatever x and y, the square root has a positive argument and hence the equation is hyperbolic. Pedantically if $x = y = 0$, then locally the equation is parabolic.

(ii) Now the characteristics satisfy

$$\frac{dy}{dx} = \frac{1 \pm \sqrt{-x^2 - y^2 - x^2y^2}}{2(1+x)}$$

which now has the square root being always negative. Hence now the equation is elliptic (except at $x = y = 0$).

(iii) This example factorises to

$$\left(x\frac{dy}{dx} - 1\right)^2 = 0$$

and the equation is of the parabolic type.

1.9 The lines of parabolicity occur when $S^2 = 4RT$ and the regions bounded by these lines can be categorised by spot checks. Hence we need to sketch $9x^2y^2 = 4(x + y)$ which is well exposed by the transformation $x = r\cos\theta$ and $y = r\sin\theta$, to leave the polar equation:

$$r^3 = \frac{4(\cos\theta + \sin\theta)}{9\cos^2\theta\sin^2\theta}$$

from which asymptotes at 0, $\pi/2$, π and $3\pi/2$ are immediately clear, and $r = 0$ at $\theta = 3\pi/4$ and $\theta = 7\pi/4$. The result is shown in Figure A.1.

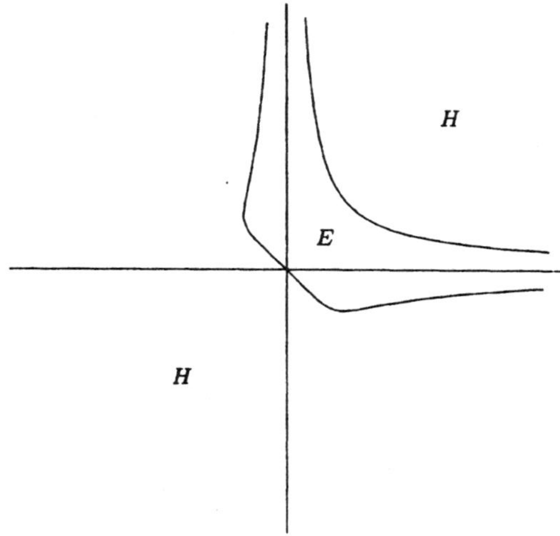

Fig. A.1.

1.10 For this partial differential equation the characteristic equation is

$$\begin{aligned}\frac{dy}{dx} &= \frac{2(x+1/y) \pm \sqrt{4(x+1/y)^2 - 16x/y}}{2} \\ &= \begin{cases} 2x \\ 2/y. \end{cases}\end{aligned}$$

Hence the characteristics are $y = x^2 + A$ and $y^2 = 4x + B$.

1.11 The results of 1.2.17 are used here but the example will be worked from first principles. The characteristics satisfy:

$$\left(\frac{dy}{dx}\right)^2 - 2\frac{dy}{dx} + 3 = 0.$$

Hence

$$\frac{dy}{dx} = 1 \pm i\sqrt{2}$$

and the required transformation is $\xi = y - (1 + i\sqrt{2})x$ and $\eta = y - (1 - i\sqrt{2})x$. Hence the required partial derivatives are

$$\begin{aligned}\frac{\partial z}{\partial x} &= -\frac{\partial z}{\partial \xi}(1 + i\sqrt{2}) - \frac{\partial z}{\partial \eta}(1 - i\sqrt{2}) \\ \frac{\partial z}{\partial y} &= \frac{\partial z}{\partial \xi} + \frac{\partial z}{\partial \eta}\end{aligned}$$

$$\begin{aligned}\frac{\partial^2 z}{\partial x^2} &= \frac{\partial^2 z}{\partial \xi^2}(-1+2i\sqrt{2})+6\frac{\partial^2 z}{\partial\xi\partial\eta}+\frac{\partial^2 z}{\partial\eta^2}(-1-2i\sqrt{2})\\ \frac{\partial^2 z}{\partial x\partial y} &= -\frac{\partial^2 z}{\partial \xi^2}(1+i\sqrt{2})-\frac{\partial^2 z}{\partial\xi\partial\eta}(1-i\sqrt{2})-\frac{\partial^2 z}{\partial\xi\partial\eta}(1+i\sqrt{2})\\ &\quad -\frac{\partial^2 z}{\partial\eta^2}(1-i\sqrt{2})\\ \frac{\partial^2 z}{\partial y^2} &= \frac{\partial^2 z}{\partial \xi^2}+2\frac{\partial^2 z}{\partial\xi\partial\eta}+\frac{\partial^2 z}{\partial\eta^2}\end{aligned}$$

and substitution into the original equation results in cancellation to

$$8\frac{\partial^2 z}{\partial\xi\partial\eta}=z.$$

To obtain a real form set $\xi=\alpha+i\beta$ and $\eta=\alpha-i\beta$. Then $\alpha=(\xi+\eta)/2$ and $\beta=(\xi-\eta)/2i$ and

$$\begin{aligned}\frac{\partial z}{\partial\xi} &= \frac{\partial z}{\partial\alpha}\frac{1}{2}+\frac{\partial z}{\partial\beta}\frac{1}{2i}\\ \frac{\partial^2 z}{\partial\xi\partial\eta} &= \frac{\partial^2 z}{\partial\alpha^2}\frac{1}{4}+\frac{\partial^2 z}{\partial\alpha\partial\beta}\left(-\frac{1}{4i}+\frac{1}{4i}\right)\frac{\partial^2 z}{\partial\beta^2}\frac{1}{4}.\end{aligned}$$

Hence the real form of the equation is

$$2\left(\frac{\partial^2 z}{\partial\alpha^2}+\frac{\partial^2 z}{\partial\beta^2}\right)=z.$$

1.12 For set (a):

$$\int_0^\pi \cos nx\cos mx\,dx=\frac{1}{2}\int_0^\pi \cos(n+m)x\,dx+\frac{1}{2}\int_0^\pi\cos(n-m)x\,dx=0$$

as long as $n\neq m$. In the case $n=m$,

$$\int_0^\pi \cos^2 nx\,dx=\int_0^\pi\frac{1+\cos 2nx}{2}\,dx=\frac{\pi}{2}.$$

For set (b) the same approach applies with

$$\sin n\pi x\sin m\pi x=(\cos(n-m)\pi x\cos(n+m)\pi x)/2$$

and the range $[-1,1]$. When $n=m$,

$$\begin{aligned}\int_0^\pi \sin^2 nx\,dx &= \int_0^\pi\frac{1-\cos 2nx}{2}\,dx\\ &= 1.\end{aligned}$$

To prove (c), the three integrals:

$$\int_0^\infty (1-x)e^{-x}\,dx$$
$$\int_0^\infty (1-2x+x^2/2)e^{-x}\,dx$$
$$\int_0^\infty x(1-2x+x^2/2)e^{-x}\,dx$$

need to be shown to be zero.

Use

$$I_n = \int_0^\infty x^n e^{-x}\,dx = n\int_0^\infty x^{n-1}e^{-x}\,dx = I_{n-1}$$

by integration by parts, then

$$\begin{aligned}\int_0^\infty (1-x)e^{-x}\,dx &= 1-I_0 = 0\\ \int_0^\infty (1-2x+x^2/2)e^{-x}\,dx &= 1-2+\frac{1}{2}(2.1) = 0\\ \int_0^\infty x(1-2x+x^2/2)e^{-x}\,dx &= 1-2(2.1)+\frac{1}{2}(3.2.1) = 0.\end{aligned}$$

1.13 For f and g to be orthogonal

$$\int_{-1}^1 x\,dx = \left.\frac{x^2}{2}\right|_{-1}^1 = 0$$

and α and β need to satisfy:

$$\begin{aligned}\int_{-1}^1 (1+\alpha x+\beta x^2)\,dx &= 2+\frac{2\beta}{3} = 0\\ \int_{-1}^1 x(1+\alpha x+\beta x^2)\,dx &= \frac{\alpha}{2} = 0\end{aligned}$$

and hence $\alpha = 0$ and $\beta = -3$.

1.14 Consider

$$x-2 = a_0 + \sum_{n=1}^\infty a_n \cos nx$$

then

$$a_n = \int_0^\pi (x-2)dx\Big/\int_0^1 1^2\,dx = (\pi/2-2)$$

and

$$\begin{aligned} a_n &= \int_0^\pi (x-2)\cos nx\,dx \Big/ \int_0^\pi \cos^2 nx\,dx \\ &= -\frac{2}{n\pi}\left[-\frac{\cos nx}{n}\right]_0^\pi = -\frac{4}{n^2\pi}. \end{aligned}$$

1.15 Put $x = \cos\theta$ then $\cos 3\theta = 4\cos^3\theta - 3\cos\theta$. This gives immediately $T_3(x) = 4x^3 - 3x$ and the orthogonality condition reduces to investigating the integral:

$$\int_0^\pi \cos r\theta \cos s\theta d\theta.$$

The case $r \neq s$ and $r = s \neq 0$ is now completed as in 1.4.1. When $r = s = 0$, the integral is just that of unity to yield π.

1.16 These exercises use the results

$$(2n+1)xP_n(x) = (n+1)P_{n+1}(x) + nP_{n-1}(x)$$

and the orthogonality condition 1.5.18 extensively. Hence

$$\begin{aligned} &\int_{-1}^1 xP_l(x)P_{l-1}(x)\,dx \\ &= \int_{-1}^1 \left[\frac{l+1}{2l+1}P_{l+1}(x) + \frac{l}{2l+1}P_{l-1}(x)\right] P_{l-1}(x)\,dx \\ &= \frac{l}{2l+1}\int_{-1}^1 P_{l-1}^2\,dx = \frac{2l}{4l^2-1}. \end{aligned}$$

(The first integral is zero by orthogonality.)

1.17 Start with the differential equation

$$\frac{d}{dx}\left[(1-x^2)\frac{dP_l}{dx}\right] + l(l+1)P_l(x) = 0$$

and multiply throughout by $P_m(x)$ and integrate from -1 to 1. Then after integrating the first integral by parts (trivially),

$$\begin{aligned} &\int_{-1}^1 (1-x^2)P_l'(x)P_m'(x)\,dx \\ &= \int_{-1}^1 l(l+1)P_l(x)P_m(x)\,dx = l(l+1)\frac{2}{2l+1}\delta_{ml} \end{aligned}$$

1.18 By the usual Fourier coefficient formula,

$$c_r = \frac{\int_{-1}^1 f(x)P_r(x)\,dx}{\int_{-1}^1 P_r^2(x)\,dx}.$$

The numerator splits into

$$\int_{-1}^{0}\left(-\frac{1}{2}\right)P_r(x)\,dx+\int_{0}^{1}\left(\frac{1}{2}\right)P_r(x)\,dx.$$

For r even, put $x=-y$ in the first integral which becomes minus the second to give $c_r=0$. For r odd, use 1.5.16, which integrates trivially with the denominator being $2/(2r+1)$ to give the result.

1.19 $$nu_n+(n-1)u_{n-1}$$
$$=\int_{-1}^{1}nx^{-1}P_n(x)P_{n-1}(x)+(n-1)x^{-1}P_{n-1}(x)P_{n-2}(x)\,dx.$$

However

$$\frac{lP_l(x)}{x}=P_l'(x)=\frac{P_{l-1}'}{x}.$$

Use $l=n$ in the first term to eliminate $nx^{-1}P_n(x)$ and $l=n-1$ in the second term to give

$$\int_{-1}^{1}\left[P_{n-1}\left(P_n'-\frac{P_{n-1}'}{x}\right)+P_{n-1}\left(\frac{P_{n-1}'}{x}-P_{n-2}'\right)\right]dx$$
$$=\int_{-1}^{1}P_{n-1}\,d(P_n)+\int_{-1}^{1}P_{n-1}d(P_{n-2})\,dx=1+1=2.$$

However, $u_1=2$ and $u_2=0$ from the power series forms of P_0, P_1 and P_2, from 1.5.13. Then assume $u_n=2/n$ for n odd and n with the value $n-1$. Then by induction, $nu_n=2-0=2/n$ as required. For n even we obtain $nu_n=2-(n-1)2/(n-1)=0$ as required.

1.20 Use the recurrence relation

$$-(2r+1)xP_r(x)=-(r+1)P_{r+1}(x)-rP_{r-1}(x)$$

to give

$$\begin{aligned}
&(1-x)\sum_{r=0}^{n}(2r+1)P_r(x)\\
=&\sum_{r=0}^{n}(2r+1)P_r(x)-(r+1)P_{r+1}(x)-rP_{r-1}(x)\\
=&\;P_0(x)-P_1(x)\\
&+3P_1(x)-2P_2(x)-P_0(x)\\
&+5P_2(x)-3P_3(x)-2P_1(x)\\
&+7P_3(x)-4P_4(x)-3P_3(x)\\
&+\ldots\ldots\\
&+(2n-1)P_{n-1}(x)-nP_n(x)-(n-1)P_{n-2}(x)\\
&+(2n+1)P_n(x)-(n+1)P_{n+1}(x)-nP_{n-1}(x).
\end{aligned}$$

Cancellation occurs through this expansion, in that any first term in a line plus the second term in the previous line plus the third term in the next line sum to zero. All that is left is

$$(2n+1)P_n(x) - nP_n(x) - (n+1)P_{n+1}(x)$$

to give the result.

1.21 Use the previous result to give

$$\begin{aligned}\sum_{r=0}^{n}(2r+1)P_r(x) &= \frac{(n+1)[P_n(x)-P_{n+1}(x)]}{(1-x)} \\ &= \frac{P'_{n+1}(x)-xP'_n(x)}{1-x} - \frac{xP'_{n+1}(x)-P'_n(x)}{1-x} \\ &= P_{n+1}(x) - P'_n(x)\end{aligned}$$

using both 1.5.16, (ii) and (iii).

1.22
$$\begin{aligned}\int_{-1}^{1} P_n(x)(1-2xh+h^2)^{-1/2}\,dx &= \int_{-1}^{1} P_n(x)\sum_{l=0}^{\infty} t^l P_l(x)\,dx \\ &= \int_{-1}^{1} h^n P_n^2(x)\,dx = \frac{2h^n}{2n+1}\end{aligned}$$

by using the generating function and orthogonality.

1.23 The two major formulae being manipulated here are

$$\begin{aligned}x(2l+1)P_l(x) &= (l+1)P_{l+1}(x) + lP_{l-1}(x) &(1)\\ P_l(x) &= P'_{l+1}(x) - 2xP'_l(x) + P'_{l-1}(x). &(2)\end{aligned}$$

(i) Differentiate (1):

$$(2l+1)P_l(x) + (2l+1)xP'_l(x) = (l+1)P'_{l+1}(x) + lP'_{l+1}(x) \quad (3)$$

$$2P_l(x) = -2xP'_l(x) + \frac{2(l+1)}{2l+1}P'_{l+1}(x) + \frac{2l}{2l+1}P'_{l-1}(x). \quad (4)$$

Subtract (2) and (4)

$$P_l(x) = \frac{2(l+1)}{(2l+1)}P'_{l+1}(x) + \frac{2l}{2l+1}P'_{l-1}(x) - P'_{l+1}(x) - P'_{l-1}(x)$$

$$\begin{aligned}(2l+1)P_l(x) &= P'_{l-1}(x)[2l+1-(2l-1)] + P'_{l-1}(x)[2l-(2l+1)] \\ &= P'_{l+1}(x) - P'_{l-1}(x).\end{aligned}$$

(ii) Multiply (2) by $l+1$ to give

$$(l+1)P_l(x) = (l+1)P'_{l+1}(x) - 2x(l+1)P'_l(x) + (l+1)P'_{l-1}(x). \quad (5)$$

Subtract (3) from (5) to give

$$xP'_l(x) - P'_{l-1}(x) = lP_l(x). \quad (6)$$

(iii) From (6)

$$xP'_{l-1}(x) = P'_{l-2}(x) + (l-1)P_{l-1}(x).$$

Substitute out $P'_{l-2}(x)$ from (2):

$$\begin{aligned} xP'_{l-1}(x) &= P_{l-1}(x) + 2xP'_{l-1}(x) - P'_l(x) + (l-1)P_{l-1}(x) \\ &= lP_{l-1}(x) + 2xP'_{l-1}(x) - P'_l(x). \end{aligned}$$

Hence

$$P'_l(x) - xP'_{l-1}(x) = lP_{l-1}(x). \tag{7}$$

(iv) Multiply (6) by x to give

$$x^2P'_l(x) = xP'_{l-1}(x) = xlP_l(x)$$

and from (7)

$$x^2P'_l(x) + lP_{l-1}(x) - P'_l(x) = xlP_l(x)$$

and hence

$$(x^2-1)P'_l(x) = lxP_l(x) - lP_{l-1}(x).$$

1.24 From 1.6.29

$$J_{1/2}(x) = \sum_{s=0}^{\infty} \frac{(-1)^s}{\Gamma\left(\frac{3}{2}+s\right)s!}\left(\frac{x}{2}\right)^{2s+1/2} = \frac{2^{1/2}}{x^{1/2}}\sum_{s=0}^{\infty}\frac{(-1)^s x^{2s+1}}{2^{2s+1}s!\Gamma\left(\frac{3}{2}+s\right)}.$$

However, from Abramowitz and Stegun (1964, equation 6.1.12),

$$\begin{aligned} \Gamma\left(n+\frac{1}{2}\right) &= \frac{1.3.5.7\ldots(2n-1)}{2^n}\Gamma\left(\frac{1}{2}\right) \\ &= \frac{1.3.5.7\ldots(2n-1)}{2^n}\pi^{1/2}. \end{aligned}$$

Hence

$$\Gamma\left(s+1+\frac{1}{2}\right) = \frac{1.3.5.7\ldots(2s+1)}{2^{s+1}}\pi^{1/2},$$

and

$$\begin{aligned} J_{1/2}(x) &= \left(\frac{2}{\pi x}\right)^{1/2}\sum_{s=0}^{\infty}\frac{(-1)^s x^{2s+1}}{2^s s!(1.3.5\ldots(2s+1))} \\ &= \left(\frac{2}{\pi x}\right)^{1/2}\sin x. \end{aligned}$$

1.25
$$J_{-1/2}(x) = \frac{2^{1/2}}{x^{1/2}}\sum_{s=0}^{\infty}\frac{(-1)^s x^{2s}}{2^{2s}\Gamma(1/2+s)s!}$$
$$= \frac{2^{1/2}}{x^{1/2}\pi^{1/2}}\sum_{s=0}^{\infty}\frac{(-1)^s x^{2s}}{2^{2s}(1.3.5.7\ldots(2s-1)s!}$$
$$= \left(\frac{2}{x\pi}\right)^{1/2}\cos x$$

1.26 $E = \frac{d}{dx}(xJ_nJ_{n+1}) = J_nJ_{n+1} + xJ_n'J_{n+1} + xT_nJ'n + 1.$

Use $J_n' = (J_{n-1} - J_{n+1})/2$ to give

$$E = J_nJ_{n+1} + \frac{1}{2}xJ_{n+1}J_{n-1} - \frac{1}{2}xJ_{n+1}^2 + \frac{1}{2}xJ_n^2 - \frac{1}{2}xJ_nJ_{n+2}.$$

Now use $xJ_{n-1} - 2nJ_n - xJ_{n+1}$ and $xJ_{n+2} = 2(n+1)J_{n+1} - xJ_n$ to obtain

$$E = J_nJ_{n+1} + nJ_nJ_{n+1} - \frac{1}{2}xJ_{n+1}^2 + \frac{1}{2}xJ_n^2 + \frac{1}{2}xJ_n^2 - (n+1)J_nJ_{n+1}$$
$$= x(J_n^2 - J_{n+1}^2).$$

1.27
$$2J_n' = J_{n-1} - J_{n+1} \quad 2J_n'' = J_{n-1}' - J_{n+1}'$$
$$4J_n'' = J_{n-2} - J_n - J_n + J_{n+1} \quad 4J_n''' = J_{n-2}' - 2J_n' + J_{n+1}'$$
$$8J_n''' = J_{n-3} - 3J_{n-1} + 3J_{n+1} - J_{n+3}.$$

1.28 Start with $J_0' = -J_1$ and hence $J_0'' = -J_1'$ and $J_0''' = -J_1''$. However $2J_1' = J_0 - J_2$, and hence $2J_1'' = J_0' - J_2' = J_0' - \frac{1}{2}(-J_3 + J_1)$ to give $4J_0''' + 3J_0' + J_3 = 0$.

1.29 This follows from equation 1.6.9 in the text.

1.30
$$\exp\left[\frac{x}{2}\left(t-\frac{1}{t}\right)\right]\exp\left[-\frac{x}{2}\left(t-\frac{1}{t}\right)\right]$$
$$= \sum_{n=-\infty}^{\infty} t^nJ_n(x)\sum_{n=-\infty}^{\infty} s^nJ_n(-x)$$
$$= \sum_{n=-\infty}^{\infty} t^nJ_n(x)\sum_{n=-\infty}^{\infty} s^n(-1)^nJ_n(x)$$
$$= \left[J_0 + (tJ_1 + t^2J_2 + \ldots) + \left(\frac{J_1}{t} + \frac{J_2}{t^2} + \ldots\right)\right]$$
$$+ \left[[J_0 + (-sJ_1 + s^2J_2 - \ldots) + \left(-\frac{J_1}{s} + \frac{J_2}{s^2} - \ldots\right)\right]$$

$$
\begin{aligned}
&= J_0^2 + J_0(tJ_1 + t^2J_2 + \ldots) + J_0\left(\frac{J_1}{t} + \frac{J_2}{t^2} + \ldots\right) \\
&\quad + J_0(-sJ_1 + s^2J_2 + \ldots) + J_0\left(-\frac{J_1}{s} + \frac{J_2}{s^2} + \ldots\right) \\
&\quad + (tJ_1 + t^2J_2 + \ldots)(-sJ_1 + s^2J_2 + \ldots) \\
&\quad + (-sJ_1 + s^2J_2 + \ldots)\left(-\frac{J_1}{s} + \frac{J_2}{s^2} - \ldots\right) \\
&\quad + \left(\frac{J_1}{t} + \frac{J_2}{t^2} + \ldots\right)\left(-\frac{J_1}{s} + \frac{J_2}{s^2} - \ldots\right).
\end{aligned}
$$

To obtain the required formula, pick out the coefficients of unity and put $s = t$ to give just the terms

$$1 = J_0^2 + 2J_1^2 + \ldots.$$

1.31
$$
\begin{aligned}
I &= \frac{d}{dx}\left(\frac{x^2}{2}(J_n^2 - J_{n-1}J_{n+1})\right) \\
&= x(J_n^2 - J_{n-1}J_{n+1}) + \frac{x^2}{2}(2J_nJ_n' - J_{n-1}'J_{n+1} - J_{n-1}J_{n+1}') \\
&= xJ_n^2 - xJ_{n-1}J_{n+1} - \frac{x^2}{4}J_{n+1}J_{n-2} \\
&\quad + \frac{x^2}{4}J_{n-1}J_{n+2} + \frac{x^2}{4}J_nJ_{n-1} - \frac{x^2}{4}J_nJ_{n+1}.
\end{aligned}
$$

However

$$
\begin{aligned}
\frac{x}{2}J_{n-2} + \frac{x}{2}J_n &= (n-1)J_{n-1} \\
\frac{x}{2}J_{n+2} + \frac{x}{2}J_n &= (n+1)J_{n+1}
\end{aligned}
$$

to give

$$
\begin{aligned}
I &= xJ_n^2 - xJ_{n-1}J_{n+1} - \frac{x}{2}J_{n+1}\left((n-1)J_{n-1} - \frac{x}{2}J_n\right) \\
&\quad + \frac{x}{2}J_{n-1}\left((n+1)J_{n+1} - \frac{x}{2}J_n\right) = xJ_n^2.
\end{aligned}
$$

Integrate to give the required result.

1.32 The three-term recurrence gives

$$\frac{J_n(x)}{x} = \frac{1}{2n}[J_{n+1} + J_{n-1}].$$

Hence

$$\int_0^\infty \frac{J_n(x)}{x}\,dx = \frac{1}{2n} + \frac{1}{2n} = \frac{1}{n}.$$

1.33 Use induction: for $r = 1$, see equation 1.6.22. Assume true for $r-1$, that is:

$$\left(\frac{1}{x}\frac{d}{dx}\right)^{r-1}[x^n J_n(x)] = x^{n-r}J_{n-r}(x).$$

Then

$$\begin{aligned}
&\left(\frac{1}{x}\frac{d}{dx}\right)^{r}[x^n J_n(x)] \\
&= \frac{1}{x}\frac{d}{dx}\left[\left(\frac{1}{x}\frac{d}{dx}\right)^{r-1}[x^n J_n(x)]\right] \\
&= \frac{1}{x}\frac{d}{dx}(x^{n-r+1}J_{n-r+1}) \\
&= \frac{1}{x}[(n-r+1)x^{n-r}J_{n-r+1} + x^{n-r+1}J'_{n-r+1}] \\
&= \frac{1}{x}[(n-r+1)x^{n-r}J_{n-r+1} + x^{n-r+1}(J_{n-r} - J_{n-r+2})/2] \\
&= x^{n-r}J_{n-r}(x).
\end{aligned}$$

It is a similar exercise for the second requirement:

$$\begin{aligned}
&\left(\frac{1}{x}\frac{d}{dx}\right)^{r}[x^{-n} J_n(x)] \\
&= \frac{1}{x}\frac{d}{dx}((-1)^{r-1}x^{-n-r+1}J_{n+r-1}) \\
&= \frac{(-1)^{r-1}}{x}[(-n-r+1)x^{-n-r}J_{n+r-1} + x^{-n-r+1}J'_{n+r-1}] \\
&= \frac{(-1)^{r-1}}{x}[(-n-r+1)x^{-n-r}J_{n+r-1} \\
&\quad + x^{-n-r+1}(J_{n+r-2} - J_{n+r})/2] \\
&= (-1)^r x^{-n-r}J_{n+r}(x).
\end{aligned}$$

1.34 The Laurent series for a function with a pole of order m is

$$f(z) = \frac{b_m}{(z-a)^m} + \cdots + \frac{b_1}{(z-a)} + a_0 + a_1(z-a) + \cdots.$$

Multiply by $(z-a)^m$ and differentiate $m-1$ times to leave the first term b_1 plus terms of order $(z-a)$ and higher. These all tend to zero as $z \to a$ to leave the required residue b_1.

1.35 The imaginary part of the integral along the real axis will tend to the required integral as $R \to \infty$ and $r \to 0$. Around the large semicircle the integral is

$$\int_0^{\pi} ie^{iR(\cos\theta+i\sin\theta)}\,d\theta = \int_0^{\pi} ie^{-R\sin\theta}e^{iR\cos\theta}\,d\theta \to 0 \quad \text{as } R\to\infty$$

where $R = e^{i\theta}$, and around the small semicircle:

$$\int_{\pi}^{0} ie^{ir(\cos\theta+i\sin\theta)}\,d\theta = -i\pi$$

as $r \to 0$ to give

$$\int_{-\infty}^{\infty} \frac{\sin x}{x}\,dx = \pi$$

as there are no poles inside the contour. Hence

$$\int_0^{\infty} \frac{\sin x}{x}\,dx = \pi/2$$

as the integral on the negative half range is equal to that on the positive half range.

1.36 Consider

$$I = \int_{\Gamma} \frac{e^{iz}}{z^2+a^2}\,dz$$

where Γ lies along the real axis from $-R$ to R, then round the semicircle radius R. For large R there is one pole at ia with residue $e^{-a}/(2ai)$, so that $I = \pi e^{-a}/a$. Again use $z = Re^{i\theta}$ on the semicircle and the integral on the semicircle is

$$\int_0^{\pi} iRe^{i\theta}\frac{e^{iR\cos\theta-R\sin\theta}}{R^2e^{2i\theta}+a^2}\,d\theta$$

which is less than $\pi R/(R^2-a^2)$ and therefore tends to zero as $R\to\infty$. Hence the real part of I gives

$$\int_{-\infty}^{\infty} \frac{\cos x}{x^2+a^2}\,dx = \frac{\pi e^{-a}}{a}.$$

1.37

$$\frac{1}{2\pi}\int_{-n}^{n} e^{ikx}\,dx = \frac{n}{\pi}\mathrm{sinc}\,(nx).$$

Hence

$$\lim_{n\to\infty} \frac{n}{\pi}\,\mathrm{sinc}(nx) = \frac{1}{2\pi}\int_{-\infty}^{\infty} e^{ikx}\,dx.$$

However

$$\lim_{n\to\infty}\left\{\frac{n}{\pi}\mathrm{sinc}(nx)\right\} = \delta(x)$$

to give

$$\delta(x) = \frac{1}{2\pi}\int_{-\infty}^{\infty} e^{ikx}\,dx \qquad \text{and} \qquad \int_{-\infty}^{\infty} e^{-ikx}\delta(x)\,dx = 1.$$

Finally

$$\begin{aligned}\int_{-\infty}^{\infty} \cos x e^{-ikx}\,dx &= \frac{1}{2}\left[\int_{-\infty}^{\infty} e^{ix(1-k)}\,dx + \int_{-\infty}^{\infty} e^{-ix(1+k)}\,dx\right] \\ &= \pi[\delta(1-k)+\delta(1+k)]\end{aligned}$$

and similarly the sin case gives $i\pi[\delta(1+k)-\delta(1-k)]$.

1.38 Let

$$k(x) = \left(\frac{\sin x}{x}\right)^2.$$

Then

$$\int_{-\infty}^{\infty} \frac{\sin^2 x}{x^2}\,dx = \frac{1}{2}\,\text{Re}\int_{-\infty}^{\infty} \frac{1-e^{2ix}}{x^2}\,dx.$$

However $k(z) = (1-e^{2iz})/z^2$ has a simple pole at $z=0$ with residue

$$\lim_{z\to 0} z\left\{\frac{1-e^{2iz}}{z^2}\right\} = -2i.$$

Using a contour in the upper half plane indented at the origin gives 2π, and hence $k=\pi$ and the required limit is $\pi\delta(x)$.

$\sin(n!x)$ converges to 0 only for points of the form $x = p\pi/q$. Hence it is non-convergent almost everywhere. However, for any infinitely smooth f of bounded support

$$\int_{\infty}^{\infty} \sin(n!x)f(x)\,dx = \frac{1}{n!}\int_{-\infty}^{\infty} f'(x)\cos(n!x)\,dx.$$

Hence

$$\begin{aligned}\left|\int_{-\infty}^{\infty} \sin(n!x)f(x)\,dx\right| &\le \frac{1}{n!}\int_{-\infty}^{\infty} |f'(x)|\,dx \\ &\le \frac{(b-a)}{n!}\sup|f'(x)| \to 0.\end{aligned}$$

Hence the required limit is $0\delta(x) = 0$.

1.39 Proofs involving the δ function will only be meaningful if both sides are expressed as an integrand in the basic sampling property.

(i)
$$\begin{aligned}\int_{-\infty}^{\infty} f(x)\delta(x-\alpha)\,dx &= f(\alpha) \\ \int_{-\infty}^{\infty} f(\alpha)\delta(x-\alpha)\,dx &= f(\alpha)\int_{-\infty}^{\infty} \delta(x-\alpha)\,dx = f(\alpha)\end{aligned}$$

(ii) $$\int_{-\infty}^{\infty} xf(x)\delta(x)\,dx = [xf(x)]_{x=0} = 0.$$

(iii) $$\begin{aligned}\int_{-\infty}^{\infty} \delta(\alpha - x)f(x)\,dx &= \int_{-\infty}^{\infty} \delta[y-(-\alpha)]f(-y)\,dy \\ &= f[-(-\alpha)] = \int_{-\infty}^{\infty} \delta(x-\alpha)f(x)\,dx.\end{aligned}$$

Note that $\delta(x)$ is even as $\delta(-x) = \delta(x)$.

(iv) Let

$$I = \int_{-\infty}^{\infty} \delta(\alpha x)f(x)\,dx.$$

Case 1: $\alpha > 0$; $y = |\alpha|x$, $dy = |\alpha|dx$

$$I = \frac{1}{|\alpha|}\int_{-\infty}^{\infty} \delta(y)f\left(\frac{y}{|\alpha|}\right)dy = \frac{1}{|\alpha|}f(0) = \int_{-\infty}^{\infty}\left[\frac{\delta(x)}{|\alpha|}\right]f(x)\,dx$$

to leave $\delta(\alpha x) = \delta(x)/|\alpha|$ for $\alpha > 0$, $\alpha \neq 0$.

Case 2: $\alpha < 0$; $y = -|\alpha|x$, $dy = -|\alpha|dx$

$$I = \frac{1}{|\alpha|}\int_{-\infty}^{\infty} \delta(y)f\left(\frac{-y}{|\alpha|}\right)(-dy) = \frac{1}{|\alpha|}f(0) = \int_{-\infty}^{\infty}\left[\frac{\delta(x)}{|\alpha|}\right]f(x)\,dx$$

to leave $\delta(\alpha x) = \delta(x)/|\alpha|$ for $\alpha < 0$, $\alpha \neq 0$.

(v) $$(f\delta)' = f'\delta + f\delta'$$

Hence

$$\begin{aligned}\int_{-\infty}^{\infty} f(x)\delta'(x)\,dx &= \int_{-\infty}^{\infty}(f\delta)'\,dx - \int_{-\infty}^{\infty}\delta f'\,dx \\ &= [f\delta]_{-\infty}^{\infty} - f'(0) = -f'(0).\end{aligned}$$

(vi) $$\begin{aligned}\int_{-\infty}^{\infty} f(x)\delta''(x)\,dx &= \int_{-\infty}^{\infty}(f\delta')'\,dx - \int_{-\infty}^{\infty}\delta' f'\,dx \\ &= -\left[\int_{-\infty}^{\infty}(f'\delta)'\,dx - \int_{-\infty}^{\infty}\delta f''\,dx\right] = f''(0).\end{aligned}$$

Similarly

$$\int_{-\infty}^{\infty} f(x)\delta'''(x)\,dx = -f'''(0),$$

and hence the result follows by induction.

(vii) Observe that $\delta(x^2-\alpha^2)=\delta[(x-\alpha)(x+\alpha)]$ and hence $\delta(x^2-\alpha^2)=0$ except at $x=\pm\alpha$. Hence

$$\int_{-\infty}^{\infty}\delta(x^2-\alpha^2)f(x)\,dx$$

$$=\int_{-\alpha-\epsilon}^{-\alpha+\epsilon}\delta[(x+\alpha)(x-\alpha)]f(x)\,dx+\int_{\alpha-\epsilon}^{\alpha+\epsilon}\delta[(x+\alpha)(x-\alpha)]f(x)\,dx$$

for $\alpha>0$, and where $0<\epsilon<2\alpha$ and ϵ is arbitrarily small. In the neighbourhood of $x=-\alpha$, the factor $(x-\alpha)$ may be replaced by -2α to give

$$\begin{aligned}\int_{-\alpha-\epsilon}^{-\alpha+\epsilon}\delta[(x-\alpha)(x+\alpha)]f(x)\,dx &= \int_{-\alpha-\epsilon}^{-\alpha+\epsilon}\delta[(-2\alpha)(x-\alpha)]f(x)\,dx\\ &= \int_{-\infty}^{\infty}\frac{1}{2\alpha}\delta(x+\alpha)f(x)\,dx\end{aligned}$$

as $\delta(-y)=\delta(y)$ and $\delta(\alpha y)=\delta(y)/\alpha$ for $\alpha>0$ and $\alpha\neq 0$. Similarly

$$\int_{\alpha-\epsilon}^{\alpha+\epsilon}\delta[(x-\alpha)(x+\alpha)]f(x)\,dx=\int_{-\infty}^{\infty}\frac{1}{2\alpha}\delta(x-\alpha)f(x)\,dx$$

for $\alpha>0$ and $\alpha\neq 0$, leaving the required result.

(viii) $\delta(\sin x)=0$ except at points $x=n\pi$, $n=0,\pm1,\pm2,\dots$ where $\sin x=0$. Hence

$$\begin{aligned}\int_{-\infty}^{\infty}\delta(\sin x)f(x)\,dx &= \sum_{n=-\infty}^{\infty}\int_{-\infty}^{\infty}\delta(x-n\pi)f(x)\,dx\\ &= \int_{-\infty}^{\infty}\sum_{n=-\infty}^{\infty}\delta(x-n\pi)f(x)\,dx\end{aligned}$$

or

$$\delta(\sin x)=\sum_{n=-\infty}^{\infty}(x-n\pi).$$

Similarly

$$\delta(\cos x)=\sum_{\substack{n=-\infty\\ n\neq 0}}\delta(x-n\pi/2).$$

1.40 $$\lim_{n\to\infty}\langle n\text{gauss}(nx),e^{-|x|}\rangle\equiv\lim_{n\to\infty}\frac{n}{\sqrt{\pi}}\int_{-\infty}^{\infty}e^{-n^2x^2}e^{-|x|}\,dx$$

and

$$|x|+n^2x^2=-\frac{1}{4n^2}+n^2\left(|x|+\frac{1}{2n^2}\right)^2.$$

Hence

$$\lim_{n\to\infty}\langle n\text{gauss}(nx), e^{-|x|}\rangle = \lim_{n\to\infty}\frac{n}{\sqrt{\pi}}e^{\frac{1}{4n^2}}\int_{-\infty}^{\infty} e^{-n^2\left(|x|+\frac{1}{2n^2}\right)^2}\,dx.$$

Let $y = n(|x| + 1/2n^2)$ which reduces the integral to $\sqrt{\pi}/n$, and hence

$$\lim_{n\to\infty}\langle n\text{gauss}(nx), e^{-|x|}\rangle = \lim_{n\to\infty}\frac{n}{\sqrt{\pi}}e^{\frac{1}{4n^2}}\frac{\sqrt{\pi}}{n} = 1.$$

However

$$\lim_{n\to\infty}\{n\text{gauss}(nx)\} = \delta(x)$$

to give the result.

1.41 (i)

$$\int_{-\infty}^{\infty}\delta(x)e^{-iux}\,dx = e^0 = 1.$$

Therefore the Fourier inverse:

$$\delta(x) = \frac{1}{2\pi}\int_{-\infty}^{\infty} 1e^{iux}\,du.$$

Hence

$$\delta(x) \leftrightarrow 1.$$

(ii)

$$\frac{1}{2\pi}\int_{-\infty}^{\infty}\delta(u)e^{iux}\,du = \frac{1}{2\pi}.$$

Hence

$$\int_{-\infty}^{\infty}\frac{1}{2\pi}e^{-iux}\,dx = \delta(u)$$

or $1 \leftrightarrow 2\pi\delta(u)$.

(iii)

$$\begin{aligned}\int_{-\infty}^{\infty}\delta(x-\alpha)e^{-iux}\,dx &= \int_{-\infty}^{\infty}\delta(y)e^{-iu(y+\alpha)}\,dy\\ &= e^{-iu\alpha}\int_{-\infty}^{\infty}\delta(y)e^{-iuy}\,dy = e^{-iu\alpha}.\end{aligned}$$

(iv)

$$\begin{aligned}\int_{-\infty}^{\infty}e^{i\alpha u}\,dx &= \frac{1}{2\pi}\int_{-\infty}^{\infty}2\pi e^{ix(\alpha-u)}\,dx\\ &= 2\pi\delta(\alpha-u) = 2\pi\delta(u-\alpha).\end{aligned}$$

(v)

$$\begin{aligned}\int_{-\infty}^{\infty}\delta(\alpha x+\beta)e^{-iux}\,dx &= \int_{-\infty}^{\infty}\delta(y)e^{-iu\left(\frac{y-\beta}{\alpha}\right)}\frac{dy}{\alpha}\\ &= \frac{e^{iu\beta/\alpha}}{\alpha}\int_{-\infty}^{\infty}\delta(y)e^{\frac{-iuy}{\alpha}}\,dy = \frac{e^{-iu\beta/\alpha}}{\alpha}.\end{aligned}$$

(vi)
$$\delta(x) = \frac{1}{2\pi}\int_{-\infty}^{\infty} e^{iux}\,du.$$

Hence
$$\delta^{(m)}(x) = \frac{1}{2\pi}\int_{-\infty}^{\infty} (iu)^m e^{iux}\,du.$$

Inverting
$$(iu)^m = \int_{-\infty}^{\infty} \delta^{(m)}(x)e^{-iux}\,dx.$$

(vii)
$$2\pi\delta(u) = 2\pi\delta(-u) = \int_{-\infty}^{\infty} e^{-iux}\,dx.$$

Hence
$$2\pi\delta^{(m)}(u) = \int_{-\infty}^{\infty} (-ix)^m e^{-iux}\,dx$$

which gives
$$\int_{-\infty}^{\infty} x^m e^{-iux}\,dx = \frac{2\pi}{(-i)^m}\delta^{(m)}(u) = 2\pi i^m \delta^{(m)}(u).$$

1.42
$$\int_{-\infty}^{\infty} \operatorname{sgn}(x)e^{-iux}\,dx$$
$$= \lim_{\epsilon\to 0}\lim_{a\to\infty}\left[\int_0^a e^{-\epsilon|x|}e^{-iux}\,dx - \int_{-a}^{0} e^{-\epsilon|x|}e^{-iux}\,dx\right]$$
$$= \lim_{\substack{\epsilon\to 0\\ a-\infty}}\left[\frac{e^{-a(\epsilon-iu)}}{\epsilon-iu} - \frac{e^{-a(\epsilon+iu)}}{\epsilon+iu} + \frac{1}{\epsilon+iu} - \frac{1}{\epsilon-iu}\right]$$
$$\lim_{\epsilon\to 0}\left[\frac{-2iu}{(\epsilon+iu)(\epsilon-iu)}\right] = \frac{2}{iu}.$$

Hence $\operatorname{sgn}(x) \leftrightarrow \frac{2}{iu}$.
Noting that $H(x) = \frac{1}{2}[1+\operatorname{sgn}(x)]$, then

$$\begin{aligned}\int_{-\infty}^{\infty} H(x)e^{-iux}\,dx &= \frac{1}{2}\int_{-\infty}^{\infty} 1e^{-iux}\,dx + \frac{1}{2}\int_{-\infty}^{\infty} \operatorname{sgn}(x)e^{-iux}\,dx\\ &= \pi\delta(-u) + \frac{1}{iu} = \pi\delta(u) - \frac{i}{u}.\end{aligned}$$

Chapter 2

2.1 The separated solution has the form

$$u = \sum_p (A_p \sin px + B_p \cos px)(C_p \sin pct + D_p \cos pct)$$

from which

$$\frac{\partial u}{\partial t} = \sum_p (A_p \sin px + B_p \cos px)(C_p pc \cos pct - D_p pc \sin pct).$$

Then $u(0,t) = 0$ gives $B_p = 0$ and $u(L,t) = 0$ gives $p = n\pi/L$ for integer n. Further, $u_t(x,0) = 0$ gives $C_p = 0$ and hence the relevant part of the solution is then:

$$u = \sum_{n=1}^{\infty} D_p \cos\frac{n\pi ct}{L} \sin\frac{n\pi x}{L}$$

and the boundary condition at $t = 0$ implies that there are just two terms with $n = 1$ and $n = 2$ to give

$$u = \sum_{n=1}^{2} D_p \cos\frac{n\pi ct}{L} \sin\frac{n\pi x}{L}.$$

2.2 The general separated solution is

$$u = \sum_p (A_p \sin px + B_p \cos px)\left(C_p \sin\frac{pt}{a} + D_p \cos\frac{pt}{a}\right).$$

The condition $u(0,t) = 0$ gives $B_p = 0$, $u(1,t) = 0$ forces $p = n\pi$ for integer n, and $u_t(x,0) = 0$ yields $D_p = 0$ to reduce the solution to

$$u = \sum_{n=1}^{\infty} C_n \sin n\pi x \cos\frac{n\pi t}{a}.$$

At $t = 0$ the Fourier expansion:

$$\sum_{n=1}^{\infty} C_n \sin n\pi x = f(x) = \begin{cases} 3\delta x & 0 \le x \le \frac{1}{3} \\ \frac{3\delta}{2}(1-x) & \frac{1}{3} \le x \le 1 \end{cases}$$

and using the usual Fourier coefficient for C_n gives

$$\begin{aligned} C_n &= \frac{\int_0^1 f(x)\sin n\pi x\,dx}{\int_0^1 \sin^2 n\pi x\,dx} \\ &= \left[3\delta\left\{-\frac{\cos\frac{\pi n}{3}}{3n\pi} + \frac{\sin\frac{n\pi}{3}}{n^2\pi^2}\right\} - \frac{3\delta}{2}\left\{-\frac{\cos\frac{n\pi}{3}}{n\pi}\right\}\right. \\ &\quad + \frac{3\delta}{2}\left\{-\frac{\cos\frac{\pi n}{3}}{3n\pi} + \frac{\sin\frac{n\pi}{3}}{n^2\pi^2}\right\} + \frac{3\delta}{2}\left\{-\frac{\cos n\pi}{n\pi}\right\} \\ &\quad \left. - \frac{3\delta}{2}\left\{-\frac{\cos n\pi}{n\pi}\right\}\right] \Big/ \frac{1}{2} \\ &= \frac{9\delta \sin\frac{n\pi}{3}}{n^2\pi^2}. \end{aligned}$$

2.3 The separated solution is again

$$u = \sum_p (A_p \sin px + B_p \cos px)(C_p \sin pct + D_p \cos pct)$$

and the condition $u(0,t) = 0$ gives $B_p = 0$, $u(L,t) = 0$ forces $p = n\pi/L$ for integer n, and $u(x,0) = 0$ yields $D_p = 0$ to reduce the solution to

$$u = \sum_{n=1}^{\infty} C_n \sin\frac{n\pi x}{L} \sin\frac{n\pi ct}{L}.$$

Enforcing $u_t(x,0) = sin(n\pi/L)$ gives simply $C_n = L/n\pi c$ and all other C's are zero. Hence

$$u = \frac{L}{n\pi c}\sin\frac{n\pi ct}{L}\sin\frac{n\pi x}{L}.$$

2.4 The same general separable solution again applies as in Exercise 2.2. The condition $u_x(0,t) = 0$ gives $A_p = 0$, $u_x(2,t) = 0$ gives $2p = n\pi$, and $u_t(x,0) = 0$ gives $C_n = 0$. The problem then reduces to a Fourier series expansion:

$$\sum_{n=1}^{\infty} D_n \cos\frac{n\pi x}{2} = f(x) = \begin{cases} kx & 0 \le x \le 1 \\ k(2-x) & 1 \le x \le 2. \end{cases}$$

Hence the Fourier coefficient is

$$\begin{aligned} D_n &= \int_0^1 kx \cos\frac{n\pi x}{2}\,dx + \int_1^2 k(2-x)\cos\frac{n\pi x}{2}\,dx \\ &= \frac{8k}{n^2\pi^2}\cos\frac{n\pi}{2} - \frac{4k}{n^2\pi^2}\cos n\pi - \frac{4k}{n^2\pi^2}. \end{aligned}$$

2.5 Writing $u(x,t) = X(x)T(t)$ gives:

$$\frac{c^2 X''}{X} = -p^2 = \frac{T''}{T} + \mu\frac{T'}{T}.$$

The auxiliary equation for T has roots

$$m = \left(-\mu \pm \sqrt{\mu^2 - 4p^2}\right)/2$$

and hence the separated solution has the general form

$$u(x,t) = \sum_{n=1}^{\infty} \sin\frac{n\pi x}{L} e^{\frac{\mu t}{2}}(C_p \sin\omega t + D_p \cos\omega t)$$

where

$$\omega = \frac{1}{2}\sqrt{4p^2 - \mu^2}$$

and the boundary conditions $u(0,t) = u(L,t) = 0$ force the sin solution with $pL = n\pi c$. The condition $u(x,0) = \sin(\pi x/L)$ implies that the solution has just the one term with $n = 1$, and $u_t(x,0) = 0$ gives $D_p = 0$. Hence the solution

$$u = e^{\frac{-\mu t}{2}} \sin\frac{\pi x}{L} \cos\frac{1}{2}\sqrt{\frac{4\pi^2 c^2}{L^2} - \mu^2}t.$$

2.6 The general separable solution is

$$u = \sum_p A_p e^{-\frac{p^2}{a^2}t}(B_p \sin px + C_p \cos px)$$

choosing the negative exponential to ensure that $u \to 0$ as $t \to \infty$. The conditions that $u = 0$ when $x = 0$ and $x = \pi$ force $p = n$, and hence the solution for (i) is:

$$u = \sum_{n=1}^{\infty} A_p e^{-\frac{p^2}{a^2}t} \sin nx$$

and for (ii) is:

$$u = \sum_{n=1}^{\infty} A_p e^{-\frac{p^2}{a^2}t} \cos nx.$$

2.7 From Exercise 2.6,

$$u = \sum_{n=1}^{\infty} A_p e^{-\frac{p^2}{a^2}t} \sin nx$$

and the required Fourier series expansion is

$$1 + x = \sum_{n=1}^{\infty} A_n e^{-\frac{n^2}{a^2}t} \sin nx.$$

Hence the Fourier coefficient is given by

$$\begin{aligned} A_n &= \frac{\int_0^\pi (1+x)\sin nx\,dx}{\int_0^\pi \sin^2 nx\,dx} \\ &= -\frac{2}{\pi n}[(1+\pi)\cos n\pi - 1] + \frac{2}{\pi n^2}\sin nx\Big|_0^\pi . \end{aligned}$$

The final form is then

$$u = \frac{2}{\pi}\sum_{n=1}^{\infty}\frac{1}{n}[1 - (1+\pi)\cos n\pi]e^{-\frac{n^2}{a^2}t}\sin nx.$$

2.8 The general separated solution is

$$\theta = \sum_p e^{-c^2p^2t}(A_p \sin px + B_p \cos px)$$

and

$$\frac{\partial\theta}{\partial x} = \sum_p e^{-c^2p^2t}(A_p p \cos px - B_p p \sin px).$$

Hence for

$$\frac{\partial\theta}{\partial x}(0,t) = \theta(a,t) = 0$$

then $A_p = 0$ and

$$pa = \frac{(2n-1)\pi}{2}$$

to leave

$$\theta = \sum_n B_n e^{-c^2p^2t} \cos \frac{(2n-1)\pi}{2a} x.$$

At $t = 0$ the following Fourier series is required:

$$\theta_0 = \sum_n B_n \cos \frac{(2n-1)\pi}{2a} x$$

with coefficient

$$\begin{aligned} B_n &= \frac{\int_0^a \theta_0 \cos \frac{(2n-1)\pi}{2a} x \, dx}{\int_o^a \cos \frac{(2n-1)\pi}{2a} x \, dx} \\ &= \frac{4\theta_0}{(2n-1)\pi} \sin \frac{(2n-1)\pi}{2}. \end{aligned}$$

2.9 With $u = 0$ at both $x = 0$ and $x = 1$ for all t the split variable solution can be written directly as

$$u = \sum_{n=1}^{\infty} A_n e^{-n^2\pi^2t} \sin n\pi x$$

and the Fourier coefficient splits into the two integrals from 0 to $\frac{1}{2}$, and from $\frac{1}{2}$ to 1. Hence

$$\begin{aligned} A_n &= \frac{\int_0^{\frac{1}{2}} 2x \sin n\pi x \, dx + \int_{\frac{1}{2}}^1 2(1-x) \sin n\pi x \, dx}{\int_0^1 \sin^2 n\pi x \, dx} \\ &= \frac{8}{n^2\pi^2} \sin \frac{n\pi}{2} \end{aligned}$$

as required.

2.10 The separated solution is again

$$u = \sum_p e^{-p^2 t}(A_p \sin px + B_n \cos px)$$

and with non-homogeneous boundary conditions, the x and t independent solutions need to also be brought into play. These have the form

$$u = a + bt + cx + dx^2$$

and as $u_t = b$ and $u_{xx} = 2d$ then $b = 2d$ for the partial differential equation to be satisfied. By the linearity of the partial differential equation the solution is the sum of these two solutions. Hence

$$u_x = c + 2dx + \sum_p e^{-p^2 t}(A_p p \cos px - B_p p \sin px)$$

which is zero at $x = 0$ if $c = A_p = 0$. To force $u_x = 1$ at $x = \frac{1}{2}$ gives $d = 1$ and to make the summation zero for all t gives $p = 2n\pi$. The remaining condition is for $u = 0$ at $t = 0$ for the range of x. Hence

$$0 = a + x^2 + \sum_{n=1}^{\infty} B_n \cos 2n\pi x$$

to yield the Fourier fit:

$$x^2 = a + \sum_{n=1}^{\infty} B_n \cos 2n\pi x$$

and the coefficients follow in the usual way:

$$\begin{aligned} a &= \frac{-\int_0^{1/2} x^2\, dx}{\int_0^{1/2} 1 dx} \\ &= -\frac{1}{12} \end{aligned}$$

and the general term is

$$\begin{aligned} B_n &= \frac{-\int_0^{1/2} x^2 \cos 2n\pi x\, dx}{\int_0^{1/2} \cos^2 2n\pi x\, dx} \\ &= -\frac{2}{n\pi}[x^2 \sin 2n\pi x]_0^{\frac{1}{2}} + \frac{4}{n\pi}\int_0^{\frac{1}{2}} x \sin 2n\pi x\, dx \\ &= -\frac{2}{n^2\pi^2}[x \cos 2\pi n x]_0^{1/2} + \frac{2}{n^2\pi^2}\int_0^{1/2} \cos 2n\pi x\, dx \\ &= -\frac{(-1)^n}{n^2\pi^2} \end{aligned}$$

to give the required form.

2.11 For Laplace's equation the separated solution is

$$u = \sum_p (A_p e^{px} + B_p e^{-px})(C_p \sin py + D_p \cos py)$$

and forcing $u = 0$ on $y = 0$ and $y = a$, makes $D_p = 0$ and $pa = n\pi$. Further, to force $u = 0$ when $x = a$ gives the equation

$$0 = A_p e^{pa} + B_p e^{-pa}$$

and the solution takes the form

$$u = \sum_{n=1}^{\infty} A_n \sin \frac{n\pi y}{a} \left(e^{\frac{n\pi y}{a}} - e^{\frac{2n\pi}{a}} e^{-\frac{n\pi y}{a}} \right).$$

Hence when $x = 0$ the Fourier series expansion

$$f(y) = \sum_{n=1}^{\infty} A_n (1 - e^{\frac{2n\pi}{a}}) \sin \frac{n\pi y}{a}$$

must be achieved. The usual orthogonality condition gives the Fourier coefficient

$$B_n = \frac{\int_0^{\frac{a}{2}} y \sin \frac{n\pi y}{a}\, dy + \int_{\frac{a}{2}}^{a} (a-y) \sin \frac{n\pi y}{a}\, dy}{\int_0^a \sin^2 \frac{n\pi y}{a}\, dx} = \frac{4a}{n^2\pi^2} \sin \frac{n\pi}{2}.$$

2.12 Using the separated solution

$$u = \sum_k (A_k e^{kx} + B_k e^{-kx}) \left(C_k \sin \frac{k}{c} y + D_k \cos \frac{k}{c} y \right)$$

the condition that $u \to 0$ as $x \to \infty$ puts $a_k = 0$, $u = 0$ at $y = 0$ forces $D_k = 0$ and $u_y = 0$ at $y = l$ requires

$$\frac{kl}{c} = \frac{(2n-1)\pi}{2}.$$

Fourier series expansion then makes $u = 2y$ at $t = 0$ for the solution now in the form

$$u = \sum_{n=1}^{\infty} B_n e^{-\frac{(2n-1)c\pi x}{2l}} \sin \frac{(2n-1)\pi y}{2l}$$

from which the Fourier coefficient follows as

$$B_n = \frac{\int_0^l 2y \sin \frac{(2n-1)\pi y}{2l}\, dy}{\int_0^l \sin^2 \frac{(2n-1)\pi y}{2l}\, dy} = \frac{16l}{(2n-1)^2\pi^2} \sin \frac{(2n-1)\pi}{2}.$$

2.13 The separation of variables gives

$$\frac{X''}{X} = -\frac{Y''}{Y} + 1 = \pm k^2$$

and hence

$$X'' = \pm k^2 X$$

and

$$Y'' = (1 \mp k^2)Y$$

with solution

$$u = \sum_k (A_k \sin kx + B_k \cos kx)\left(C_k e^{(1+k^2)^{1/2}y} + D_k e^{-(1+k^2)^{1/2}y}\right).$$

Hence to make $u = 0$ for $x = 0$ and $x = \pi$ causes $B_k = 0$ and $k = n\pi$. Making $u = 0$ when $y = 0$ gives the equation

$$0 = C_n + D_n$$

and $u = 1$ when $y = a$ requires the Fourier fit

$$1 = \sum_{n=1}^{\infty} C_n \sin n\pi x \left(e^{(1+k^2)^{1/2}a} - e^{-(1+k^2)^{1/2}a}\right).$$

Hence

$$C_n\left(e^{(1+k^2)^{1/2}a} - e^{-(1+k^2)^{1/2}a}\right) = H_n, \text{say}$$

and

$$H_n = \frac{\int_0^\pi \sin n\pi x\, dx}{\int_0^\pi \sin^2 n\pi x\, dx} = \frac{4}{n\pi}.$$

2.14 For a three-dimensional problem try the separation

$$\phi = X(x)Y(y)Z(z)$$

which gives

$$-\frac{X''}{X} = \frac{Y''}{Y} + \frac{Z''}{Z} = \pm k^2.$$

Hence

$$X'' = \mp k^2 X$$

and

$$\frac{Y''}{Y} \mp k^2 = -\frac{Z''}{Z} = \pm l^2$$

to give

$$Y'' = \pm l^2 Y$$

and

$$-Z'' = (\mp k^2 \pm l^2)Z.$$

To make $\phi = 0$ on $x = 0$ and $x = a$ requires $-k^2$ and the cos solution is zero to leave

$$X = \sin\frac{n\pi x}{a}$$

in the usual way. Similarly for y, to give

$$Y = \sin\frac{n\pi y}{b}.$$

The Z equation is then

$$Z'' = (k^2 + l^2)Z$$

and

$$Z = Ae^{(k^2+l^2)^{1/2}z} + Be^{-(k^2+l^2)^{1/2}z}.$$

Hence the full solution is

$$\phi = \sum_{n=1}^{\infty}\sum_{m=1}^{\infty} \sin\frac{n\pi x}{a}\sin\frac{m\pi y}{b}\left(A_{nm}e^{(k^2+l^2)^{1/2}z} + B_{nm}e^{-(k^2+l^2)^{1/2}z}\right)$$

and $\phi = 0$ on $z = 0$ gives

$$A_{nm} + B_{nm} = 0$$

and this leaves a double Fourier series for the final boundary condition to give

$$f(x) = \sum_{n=1}^{\infty}\sum_{m=1}^{\infty} \sin\frac{n\pi x}{a}\sin\frac{m\pi y}{b}\left(A_{nm}e^{(k^2+l^2)^{1/2}c} + B_{nm}e^{-(k^2+l^2)^{1/2}c}\right).$$

Write

$$H_{nm} = A_{nm}\left(A_{nm}e^{(k^2+l^2)^{1/2}c} + B_{nm}e^{-(k^2+l^2)^{1/2}c}\right)$$

then

$$\begin{aligned} H_{nm} &= \frac{4}{ab}\int_0^a\int_0^b f(x)\sin\frac{n\pi x}{a}\sin\frac{m\pi y}{b}\,dx\,dy \\ &= \frac{8}{m\pi a}\int_0^a \sin\frac{n\pi x}{a}\,dx. \end{aligned}$$

2.15 The sum of the four separate solutions of Laplace's equation trivially satisfies Laplace's equation. The first boundary condition of 2.5.63 gives

$$\alpha_1 u(0,y) + \beta_1\frac{\partial u}{\partial x}(0,y) = \sum_{i=1}^{4}\alpha_i u_i(0,y) + \beta_i\frac{\partial u}{\partial x}(0,y) = f_1(y)$$

from the first of 2.5.65–2.5.68. Similar analysis using the second equations, the third and the fourth give the other three results.

2.16 For constant non-homogeneous solutions the separated solution needs to be enhanced with the solution in x only:

$$u = ax + b + \sum_k (A_k \sin kx + B_k \cos kx)e^{-c^2k^2t}$$

where just the negative exponential solution is taken to ensure $u \to 0$ as $t \to \infty$. The condition $u = 0$ at $x = 0$ gives $B_n = 0$ and $u = 1$ at $x = 1$ gives $a = 1$ and $k = n\pi$, to reduce the solution to

$$u = x + \sum_{n=1}^{\infty} A_n \sin n\pi x e^{-n^2\pi^2c^2t}.$$

A Fourier series expansion is now used to obtain $u = 0$ at $t = 0$ by setting

$$-x = \sum_{n=1}^{\infty} A_n \sin n\pi x$$

and the Fourier coefficient is

$$\begin{aligned} a_n &= \frac{-\int_0^1 x \sin n\pi x \, dx}{\int_0^1 \sin^2 \pi n x \, dx} \\ &= \frac{2}{n\pi} \left. \frac{\sin n\pi x}{n\pi} \right|_0^1 + \frac{2}{n\pi}(-1)^n \\ &= \frac{(-1)^n 2}{n\pi}. \end{aligned}$$

2.17 This is an inhomogeneous equation, so the trick is to solve the homogeneous problem first and leave the t variable free. By substituting this solution into the full equation an ordinary differential equation arises for the t part of the solution. This will require expanding the inhomogeneous part of the original equation as a Fourier series. Hence the solution to the homogeneous equation has the form

$$u = \sum_{n=1}^{\infty} w_n(t) \sin n\pi x$$

which satifies $u(0, t) = u(1, t) = 0$. This is a good time to expand $2x$ as a Fourier series:

$$2x = \sum_{n=1}^{\infty} f_n \sin n\pi x$$

with the Fourier coefficient

$$f_n = \frac{2\int_0^1 x \sin n\pi x \, dx}{\int_0^1 \sin^2 n\pi x \, dx} = -\frac{4}{n\pi}(-1)^n.$$

Substitute into the original equation for u and equate coefficients of $\sin n\pi x$:

$$\frac{dw_n}{dt} = n^2\pi^2 w_n - \frac{2}{n\pi}(-1)^n.$$

To find boundary conditions for w_n, put $t = 0$ to give

$$x - x^2 = \sum_{n=1}^{\infty} w_n(0) \sin n\pi x$$

and again the Fourier coefficient is

$$w_n(0) = 2\int_0^1 (x - x^2) \sin n\pi x \, dx = \frac{8}{n^3\pi^3}.$$

The differential equation for w_n has the solution

$$w_n(t) = a_n e^{-n^2\pi^2 t} + \frac{2(-1)^n}{n^3\pi^3}$$

and using $w_n(0)$ from above gives

$$A_n = \frac{2}{n^3\pi^3}(4 - (-1)^n)$$

and

$$u = \sum_{n=1}^{\infty} \left[\frac{2}{n^3\pi^3}(4 - (-1)^n)e^{-n^2\pi^2 t} + \frac{2(-1)^n}{n^3\pi^3}\right] \sin n\pi x.$$

2.18 In this problem there is a combination of inhomogeneous boundary conditions and equations. As in the previous example, solve the homogeneous equation by separation to give

$$u = \sum_{n=1}^{\infty} w_n(t)[A_n \sin kx + B_n \cos kx].$$

Hence

$$\frac{\partial u}{\partial x} = \sum_{n=1}^{\infty} w_n(t)[A_n k \cos kx - B_n k \sin kx]$$

and $u_x = 0$ when $x = 0$ forces $A_n = 0$. To cope with constant inhomogeneous boundary conditions requires the extra solution which is just a function of x, hence

$$u = \sum_{n=1}^{\infty} w_n(t)[B_n \cos kx] + bx + c$$

and to make $u(10,t) = 20$ requires $c = 20$ and $20k = (2n-1)\pi$. Hence the trial solution for u is

$$u = 20 + \sum_{n=1}^{\infty} w_n(t)\cos\frac{(2n-1)\pi x}{20}$$

and the ordinary differential equation for $w_n(t)$ will require expanding the equation on the right-hand side as a Fourier series:

$$1 = \sum_{n=1}^{\infty} f_n \cos\frac{(2n-1)\pi x}{20}$$

for which the Fourier coefficients are

$$f_n = \frac{\int_0^{10}\cos\frac{(2n-1)\pi x}{20}\,dx}{\int_0^{10}\cos^2\frac{(2n-1)\pi x}{20}\,dx} = \frac{100}{(2n-1)\pi}\sin(2n-1)\frac{\pi}{2}.$$

Substituting for u gives the ordinary differential equation

$$\frac{dw_n}{dt} = -\frac{1}{4}\left(\frac{(2n-1)\pi}{20}\right)^2 w_n + \frac{100}{(2n+1)\pi}\sin[(2n-1)\pi/2].$$

The initial condition follows by using the condition $u(x,0) = 50$ to give

$$50 = 20 + \sum_{n=1}^{\infty} w_n(0)\cos\frac{(2n-1)\pi x}{20}$$

and the Fourier coefficient is

$$w_n(0) = \frac{30\int_0^{10}\cos\frac{(2n-1)\pi x}{20}\,dx}{\int_0^{10}\cos^2\frac{(2n-1)\pi x}{20}\,dx} = \frac{30.5.20}{(2n-1)\pi}\sin\frac{(2n-1)\pi}{2}$$

and hence

$$w_n(t) = D_n e^{-\frac{(2n-1)^2\pi^2}{1600}t} + \frac{100}{(2n-1)\pi}\sin(2n-1)\pi/2$$

and using $w_n(0)$ gives

$$D_n = \frac{2900}{(2n-1)\pi}\sin(2n-1)\pi/2$$

leaving the final solution:

$$\begin{aligned} u &= 20 + \sum_{n=1}^{\infty}\left[\frac{2900}{(2n-1)\pi}\sin(2n-1)\pi/2e^{-\frac{(2n-1)^2\pi^2}{1600}t} + \right. \\ &\left. \frac{100}{(2n-1)\pi}\sin(2n-1)\pi/2\right]\cos\frac{(2n-1)\pi x}{20}. \end{aligned}$$

2.19 For this problem, follow the lines of 2.5.18, and set

$$u(x,t) = v(x,t) + w(x,t)$$

where in this case $v = xt$. Substituting into the partial differential equation gives

$$\frac{\partial v}{\partial t} + \frac{\partial w}{\partial t} = \frac{\partial^2 w}{\partial x^2} + \frac{\partial^2 w}{\partial x^2}$$

and hence

$$\frac{\partial^2 w}{\partial x^2} = \frac{\partial w}{\partial t} + t$$

with $w(0,t) = 0$, $w(1,t) = 0$ and $w(x,0) = 0$. Hence the separable solution has the form

$$w = \sum_{n=1}^{\infty} w_n(t) \sin n\pi x$$

and

$$t = \sum_{n=1}^{\infty} f_n \sin n\pi x$$

with Fourier coefficients

$$f_n(t) = 2t \int_0^1 \sin n\pi x \, dx = \frac{4t}{n\pi}.$$

Hence the ordinary differential equation is

$$\frac{dw_n}{dt} = n^2\pi^2 w_n + \frac{4t}{n\pi}$$

with solution

$$w_n = A_n E^{-n^2\pi^2 t} + \text{particular part.}$$

To find the particular part: try $w_n = at + b$ then

$$a = n^2\pi^2(at + b) + \frac{4t}{n\pi}.$$

Hence comparing coefficients gives $a = n^2\pi^2 b$. Hence the full solution is

$$w_n = A_n e^{-n^2\pi^2 t} - \frac{4t}{n^3\pi^3} - \frac{4}{n^5\pi^5}.$$

The condition $w(x,0) = 0$ gives

$$0 = A_n - \frac{4}{n^5\pi^5}$$

and hence the full solution is

$$u = xt + \sum_{n=1}^{\infty} \left[\frac{4}{n^5\pi^5} e^{-n^2\pi^2 t} - \frac{4t}{n^3\pi^3} - \frac{4}{n^5\pi^5} \right] \sin n\pi x.$$

2.20 This problem is a special case of problem 2.5.64. Here, only two separate problems arise, so that $u(x,y) = u_1(x,y) + u_2(x,y)$ with $u_1 = 0$ on $x = 0$, $y = 0$ and $y = 1$, and $\frac{\partial u_1}{\partial x} = 1$ on $x = 1$; and for the second problem $u_2 = 0$ on $x = 0$ and $y = 0$, $u_2 = x$ on $y = 1$ and $\frac{\partial u_2}{\partial x} = 0$ on $x = 1$.

Hence separation of variables in the usual way with an eye on the required boundary conditions gives

$$u_1 = \sum_{n=1}^{\infty} (C_n e^{-kx} + D_n e^{kx}) \sin n\pi y$$

choosing $k = n\pi$ and no cos term to satisfy $u_1 = 0$ on both $y = 0$ and $y = 1$. For $u_1 = 0$ on $x = 0$ gives

$$0 = C_n + D_n$$

and the derivative condition requires

$$1 = \sum_{n=1}^{\infty} C_n[-ke^{-k} - ke^{k}] \sin n\pi y$$

using $D_n = -C_n$. Hence

$$C_n[-ke^{-k} - ke^{k}] = 2\int_0^1 \sin n\pi y \, dy$$

is the required Fourier coefficient, yielding

$$\frac{2}{n\pi}[\cos n\pi - 1]$$

and

$$u_1 = \sum_{n=1}^{\infty} \frac{\frac{2}{n\pi}[\cos n\pi - 1]}{[-ke^{-k} - ke^{k}]}(e^{-kx} - ne^{kx}) \sin n\pi y.$$

For u_2 the separated solution is

$$u_2 = \sum_{n=1}^{\infty} (C_n e^{-ky} + D_n e^{ky}) \sin \frac{(2n+1)\pi x}{2}$$

so accomodating the boundary conditions at $x = 0$ and $x = 1$. The condition at $y = 0$ gives

$$0 = C_n + D_n$$

and the Fourier series expansion on the interval (0,1) requires

$$x = \sum_{n=1}^{\infty} C_n(e^{-k} - e^{k}) \sin \frac{(2n-1)\pi x}{2}.$$

Hence

$$\begin{aligned} C_n(e^{-k} - e^k) &= 2\int_0^1 x \sin\frac{(2n-1)\pi x}{2}\,dx \\ &= \frac{8}{(2n-1)^2\pi^2}\sin\frac{(2n-1)\pi}{2} \end{aligned}$$

to leave

$$u_2 = \frac{8}{(2n-1)^2\pi^2}\frac{\sin\frac{(2n-1)\pi}{2}}{(e^{-k} - e^k)}(e^{-ky} - e^{ky})\sin\frac{(2n-1)\pi x}{2}.$$

2.21 Try the separated solution

$$e = X(x)T(t)$$

in

$$\frac{\partial^2 e}{\partial x^2} = LC\frac{\partial^2 e}{\partial t^2} + (RC + GL)\frac{\partial e}{\partial t} + RGe$$

to give

$$\frac{X''}{X} = LC\frac{T''}{T} + (RC + GL)\frac{T'}{T} + RG = \pm k^2.$$

Then either

$$X = A\cos kx + B\sin kx$$

or

$$X = Ae^{kx} + Be_{k}x$$

and

$$LCT'' + (RC + GL)T' + (RG \mp k^2)T = 0.$$

Hence the T solution will either be of the form

$$T = e^{\alpha t}(A\sin\beta t + B\cos\beta t)$$

if the auxiliary equation has complex roots or

$$T = Ae^{\gamma t} + Be^{\delta t}$$

if the auxiliary equation has real roots. Hence there is no separated solution of the form $E_0\cos\omega t$ at $x = 0$. Hence the separated solution breaks down.

However trying

$$e = E_0 e^{-\alpha x}\cos(\omega t + bx)$$

gives

$$\begin{aligned}
\frac{\partial e}{\partial t} &= -E_0 e^{-\alpha x} \omega \sin(\omega t + bx) \\
\frac{\partial e}{\partial x} &= -E_0 \alpha e^{-\alpha x} \cos(\omega t + bx) - E_0 b e^{-\alpha x} \sin(\omega t + bx) \\
\frac{\partial^2 e}{\partial t^2} &= -E_0 e^{-\alpha x} \omega^2 \cos(\omega t + bx) \\
\frac{\partial^2 e}{\partial x^2} &= E_0 \alpha^2 e^{-\alpha x} \cos(\omega t + bx) + 2E_0 \alpha b e^{-\alpha x} \sin(\omega t + bx) \\
&\quad - E_0 b^2 e^{-\alpha x} \cos(\omega t + bx).
\end{aligned}$$

Substituting into the original equation gives:

$$\begin{aligned}
&E_0 \alpha^2 e^{-\alpha x} \cos(\omega t + bx) + 2E_0 \alpha b e^{-\alpha x} \sin(\omega t + bx) \\
&\quad - E_0 b^2 e^{-\alpha x} \cos(\omega t + bx) = -E_0 LC e^{-\alpha x} \omega^2 \cos(\omega t + bx) \\
&\quad - (RC + GL) E_0 \omega e^{-\alpha x} \sin(\omega t + bx) + RGE_0 e^{-\alpha x} \cos(\omega t + bx).
\end{aligned}$$

Pick out the coefficients of sin and cos to give

$$a^2 - b^2 = -LC\omega^2 + RG$$

and

$$2ab = -(RC + GL)\omega.$$

2.22 Try the separated solution

$$v = R(r)\Theta(\theta)$$

then

$$r^2 \frac{R''}{R} + r\frac{R'}{R} = -\frac{\Theta''}{\Theta} = \pm k^2$$

and

$$v = \sum_{n=1}^{\infty} \left(\frac{C_n}{r^n} + D_n r^n \right) (A_n \cos n\theta + B_n \sin n\theta).$$

The condition $v = 0$ on $\theta = 0$ gives $A_n = 0$, and $\frac{\partial v}{\partial \theta} = 0$ on $\theta = \pi/2$ requires $B_n = 0$ for $n = 2, 4, 6, \ldots$. For v to be finite at $r = 0$, it is necessary that $C_n = 0$ to leave the solution:

$$v = \sum_{n=1}^{\infty} D_n r^{(2n-1)} \sin(2n-1)\theta$$

with the final generalized Fourier series expansion

$$2\theta = \sum_{n=1}^{\infty} D_n \sin(2n-1)\theta$$

and

$$D_n = \frac{\int_0^{\pi/2} 2\theta \sin(2n-1)\theta\, d\theta}{\int_0^{\pi/2} \sin^2(2n-1)\theta\, d\theta} = \frac{8\sin(2n-1)\frac{\pi}{2}}{\pi(2n-1)^2}.$$

2.23 The separated solution in cylindrical polar coordinates is

$$u = \sum_{n=1}^{\infty} \left(\frac{C_n}{r^n} + D_n r^n \right) (A_n \cos n\phi + B_n \sin n\phi)$$

and the solution needs to be finite at $r = 0$ giving $C_n = 0$. Now

$$\frac{\partial u}{\partial \phi} = \sum_{n=1}^{\infty} D_n(-nA_n \sin n\phi + nB_n \cos n\phi)$$

and $\frac{\partial u}{\partial \phi} = 0$ at $\phi = 0$ gives $B_n = 0$, whereas at $\phi = \pi/2$ it implies that $A_n = 0$ for $n = 1, 3, 5, \ldots$. The solution then has the reduced form

$$u = \sum_{n=1}^{\infty} D_n r^{2n} \cos 2n\phi$$

and the boundary condition on $r = a$ gives

$$u = \sum_{n=1}^{\infty} D_n a^{2n} \cos 2n\phi$$

with the Fourier coefficient

$$A^{2n} D_n = \frac{4}{\pi} \int_0^{\pi/2} \phi \cos 2n\phi\, d\phi = \frac{1}{n^2\pi}[\cos n\pi - 1].$$

2.24 The above separated solution needs enhancing with a ϕ only solution to satisfy the constant inhomogeneous boundary conditions to give the required solution

$$u = \sum_{n=1}^{\infty} \left(\frac{C_n}{r^n} + D_n r^n \right) (A_n \cos n\theta + B_n \sin n\theta) + a\phi + b.$$

For finite u at $r = 0$, $C_n = 0$, and $u = 0$ on $\phi = 0$ gives $b = 0$ and $B_n = 0$. The condition $u = A$ on $\phi = \pi$ gives $a = A/\pi$, leaving

$$u = \frac{A}{\pi}\phi + \sum_{n=1}^{\infty} D_n r^n \sin n\phi.$$

The condition $u = a$ on $r = a$ gives the Fourier series expansion problem

$$A\left(1 - \frac{\phi}{\pi}\right) = \sum_{n=1}^{\infty} D_n a^n \sin n\phi$$

with Fourier coefficient:

$$D_n = \frac{2}{\pi}\int_0^\pi A\left(1-\frac{\phi}{\pi}\right)\sin n\phi\, d\phi = \frac{2A}{\pi n}.$$

2.25 The separated variable solution of the form $u = R(r)T(t)$ satisfies

$$\frac{1}{r}\left(\frac{\partial u}{\partial r}\right) = \frac{\partial^2 u}{\partial r^2} = \frac{1}{k}\frac{\partial u}{\partial t}$$

giving

$$\frac{1}{r}\frac{R'}{R} + \frac{R''}{R} = \frac{1}{k}\frac{T'}{T} = -s^2.$$

Hence

$$R'' + \frac{1}{r}R' + s^2 R = 0$$

with solution

$$R = AJ_0(sr) + BY_0(sr)$$

and

$$T = Ae^{-ks^2t}.$$

As $Y_0 \to \infty$ as $r \to 0$ then $B = 0$ and the solution looks like

$$u = \sum_s A_s J_0(sr)e^{-ks^2t}.$$

To make $u(a,t) = 0$ requires

$$sa = \lambda_n a$$

where λ_n is such that

$$J_0(\lambda_n a) = 0$$

and the generalised Fourier series expansion is now

$$\tau_0 = \sum_{n=1}^{\infty} A_n J_0(\lambda_n r)$$

which, using the given Fourier expansion, gives the required result.

2.26 For the external problem the relevant solution is

$$u = \sum_{n=1}^{\infty} \frac{B_n}{r^{n+1}} P_n(\cos\theta)$$

and the Fourier series expansion gives the coefficient

$$\begin{aligned}\frac{B_n}{a^{n+1}} &= \frac{(2n+1}{2}\int_{-1}^{1} f(w)P_n(w)\,dw \\ &= \frac{2n+1}{2}\int_{-1}^{0} \mu_2 P_n(w)\,dw + \frac{2n+1}{2}\int_0^1 \mu_1 P_n(w)\,dw\end{aligned}$$

which as in 2.6.130 gives

$$\begin{aligned} u &= \frac{1}{2}(\mu_1 + \mu_2)\left(\frac{a}{r}\right) - \frac{3}{4}\left(\frac{a}{r}\right)^2 P_1(\cos\theta) \\ &+ \frac{7}{16}(\mu_2 - \mu_1)\left(\frac{a}{r}\right)^3 P_2(\cos\theta) + \ldots. \end{aligned}$$

Chapter 3

3.1 $dx = dy/c$ to give $z = F(y - cx)$ as the solution for arbitrary F. When the variable x represents time, this solution is a travelling wave with propagation velocity c, and is discussed in more detail in §3.3 in the bi-directional case.

3.2 as

$$\frac{dx}{y + \frac{z}{x+y}} = \frac{dy}{x - \frac{z}{x+y}} = \frac{dz}{z}$$

having divided throughout by $x + y$ in order to leave just z terms in the last term. The denominators and numerators of the first two terms may be summed and still be equal to the above ratios so cancelling out the awkward $z/(x + y$ term to give

$$\frac{dz}{z} = \frac{d(x+y)}{x+y}$$

and hence $z = c_1(x + y)$ and the first equation reduces to

$$\frac{dx}{y + c_1} = \frac{dy}{x - c_1}$$

with solution $x^2 - y^2 - 2z = c_2$ and hence the general solution

$$F(z/(x+y), x^2 - y^2 - 2z) = 0$$

or equivalent form.

3.3 The characteristic equation is

$$\frac{dx}{y+z} = \frac{dy}{z+x} = \frac{dz}{x+y}$$

and these ratios are equal to

$$\frac{dy - dz}{z - y} = \frac{dx - dy}{y - x} = \frac{dx - dz}{z - x}$$

which yield immediately

$$(z - y) = c_1(y - x) \qquad (z - x) = c_2(y - x)$$

with general solution

$$F((z-y)/(y-x),(z-x)/(y-x)) = 0.$$

3.4

$$x = s \qquad y = 0 \qquad z = s$$

and $p(s)$ and $q(s)$ satisfy

$$p(s)q(s) = 1 \qquad p(s) = 1$$

to give $p(s) = 1$ and $q(s) = 1$. The characteristic equations are

$$\frac{dx}{dt} = \qquad \frac{dy}{dt} = p \qquad \frac{dz}{dt} = 2pq \qquad \frac{dp}{dt} = 0 \qquad \frac{dq}{dt} = 0.$$

The initial values for p and q are

$$p(0,s) = 1 \qquad \text{and} \qquad q(o,s) = 1$$

and the characteristic equations solve to give:

$$x(t,s) = t + s \qquad y(t,s) = t \qquad z(t,s) = 2t + s.$$

The first two equations give $t = y$ and $s-x-y$ and hence $z(x,y) = x+y$.

3.5

$$x = s \qquad t = 0 \qquad z = ds$$

and initially p and q satisfy

$$q(s) + p^2(s) = 0 \qquad p(s) = d$$

to give $p(s) = d$ and $q(s) = -d^2$. The characteristic equations are

$$\begin{aligned} \frac{dx}{d\tau} &= 2p \qquad \frac{dt}{d\tau} = 1 \qquad \frac{dz}{d\tau} = q + 2p^2 \\ \frac{dp}{d\tau} &= 0 \qquad \frac{dq}{d\tau} = 0 \end{aligned}$$

with solutions

$$x(\tau,s) = 2d\tau + s \qquad t(\tau,s) = \tau \qquad z(\tau,s) = d^2\tau + ds$$

and

$$p(\tau,s) = d \qquad q(\tau,s) = -d^2$$

from which $\tau = t$ and $s = x - 2at$ and hence $z(x,t) = d(x - dt)$.

3.6 Figure A.2 shows a series of left-progressing waves.

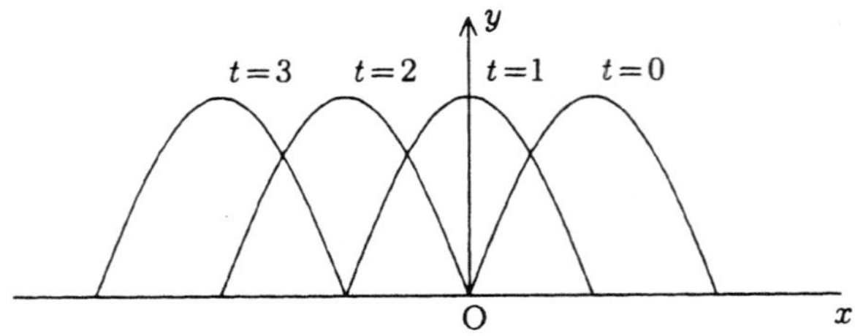

Fig. A.2.

3.7 The D'Alembert solution is

$$u = f(x - ct) + g(x + ct)$$

and the boundary conditions give

$$\begin{aligned} 0 &= f(x) + g(x) \\ \cos x &= -cf'(x) + cg'(x). \end{aligned}$$

Differentiate the first equation to give:

$$0 = cf'(x) + cg'(x)$$

and adding and subtracting gives

$$f(x) = -\frac{1}{2c} \sin x$$

and

$$g(x) = \frac{1}{2c} \sin x$$

and hence

$$u = -\frac{1}{2c} \sin(x - ct) + \frac{1}{2c} \sin(x + ct).$$

3.8 D'Alembert's general solution is

$$u(x) = \frac{1}{2}[\phi(x - ct) + \phi(x + ct)] + \frac{1}{2c} \int_{x-ct}^{x+ct} q(\zeta)\, d\zeta$$

and in this problem $c = \frac{1}{2}$, $\phi = e^{-x^2}$ and $q = xe - -x^2$. Hence

$$\begin{aligned} u &= \frac{1}{2}[e^{-(x-ct)^2} + e^{(x+ct)^2}] + \frac{1}{2} \int_{x-ct}^{x+ct} e^{-\zeta^2}\, d(\zeta^2) \\ &= \frac{1}{2}[e^{-(x-ct)^2} + e^{(x+ct)^2}] + -e^{-x^2} \Big|_{x-ct}^{x+ct} \\ &= e^{-(x-t/2)^2} \end{aligned}$$

3.9 Use the same general solution as above with $c = 1$, $\phi = \sin x$ and $q = \cos x$. Hence

$$\begin{aligned} u &= \frac{1}{2}[\sin(x - ct) + \sin(x + ct)] + \frac{1}{2}\int_{x-t}^{x+t} \cos(\zeta)\, d\zeta \\ &= \frac{1}{2}[\sin(x - ct) + \sin(x + ct)] + \frac{1}{2}\left[\sin\zeta\right]_{x-t}^{x+t} \\ &= \sin(x + t). \end{aligned}$$

3.10 The discontinuity will propagate from $x = 2$, $t = 0$ along the two characteristics which have equations $x + t = 2$ and $x - t = 2$. For any point for which the intersecting characteristics miss $x = 2$ then $\phi = 0$ and $q = 0$ in D'Alembert's solution and hence $u = 0$. If one of the intersecting characteristics hits $x = 2$, that is along the two sketched characteristics then at $x = 2$, $\phi = a$ and so the solution is $u = a/2$. Hence the spike propagates out along the two characteristics.

3.11 Here the D'Alembert solution will have $q = 0$. Figure A.3 shows the three regions of the solution.

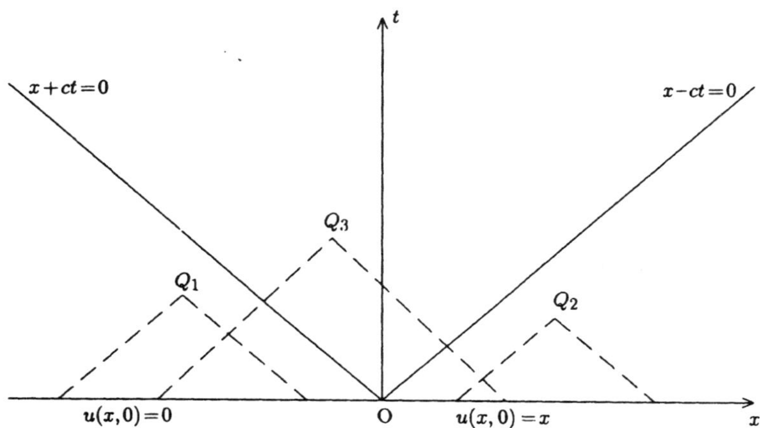

Fig. A.3.

These regions are bounded by the two characteristics $x+ct = 0$ and $x-ct = 0$. In region Q_2 the intersecting characteristics will both intersect $t = 0$ where $\phi = x$ and hence $u = \frac{1}{2}(x + x) = x$. In Q_1, both intersect $t = 0$ in the regime where $\phi = 0$, giving $u = 0$. In the middle region Q_3 one characteristic will intersect $t = 0$ where $\phi = x$ and one where $\phi = 0$ and will give $u = x/2$.

3.12 The problem is extended to an infinite range which gives the correct boundary conditions at $x = 0$ by setting

$$u(x,0) = \begin{cases} \sin x & x > 0 \\ -\sin(-x) & x < 0, \end{cases}$$

and now Figure A.3 will apply, although only the region for which $x > 0$ is now relevant. In region Q_2, using D'Alembert's solution with $q = 0$ gives

$$u = \frac{1}{2}(\sin(x - ct) + \sin(x + ct))$$

with both characteristics intersecting $t = 0$ in the $\sin x$ region with $x > 0$. In the region Q_3 for $x > 0$, one characteristic will fall in the $-\sin(-x)$ region and then

$$u = \frac{1}{2}(-\sin(-x + ct) + \sin(x + ct))$$

giving the same solution as before.

3.13 In this example $\phi = 0$ and the extended q is

$$q(x) = xe^{-x^2}$$

both for $x > 0$ and $x < 0$. Hence

$$\begin{aligned} u &= \frac{1}{2c}\int_{x-ct}^{x+ct} xe^{-x^2}\,dx \\ &= \frac{1}{4c}[e^{-(x-ct)^2} - e^{-(x+ct)^2}]. \end{aligned}$$

3.14 Again using the regions of Figure A.3, in region Q_3 for $x > 0$, D'Alembert's solution gives

$$\begin{aligned} u &= \frac{1}{2}\int_{x-t}^{0} -\sin(-x)\,dx + \frac{1}{2}\int_{0}^{x+t} \sin x\,dx \\ &= \frac{1}{2}[\cos(x - t) - \cos(x + t)]. \end{aligned}$$

Region Q_2 reduces to the same integral as $-\sin(-x) = \sin x$.

Chapter 4

4.1 Use

$$f(t) = \int_0^\infty [A(\omega)\cos\omega t + B(\omega)\sin\omega t]d\omega$$

with

$$A(\omega) = \frac{1}{\pi}\int_{-\infty}^{\infty} f(v)\cos\omega v\, dv$$

and

$$B(\omega) = \frac{1}{\pi}\int_{-\infty}^{\infty} f(v)\sin\omega v\, dv$$

which gives

$$\begin{aligned} A(\omega) &= \int_0^{\infty} e^{-v}\cos\omega v\, dv \\ &= \frac{1}{1+\omega^2}. \end{aligned}$$

The imaginary part of the same integral gives

$$B(\omega) = \frac{\omega}{1+\omega^2}.$$

4.2 In this example, $f(-t) = -f(t)$ and hence $A(\omega) = 0$. Then

$$\begin{aligned} B(\omega) &= \int_0^{\infty} e^{-v}\cos v \sin\omega v\, dv \\ &= \frac{1}{2}\frac{\omega^3}{\omega^4+4}. \end{aligned}$$

4.3 For this case f is odd to leave just

$$\begin{aligned} B(\omega) &= \int_0^{\pi} \sin v \sin\omega v\, dv \\ &= \frac{\sin\pi\omega}{1-\omega^2}. \end{aligned}$$

4.4

$$B(\omega) = \frac{2}{\pi}\int_0^{\pi} \frac{\pi}{2}\sin\omega v\, dv = \frac{1-\cos\pi v}{\omega}.$$

4.5 This case has f even and hence $B(\omega) = 0$ and

$$A(\omega) = \int_0^{\infty} e^{-v}\cos\omega v\, dv = \frac{1}{1+\omega^2}.$$

4.6

$$A(\omega) = \frac{2}{\pi}\int_0^{1} \frac{\pi}{2}\cos\omega v\, dv = \frac{\sin\omega}{\omega}$$

and at the discontinuity the integral will be half the values on each side to give $\pi/4$.

4.7 With $f(x)$ even,

$$A(\omega) = \int_0^{\pi/2} \cos v \cos \omega v \, dv = \frac{\cos \frac{\pi\omega}{2}}{1-\omega^2}.$$

4.8
$$A(\omega) = \frac{2}{\pi}\int_0^a \cos \omega v \, dv = \frac{2}{\pi\omega}\sin \omega a.$$

4.9
$$\begin{aligned} A(\omega) &= \frac{2}{\pi}\int_0^a v^2 \cos \omega v \, dv \\ &= \frac{2a^2 \sin \omega a}{\pi\omega} + \frac{4a}{\pi\omega^2}\cos \omega a - \frac{4}{\pi\omega^3}\sin \omega a. \end{aligned}$$

4.10 Starting with

$$f(x) = \frac{1}{\pi}\int_0^\infty A(\omega)\cos \omega x \, dx,$$

put $x = ay$ to give

$$f(ay) = \frac{a}{\pi}\int_0^\infty A(\omega)\cos \omega ay \, dy.$$

Then $a\omega = w$ gives

$$f(ax) = \frac{a}{\pi}\int_0^\infty A\left(\frac{w}{a}\right)\cos wy \, dy.$$

4.11 To expand $f(x) = 1$ requires the integral

$$\int_0^\infty \cos \omega x \, dx$$

which does not converge.

4.12 The separated solution has the form

$$u(x,y) = \sum_p (A_p \cos px + B_p \sin px)(Cepy + De^{-py})$$

and $u(x,a) = 0$ gives

$$Ce^{pa} = -De^{-pa}$$

to give the trial solution with continuous p

$$u(x,y) = \int_{-\infty}^{\infty} (A(p)\cos px + B(p)\sin px)(Cep(y-a) + De^{p(a-y)}) \, dp.$$

The condition $u(x,0) = f(x)$ gives the Fourier integral

$$f(x) = \int_{-\infty}^{\infty} [A(p)\cos px + B(p)\sin px] \sinh pa \, dp$$

with

$$A(p) = \frac{1}{\pi}\int_{-\infty}^{\infty} f(v)\cos pv\,dv$$

and

$$B(p) = \frac{1}{\pi}\int_{-\infty}^{\infty} f(v)\sin pv\,dv$$

using

$$\cos pv\cos px + \sin pv\sin px = \cos p(v-x)$$

gives

$$u(x,y) = \frac{1}{\pi}\int_0^{\infty}\left[\int_{-\infty}^{\infty}\frac{\sinh[p(a-y)]}{\sinh pa}f(v)\cos p(v-x)\,dv\right]dp.$$

4.13 The separated solution is as in the previous exercise and this reduces to

$$u(x,y) = \sum_p A(p)e^{-yp}\sin px$$

under the boundary conditions. Hence the Fourier integral is

$$f(x) = u(x,0) = \int_0^{\infty} A(p)\sin px\,dp$$

and

$$A(p) = \frac{2}{\pi}\int_0^{\infty} f(\zeta)\sin(\zeta p)\,dp$$

to give

$$u(x,y) = \frac{2}{\pi}\int_0^{\infty}\int_0^{\infty} e^{-py}f(\zeta)\sin px\sin p\zeta\,d\zeta dp.$$

The integral with respect to p is

$$\int_0^{\infty} e^{-py}\sin px\sin p\zeta dp = \frac{y}{2}\left[\frac{1}{y^2+(\zeta-x)^2} - \frac{1}{y^2+(\zeta+x)^2}\right],$$

as required.

4.14 The separable solution is

$$u(r,t) = \sum_p (A_pJ_0(pr) + B_pY_0(pr))e^{-p^2kt},$$

and u finite at $r=0$ implies $B_p = 0$. Hence

$$u = \int_0^{\infty} A(p)e^{-p^2kt}J_0(pr)\,dp$$

where

$$f(r) = \int_0^{\infty} A(p)J_0(pr)\,dp.$$

4.15 The separable solution with $u(0,t) = 0$ is

$$u = \sum_p A_p \sin pxe^{-p^2kt}$$

and hence for the infinite range

$$u = \int_0^\infty A(p) \sin pxe^{-p^2kt}\, dp$$

where

$$A(p) = \frac{2}{\pi}\int_0^\infty f(\zeta) \sin \zeta p\, d\zeta$$

to give the required result.

4.16 Using the boundary conditions $u_x = 0$ at $x = 0$ and $u_y = 0$ at $y = 0$ gives the separable solution

$$u(x,y) = \sum_p B_p \cos px \cosh py$$

or for continuous p

$$u(x,y) = \int_0^\infty B(p) \cos px \cosh py\, dp$$

and

$$f(x) = \int_0^\infty B(p) \cos px \cosh p\, dp$$

with

$$B(p) \cosh p = \frac{2}{\pi}\int_0^\infty f(\zeta) \cos(\zeta p)\, d\zeta$$

to give the result

$$u(x,y) = \frac{2}{\pi}\int_0^\infty \frac{\cos px \cosh py}{\cosh p}\left[\int_0^\infty f(\zeta \cos(p\zeta)\, d\zeta\right] dp.$$

When

$$f(x) = \begin{cases} 1 & 0 \le x \le 1 \\ 0 & x > 1, \end{cases}$$

we have

$$\int_0^1 \cos p\zeta\, d\zeta = \frac{\sin p}{p}$$

to give

$$u = \frac{2}{\pi}\int_0^\infty \frac{\cos px \cosh py \sin p}{p \cosh p}\, dp$$

as required.

4.17 Multiply both sides of the partial differential equation by $\cos px$ and integrate from 0 to ∞:

$$\int_0^\infty \frac{\partial^2 u}{\partial x^2} \cos px\, dx = \frac{1}{k} \int_0^\infty \frac{\partial u}{\partial t} \cos px\, dx.$$

Hence

$$\frac{\partial u}{\partial x} \cos px \bigg|_0^\infty + pu \sin px|_0^\infty - p^2 \int_0^\infty u \cos px\, dx = \frac{1}{k}\frac{dU_c}{dt}$$

and now the choice of the cosine transformation allows the boundary condition $u_x = -v$ to be used to give just

$$\frac{1}{k}\frac{dU_c}{dt} = v - p^2 U_c$$

with solution

$$U_c = Ae^{-p^2kt} + \frac{v}{p^2}.$$

Applying $U_c = 0$ at $t = 0$ gives $A = -v/p^2$ and thereby

$$U_c = \frac{v}{p^2}(1 - e^{-p^2kt}).$$

The inversion formula gives

$$u(x,t) = \frac{2}{\pi}\int_0^\infty \frac{1 - e^{-p^2kt}}{p^2} \sin px\, dp.$$

4.18 With the boundary condition $u(x, 0) = 0$ use the sine transform to give

$$\frac{dU_s}{dt} = -kp^2 U_s$$

with solution

$$U_s = Ae^{-kp^2t},$$

and the boundary condition is

$$u(x,0) = \begin{cases} 1 & 0 < x < 1 \\ 0 & 1 \le x. \end{cases}$$

Taking transforms gives

$$U_s(0) = \int_0^1 \sin px\, dx = \frac{1 - \cos p}{p}.$$

Hence

$$A = \frac{1 - \cos p}{p}$$

and

$$U_s = \frac{1 - \cos p}{p} e^{-kp^2 t}$$

with inverse

$$u(x,t) = \frac{2}{\pi} \int_0^\infty \frac{1 - \cos p}{p} e^{-p^2 t} \sin px \, dp$$

as required.

4.19 Taking the usual sine transform gives

$$\frac{dU_s}{dt} = -2p^2 U_s$$

with solution

$$U_s = Ae^{-2p^2 t}.$$

However, $u(x,0) = e^{-x}$ which transforms to

$$U_s(0) = \int_0^\infty e^{-x} \sin px \, dx = \frac{p}{1+p^2} = A.$$

Hence

$$u(x,t) = \frac{2}{\pi} \int_0^\infty \left(\frac{p}{1+p^2} \right) e^{-2p^2 t} \sin px \, dp.$$

4.20 The cosine transform gives

$$\frac{dU_c}{dt} = -p^2 t U_c$$

with solution

$$U_c = Ae^{-p^2 t}.$$

However,

$$u(x,0) = \begin{cases} x & 0 \le x \le 1 \\ 0 & 1 < x \end{cases}$$

and transforming gives

$$U_c(0) = \int_0^1 x \cos px \, dx = \frac{\sin p}{p} + \frac{\cos p - 1}{p^2}$$

to yield

$$u(x,t) = \frac{2}{\pi} \int_0^\infty \left(\frac{\sin p}{p} + \frac{\cos p - 1}{p^2} \right) e^{-p^2 t} \cos px \, dp.$$

4.21 Take the complex Fourier transform of

$$a^2\frac{\partial^4 u}{\partial x^4}+\frac{\partial^2 u}{\partial t^2}=0$$

to obtain

$$a^2\int_{-\infty}^{\infty}\frac{\partial^4 u}{\partial x^4}e^{i\omega x}\,dx+\int_0^{\infty}\frac{\partial^2 u}{\partial t^2}e^{i\omega x}\,dx=0.$$

Hence

$$\begin{aligned}-\frac{d^2U}{dt^2} &= a^2\left[e^{i\omega x}\frac{\partial^3 u}{\partial x^3}\right]_{-\infty}^{\infty}-a^2\int_{-\infty}^{\infty}\frac{\partial^3 u}{\partial x^3}(i\omega)e^{i\omega x}\,dx\\ &= -i\omega a^2\left[e^{i\omega x}\frac{\partial^2 u}{\partial x^2}\right]_{-\infty}^{\infty}+(i\omega)^2a^2\int_{-\infty}^{\infty}\frac{\partial^2 u}{\partial x^2}e^{i\omega x}\,dx\\ &= a^2(i\omega)^2\left[e^{i\omega x}\frac{\partial u}{\partial x}\right]_{-\infty}^{\infty}-(i\omega)^3a^2\int_{-\infty}^{\infty}\frac{\partial u}{\partial x}e^{i\omega x}\,dx\\ &= -\,a^2(i\omega)^3e^{i\omega x}u\Big|_{-\infty}^{\infty}+a^2(i\omega)^4U,\end{aligned}$$

using the derivative boundary conditions at each step. The transformed equation is

$$\frac{d^2U}{dt^2}+a^2\omega^4U=0.$$

The solution is

$$U=A\cos a\omega^2t+B\sin a\omega^2t.$$

However, $U_t=0$ at $t=0$ to force $B=0$ and transforming $u(x,0)=f(x)$ gives

$$U(0)=\int_{-\infty}^{\infty}f(x)e^{i\omega x}\,dx.$$

Hence

$$U(t)=\left(\int_{-\infty}^{\infty}f(x)e^{i\omega x}\,dx\right)\cos a\omega^2t=F(\omega)\cos a\omega^2t.$$

Now use

$$\mathcal{F}^{-1}(FG)=f\otimes g=\int_{-\infty}^{\infty}f(y)g(x-y)\,dy$$

with the given inversion formula to give

$$u(x,t)=\frac{1}{\sqrt{4\pi at}}\int_{-\infty}^{\infty}f(x-y)\cos\left(\frac{y^2}{4at}-\frac{\pi}{4}\right)dy.$$

4.22 The cosine transformation gives

$$\frac{d^2U_c}{dy^2} + p^2U_c = 0$$

with solution

$$U_c = Be^{-py}$$

so that U_c remains finite at ∞. At $t = 0$

$$U_c(0) = \int_0^a \cos px\, dx = \frac{\sin ap}{p} = B.$$

Hence

$$U_c(y) = \frac{\sin ap}{p} e^{-py}$$

and the inversion formula gives

$$\begin{aligned} u(x, y) &= \frac{2}{\pi} \int_0^\infty e^{-py} \frac{\sin ap}{p} \cos px\, dp \\ &= \frac{1}{\pi} \left[\arctan\left(\frac{a+x}{y}\right) + \arctan\left(\frac{a-x}{y}\right) \right]. \end{aligned}$$

4.23 The transformed equation is

$$\frac{d^2V_s}{dy^2} = p^2V_s$$

using the boundary condition $v = 0$ when $x = 0$ in the integration by parts. The solution is

$$V_s = Ae^{-py} + Be^{py}$$

and

$$V_s(0) = \int_0^\infty f(\zeta) \sin p\zeta\, d\zeta$$

to give the two equations

$$\begin{aligned} A + B &= V_s(0) \\ Ae^{-pb} + Be^{pb} &= 0 \end{aligned}$$

for the boundary conditions. Hence

$$\begin{aligned} A &= V_s(0)e^{pb}/\sinh pb \\ B &= -V_s e^{-pb}/\sinh pb \end{aligned}$$

to give the full solution

$$\begin{aligned} u(x, y) &= \frac{2}{\pi} \int_0^\infty f(\zeta) \int_0^\infty \frac{\sin \zeta p}{\sinh pb} \sin xp (e^{pb}e^{-py} - e^{-pb}e^{py})\, d\zeta\, dp \\ &= \frac{2}{\pi} \int_0^\infty f(\zeta) \int_0^\infty \frac{\sin \zeta p}{\sinh pb} \sin xp \sinh p(b-y)\, d\zeta\, dp. \end{aligned}$$

4.24 (a) The inverse is

$$I = \frac{1}{2\pi i}\int_B \frac{p}{p^2+a^2}e^{px}\,dp$$

where B is the Bromwich contour. Hence

$$\begin{aligned} I &= \sum_{\text{Res } ia,-ia} \\ &= \lim_{p\to ia}\frac{(p-ia)(ia)e^{px}}{(p-ia)(p+ia)} + \lim_{p\to -ia}\frac{(p+ia)(-ia)e^{px}}{(p-ia)(p+ia)} \\ &= \frac{ia}{2ia}e^{iax} + \frac{ia}{2ia}e^{-iax} = \cos ax. \end{aligned}$$

(b)

$$I = \frac{1}{2\pi i}\int_B \frac{1}{p^2+a^2}e^{px}\,dp$$

and

$$\begin{aligned} I &= \sum_{\text{Res } ia,-ia} \\ &= \frac{e^{iax}}{2ia} - \frac{e^{-iax}}{2ia} = \frac{\sin ax}{a}. \end{aligned}$$

(c)

$$I = \frac{1}{2\pi i}\int_B \frac{e^{px}}{(p+1)(p^2+1)}\,dp$$

gives

$$\begin{aligned} I &= \sum_{\text{Res}-1,-i,i} \\ &= \lim_{p\to -1}\frac{e^{px}}{(p^2+1)} + \lim_{p\to -i}\frac{e^{px}}{(p+1)(p-i)} + \lim_{p\to i}\frac{e^{px}}{(p+1)(p+i)} \\ &= \frac{e^{-x}}{2} + \frac{1}{2}(\sin x - \cos x). \end{aligned}$$

(d)

$$I = \frac{1}{2\pi i}\int_B \frac{e^{px}}{(p+1)^2}\,dp$$

yields

$$\begin{aligned} I &= \text{Res}_{p\to -1} \\ &= \lim_{p\to -1}\frac{d}{dp}\left[\frac{(p+1)^2e^{px}}{(p+1)^2}\right] = xe^{-x}. \end{aligned}$$

(e)

$$I = \frac{1}{2\pi i}\int_B \frac{e^{px}}{p^3(p^2+1)}\,dp$$

which gives

$$\begin{aligned} I &= \sum_{\text{Res } 0,i,-i} \\ &= \lim_{p\to 0} \frac{1}{2!}\frac{d^2}{dp^2}\left(\frac{e^{px}}{p^2+1}\right) + \lim_{p\to i} \frac{e^{px}}{p^3(p+i)} + \lim_{p\to -i} \frac{e^{px}}{p^3(p-i)} \\ &= \frac{1}{2}x^2 + \cos x. \end{aligned}$$

4.25 $$I = \frac{1}{2\pi i}\int_B \frac{e^{px}}{p(e^p+1)}\, dp.$$

Hence

$$\begin{aligned} I &= \sum_{\text{Res } 0,\pm i\pi,\pm 3i\pi\ldots} \\ &= \frac{1}{2} + \lim_{p\to i\pi} \frac{e^{px}(p-i\pi)}{p(e^p+1)} + \lim_{p\to -i\pi} \frac{e^{px}(p+i\pi)}{p(e^p+1)} \\ &\quad + \lim_{p\to 3i\pi} \frac{e^{px}(p-3i\pi)}{p(e^p+1)} + \lim_{p\to -3i\pi} \frac{e^{px}(p+3i\pi)}{p(e^p+1)} + \ldots \\ &= \frac{1}{2} - \frac{2}{\pi}\left(\sin \pi x - \frac{\sin 3\pi x}{3} + \ldots\right). \end{aligned}$$

4.26 $$I = \frac{1}{2\pi i}\int_B \frac{e^{px}}{p\cosh p}\, dp$$

with a series of poles at the zeros of $\cosh p$. Hence

$$\begin{aligned} I &= \sum_{\text{Res } 0,\pm i\pi/2,\pm 3i\pi/2\ldots} \\ &= \lim_{p\to 0} \frac{e^{px}}{\cosh p} + \lim_{p\to i\pi/2} \frac{e^{px}(p-i\pi/2)}{p\cosh p} + \lim_{p\to -i\pi/2} \frac{e^{px}(p+i\pi/2)}{p\cosh p} \\ &\quad + \lim_{p\to 3i\pi/2} \frac{e^{px}(p-3i\pi/2)}{p\cosh p} + \lim_{p\to -3i\pi/2} \frac{e^{px}(p+3i\pi/2)}{p\cosh p} + \ldots \\ &= 1 - \frac{4}{\pi}\left(\cos\frac{\pi x}{2} - \frac{1}{3}\cos\frac{3\pi x}{2} + \ldots\right). \end{aligned}$$

4.27 $$I = \frac{1}{2\pi i}\int_B \frac{e^{px}}{p^2\sinh p}\, dp$$

gives

$$\begin{aligned} I &= \sum_{\text{Res} 0,\pm i\pi,\pm 2i\pi\ldots} \\ &= \lim_{p\to 0} \frac{1}{2}\frac{d^2}{dp^2}\frac{pe^{px}}{\sinh p} \end{aligned}$$

$$+\lim_{p\to i\pi}\frac{e^{px}(p-i\pi)}{p^2\sinh p}+\lim_{p\to -i\pi}\frac{e^{px}(p+i\pi)}{p^2\sinh p}$$
$$+\lim_{p\to 2i\pi}\frac{e^{px}(p-2i\pi)}{p^2\sinh p}+\lim_{p\to -2i\pi}\frac{e^{px}(p+2i\pi)}{p^2\sinh p}+\dots$$
$$=\frac{1}{2}x^2+\frac{2}{\pi^2}\sum_{n=1}^{\infty}\frac{(-1)^n}{n^2}(1-\cos n\pi x).$$

4.28
$$I=\frac{1}{2\pi i}\int_B\frac{e^{px}}{p^3\sinh ap}\,dp.$$

The function $\sinh ap$ has zeros when $ap=n\pi i$ which makes $p=0$ a fourth-order pole. Most of the solution is devoted to computing this residue. Consider first the residues at $ap=n\pi i$.

$$\lim_{p\to n\pi i/a}\frac{e^{px}(p-n\pi i/a}{p^3\sinh ap}=\frac{e^{in\pi/a}}{ai^3n^3\pi^3\cosh in\pi}$$

and

$$\lim_{p\to -n\pi i/a}\frac{e^{px}(p+n\pi i/a}{p^3\sinh ap}=\frac{e^{-in\pi/a}}{a(-i)^3n^3\pi^3\cosh in\pi}$$

and the sum of these residues gives

$$-\frac{2a^2}{\pi^3}\frac{(-1)^n}{n^3}\sin\frac{n\pi x}{a}$$

which gives the sum in the final result. The fourth-order pole gives a residue

$$\frac{1}{3!}\frac{d^3}{dp^3}\left\{\frac{pe^{px}}{\sinh ap}\right\}$$
$$=\frac{d^2}{dp^2}\left\{\frac{e^{px}}{\sinh ap}+\frac{pxe^{px}}{\sinh ap}-\frac{ape^{px}}{\sinh^2 ap}\right\}$$
$$=\frac{d}{dp}\left\{\frac{2xe^{px}}{\sinh ap}+\frac{2ae^{px}}{\sinh^2 ap}-\frac{2apxe^{px}}{\sinh^2 ap}\frac{px^2e^{px}}{\sinh ap}+\frac{2a^2pe^{px}}{\sinh^3 ap}\right\}$$
$$=\lim_{p\to 0}\frac{3x^2e^{px}}{\sinh ap}-\frac{6axe^{px}}{\sinh^2 ap}+\frac{4ae^{px}}{\sinh^3 ap}-\frac{3apx^2e^{px}}{\sinh^2 ap}$$
$$+\frac{6a^2pxe^{px}}{\sinh^3 ap}+\frac{x^2e^{px}}{\sinh ap}+\frac{px^3e^{px}}{\sinh ap}+\frac{2a^2e^{px}}{\sinh^3 ap}-\frac{6a^3pe^{px}}{\sinh^4 ap}$$
$$=\frac{1}{6}\frac{z^3}{a}.$$

4.29 For

$$x\frac{\partial u}{\partial t}+\frac{\partial u}{\partial x}=x$$

the transformed equation is

$$xpU + \frac{dU}{dx} = \frac{x}{p}$$

which solves with an integrating factor to give

$$U = \frac{1}{p^2} + Ae^{-x^2p/2}$$

and the boundary condition $U(0) = 0$ gives

$$U = \frac{1}{p^2}(1 - e^{-x^2p/2a})$$

with inverse

$$u = t - \left(t - \frac{x^2}{2}\right) H\left(t - \frac{x^2}{2}\right)$$

using the Heaviside function result:

$$\mathcal{L}^{-1}\{f(t-T)H(t-T)\} = e^{-pT}F(p)$$

and

$$\mathcal{L}^{-1}\{1/p^2\} = t.$$

4.30 The Laplace transform is

$$\frac{d^2U}{dx^2} = \frac{p^2}{c^2}U - \frac{k\sin\pi x}{p}$$

using standard transforms. The transformed solution is

$$U = A\sin\frac{px}{c} + B\cos\frac{px}{c} + \frac{kc^2}{p(p^2 + c^2zpi^2)}\sin\pi x$$

and $U(0,t) = 0$ gives $B = 0$. Further $U(1,t) = 0$ gives $A = 0$ to leave

$$U = \frac{kc^2}{p(p^2 + c^2zpi^2)}\sin\pi x$$

with standard inverse

$$u(x,t) = \frac{k}{\pi^2}(1 - \cos c\pi t)\sin\pi x.$$

4.31 The transformed equation is

$$\frac{dU}{dx} + \frac{p}{x}U = \frac{1}{p}$$

with solution

$$U = \frac{x}{p(p+1)} + \frac{A}{x^p}$$

using the integrating factor x^p. The condition $U(0) = 0$ gives $A = 0$ and hence

$$U = \frac{x}{p(p+1)}$$

with standard inverse

$$u(x,t) = x(1 - e^{-t}).$$

4.32 For this example the transformed equation is

$$\frac{d\Phi}{dx} + xp\Phi = \frac{x^3}{p}$$

which has an integrating factor $e^{x^2p/2}$ to give after integration by parts:

$$\Phi = \frac{x^2}{p^2} - \frac{2}{p^3} + Ae^{-\frac{x^2p}{2}}.$$

The condition $\Phi(0) = 0$ gives

$$\Phi = \left(\frac{x^2}{p^2} - \frac{2}{p^3}\right)\left(1 - e^{-\frac{x^2p}{2}}\right)$$

with standard transform

$$\phi(x,t) = x^2t - t^2 + \left(t - \frac{x^2}{2}\right)\left(\frac{3x^2}{2} - t\right)H\left(t - \frac{x^2}{2}\right).$$

4.33 The transformed equation is

$$\frac{d\Phi}{dx} + xp\Phi = \frac{x}{p} + x\phi_0$$

the last term arising from the boundary condition $\phi = \phi_0$ at $t = 0$. The solution is

$$\Phi = \frac{1}{p}\left(\frac{1}{p} + \phi_0\right)\left(1 - e^{-\frac{x^2p}{2}}\right)$$

with standard transform

$$\phi = \phi_0 + t) - \left(\phi_0 + y - \frac{x^2}{2}\right)H\left(t - \frac{x^2}{2}\right).$$

4.34 The transformed equation is

$$\frac{d^2U}{dx^2} = 6p^2U - \sin \pi x$$

with solution

$$U = Ae^{p\sqrt{6}x} + Be^{-p\sqrt{6}x} - a \sin \pi x$$

and for the particular integral, differentiating and substituting into the original gives

$$-a\pi^2 = 6p^2 a - 1$$

and the boundary conditions $U(0) = 0$ and $U(2) = 0$ give both $A = 0$ and $B = 0$. Hence

$$U = -\frac{1}{6}\left[\frac{\sin \pi x}{p^2 + \pi^2/6}\right]$$

with a standard inverse

$$u(x,t) = -\frac{\sqrt{6}}{6b\pi} \sin \pi x \sin \frac{\pi t}{\sqrt{6}}.$$

4.35 The transformed equation is

$$x\frac{dY}{dx} + pY + Y = \frac{x}{p}$$

with solution

$$Y = \frac{x}{p(p+2)} + Ax^{-(1+p)}$$

and $Y(0) = 0$, hence $A = 0$. The standard inverse is then

$$y = \frac{x}{2}(1 - e^{-2t}).$$

Chapter 5

5.1 The Green's function for the infinite space solution is, by definition, given by the solution of

$$\left(\frac{\partial^2}{\partial x^2} + k^2\right) g(x|x_0, k) = -\delta(x - x_0) \qquad -\infty \le x \le \infty$$

where $0 \le x_0 \le \infty$. Take a Laplace transform to obtain

$$p^2\bar{g}(p|x_0, k) - pg(0|x_0, k) - g'(0|x_0, k) + k^2\bar{g}(p|x_0, k) = e^{-px_0}$$

where

$$g'(0|x_0,k) = \left[\frac{\partial g}{\partial x}(x|x_0,k)\right]_{x=0}$$

Hence

$$\bar{g}(p|x_0,k) = \frac{p}{p^2+k^2}g(0|x_0,k) + \frac{1}{p^2+k^2}g'(0|x_0,k) + \frac{e^{-px_0}}{p^2+k^2}.$$

Inverting

$$\begin{aligned} g(x|x_0,k) &= g(0|x_0,k)\cos kx + \frac{1}{k}g'(0|x_0,k)\sin kx \\ &\quad + \frac{1}{k}\sin[k(x-x_0)]H(x-x_0) \end{aligned}$$

where $H(x)$ is the Heaviside step function. The solution is given by

$$\begin{aligned} u(x_0,k) &= \int_0^\infty g(x|x_0,k)f(x)\,dx \\ &= \int_0^\infty g(0|x,k)\cos kx_0\,dx \\ &\quad + \frac{1}{k}\int_0^\infty g'(0|x,k)f(x)\sin kx_0\,dx \\ &\quad + \frac{1}{k}\int_0^\infty \sin[k(x_0-x)]H(x-x_0)f(x)\,dx \\ &= A(k)\cos kx_0 + B(k)\sin kx_0 \\ &\quad + \frac{1}{k}\int_0^\infty \sin[k(x_0-x)]H(x-x_0)f(x)\,dx \end{aligned}$$

where

$$A(k) = \int_0^\infty g(0|x,k)\,dx, \qquad B(k) = \frac{1}{k}\int_0^\infty g'(0|x,k)f(x)\,dx.$$

5.2 The Green's function is

$$g(x|x_0,k) = -\frac{1}{k}\sin k|x-x_0|$$

being the solution of

$$\left(\frac{\partial^2}{\partial x^2} + k^2\right)g(x|x_0,k) = -\delta(x-x_0)$$

for both right- and left-travelling waves. The homogeneous equation has solutions $\sin kx$ and $\sin k(l-x)$ which vanish at $x=0$ and $x=L$. Consider the linear combination

$$u(x,k) = -\frac{1}{k}\sin k|x-x_0| + A\sin kx + B\sin k(l-x).$$

Then

$$0 = -\frac{1}{k}\sin kx_0 + B\sin kL$$

and

$$0 = -\frac{1}{k}\sin k(L - x_0) + A\sin kL.$$

Hence

$$\begin{aligned} u(x|x_0,k) &= -\frac{1}{k}\left[\sin k|x-x_0| - \frac{\sin kx_0 \sin k(L-x)}{\sin kL}\right. \\ &\quad \left. - \frac{\sin kx \sin k(L-x_0)}{\sin kL}\right]. \end{aligned}$$

5.3 From 5.1 $g(x|x_0,k)$ gives $g(0|x_0,k) = 0$, $g'(0|x_0,k) = g(1|x_0,k)$ and

$$g(1|x+0,k) = \frac{1}{k}g(1|x_0,k)\sin k + \frac{1}{k}\sin[k(1-x_0)]H(1-x_0).$$

Solving for g gives

$$g(1|x_0,k) = \frac{\sin[k(1-x_0)]H(1-x_0)}{k - \sin k}$$

and hence

$$g(x|x_0,k) = \frac{\sin kx \sin[k(1-x_0)]}{k(k-\sin k)} + \frac{1}{k}\sin[k(x-x_0)]H(x-x_0)$$

for $0 \le x \le 1$ and $0 \le x_0 \le 1$.

5.4 Writing $\mathbf{R} = \mathbf{r} = \mathbf{r}_0$ and take the three-dimensional Fourier transform to obtain

$$G(\mathbf{u}) = \frac{1}{u^2 + \lambda}$$

where

$$G(\mathbf{u}) = \int_{-\infty}^{\infty} g(\mathbf{R})e^{-i\mathbf{u}\cdot\mathbf{R}}\, d^3\mathbf{r}$$

and $u \equiv |\mathbf{u}|$. Transforming back using spherical polars gives

$$\begin{aligned} g(R) &= \frac{1}{(2\pi)^3}\int_{-\infty}^{\infty} \frac{e^{i\mathbf{u}.\mathbf{R}}}{u^2+\lambda}\, d^3\mathbf{u} \\ &= \frac{1}{(2\pi)^3}\int_0^{\infty} u^2\, du \int_{-1}^{1} d(\cos\theta) \int_0^{2\pi} \frac{e^{iuR\cos\theta}}{u^2+\lambda}\, d\phi \\ &= \frac{1}{4\pi^2 R}\int_{-\infty}^{\infty} \frac{u\sin(uR)}{u^2+\lambda}\, du. \end{aligned}$$

However, the contour integral

$$\oint_C \frac{ze^{izR}}{z^2+\lambda}dz$$

has two simple poles at $\pm i\sqrt{\lambda}$. Choosing the contour C in the upper half plane gives the residue

$$\lim_{z\to i\sqrt{\lambda}}\left[\frac{(z-i\sqrt{\lambda})ze^{izR}}{z^2+\lambda}\right]=\frac{1}{2}e^{-\sqrt{\lambda}R}$$

and hence the solution

$$g(R)=\frac{e^{-\sqrt{\lambda}R}}{4\pi R}.$$

5.5 Rewrite the equation as

$$(\nabla^2+u^2)u(\mathbf{r},k)=-V(\mathbf{r})u(\mathbf{r},k).$$

Then

$$(\nabla^2+u^2)g(\mathbf{r}|\mathbf{r}_0,k)=-\delta^3(\mathbf{r}-\mathbf{r}_0)$$

and

$$\begin{aligned}\int(g\nabla^2u-u\nabla^2g)d^3\mathbf{r} &= -\int gVud^3\mathbf{r}+\int u(\mathbf{r},k)\delta^3(\mathbf{r}-\mathbf{r}_0)d^3\mathbf{r}\\ &= -\int gVu\,d^3\mathbf{r}+u(\mathbf{r},k).\end{aligned}$$

Hence

$$u(\mathbf{r}_0,k)=f(\mathbf{r}_0)+\int g(\mathbf{r}|\mathbf{r}_0,k)u(\mathbf{r},k)V(\mathbf{r})\,d^3\mathbf{r}$$

where

$$f(\mathbf{r}_0)=\int[g(\mathbf{r}|\mathbf{r}_0,k)\nabla^2u(\mathbf{r},k)-u(\mathbf{r},k)\nabla^2g(\mathbf{r}|\mathbf{r}_0,k)]d^3\mathbf{r}$$

which is a solution to

$$(\nabla^2+k^2)u(\mathbf{r},k)=0.$$

5.6 Let $U(\mathbf{r},t)=u(\mathbf{r},\omega)\exp(i\omega t)$, then

$$\nabla^2u+k^2u=-4\pi\rho;\quad k=\frac{\omega}{c}=\frac{2\pi}{\lambda}$$

where λ is the wavelength. The Green's function solution to this equation (at $\mathbf{r}_0$) is

$$u(\mathbf{r}_0,k)=4\pi\int\rho(\mathbf{r})g(\mathbf{r}\mid\mathbf{r}_0,k)d^3\mathbf{r}$$

where g is the "outgoing Green's function" given by

$$g(\mathbf{r} \mid \mathbf{r}_0, k) = \frac{\exp(ik \mid \mathbf{r} - \mathbf{r}_0 \mid)}{4\pi \mid \mathbf{r} - \mathbf{r}_0 \mid}$$

whose asymptotic form is

$$\frac{\exp(ikr_0)}{4\pi r_0}\exp(-ik\hat{\mathbf{n}} \cdot \mathbf{r}); \qquad \hat{\mathbf{n}} = \frac{\mathbf{r}_0}{r_0}, \qquad r_0 \gg r.$$

Hence, in the far field

$$u(\mathbf{r}_0, k) = \frac{\exp(ikr_0)}{r_0} \int \rho(\mathbf{r}) \exp(-ik\hat{\mathbf{n}} \cdot \mathbf{r}) d^3\mathbf{r}.$$

In spherical polar coordinates

$$\begin{aligned} u(r_0, k) &= \frac{\exp(ikr_0)}{r_0} \int \frac{\exp(-ikr\cos\theta)}{r^2} r^2 dr d(\cos\theta) d\phi \\ &= \frac{2\pi}{r_0} \exp(ikr_0) \int_0^\infty \frac{2\sin(kr)}{kr} dr = \frac{2\pi^2}{kr_0} \exp(ikr_0). \end{aligned}$$

Hence, $\mid u \mid = \pi\lambda/r_0$ and with λ=10 m and r_0=1000 m has a value of 0.01π.

5.7 (i) We require the solution to the following equation:

$$\left(\nabla^2 - \frac{1}{c^2}\frac{\partial^2}{\partial t^2} - \sigma^2\right) G(\mathbf{r} \mid \mathbf{r}_0, t \mid t_0) = -\delta^3(\mathbf{r} - \mathbf{r}_0)\delta(t - t_0).$$

Let

$$G(R, \tau) = \frac{1}{2\pi} \int_{-\infty}^{\infty} g(R, \omega) \exp(i\omega\tau) d\omega$$

and

$$\delta(\tau) = \frac{1}{2\pi} \int_{-\infty}^{\infty} \exp(i\omega\tau) d\omega$$

where $R = \mid \mathbf{r} - \mathbf{r}_0 \mid$ and $\tau = t - t_0$ so that the equation above becomes (in ω-space)

$$\left(\nabla^2 + \frac{\omega^2}{c^2} - \sigma^2\right) g(R, \omega) = -\delta^3(\mathbf{r} - \mathbf{r}_0).$$

Further, let

$$g(R, \omega) = \frac{1}{(2\pi)^3} \int_{-\infty}^{\infty} \tilde{g}(u, \omega) \exp(i\mathbf{u} \cdot \mathbf{R}) d^3\mathbf{u}$$

and

$$\delta^3(\mathbf{R}) = \frac{1}{(2\pi)^3} \int_{-\infty}^{\infty} \exp(i\mathbf{u} \cdot \mathbf{R}) d^3\mathbf{u}.$$

The equation then transforms to

$$\left(-u^2 + \frac{\omega^2}{c^2} - \sigma^2\right) \tilde{g} = -1$$

or

$$\tilde{g}(u, \omega) = \frac{1}{u^2 - (\omega^2/c^2) + \sigma^2}.$$

Fourier inverting, we obtain

$$G(R, \tau) = \frac{1}{(2\pi)^4} \int_{-\infty}^{\infty} \int_{-\infty}^{\infty} \frac{\exp(i\mathbf{u} \cdot \mathbf{R}) \exp(i\omega\tau)}{u^2 - (\omega^2/c^2) + \sigma^2} d^3\mathbf{u} d\omega.$$

Integrating over the angular components of $\mathbf{u}$ (using spherical polar coordinates) we have

$$G(R, \tau) = \frac{1}{2\pi} \frac{1}{4\pi^2 R} \int_{-\infty}^{\infty} \int_{-\infty}^{\infty} \frac{u \sin(uR) \exp(i\omega\tau)}{u^2 - (\omega^2/c^2) + \sigma^2} du d\omega.$$

The contour integral

$$\oint_C \frac{z \exp(iRz)}{z^2 - (\omega^2/c^2) + \sigma^2} dz$$

has simple poles at $z = \pm\sqrt{(\omega^2/c^2) - \sigma^2}$. If we consider the contour to enclose the positive pole only, then contour integration (using the residue theorem) gives the outgoing Green's function

$$G(R, \tau) = \frac{1}{2\pi} \frac{1}{4\pi R} \int_{-\infty}^{\infty} \exp(iR\sqrt{(\omega^2/c^2) - \sigma^2}) \exp(i\omega\tau) d\omega.$$

The last part of the calculation is therefore to compute the integral above.

Noting that $\sqrt{(\omega^2/c^2) - \sigma^2} = i\sqrt{\sigma^2 - (\omega^2/c^2)}$ and letting $i\omega = p$, we can write

$$G(R, \tau) = \frac{1}{4\pi R} \frac{1}{2\pi i} \int_{-i\infty}^{i\infty} \exp(-R\sqrt{\sigma^2 c^2 + p^2}/c) \exp(p\tau) dp.$$

We are thus required to evaluate the inverse Laplace transform of $\exp(-R\sqrt{\sigma^2c^2+p^2}/c)$. Noting that

$$\exp(-R\sqrt{\sigma^2c^2+p^2}/c) = -c\frac{\partial}{\partial R}\left(\frac{\exp(-R\sqrt{\sigma^2c^2+p^2}/c)}{\sqrt{\sigma^2c^2+p^2}}\right)$$

we can write

$$G(R,\tau) = -\frac{c}{4\pi R}\frac{\partial}{\partial R}[J_0[\sigma c\sqrt{\tau^2-(R^2/c^2)}]], \qquad \tau > R/c.$$

(ii) Using exactly the same approach as that used in part (i) above, we can write the outgoing Green's function as

$$G(R,\tau) = \frac{1}{2\pi}\frac{1}{4\pi R}\int\limits_{-\infty}^{\infty}\exp(iR\sqrt{(\omega^2/c^2)-i\omega\sigma})\exp(i\omega\tau)d\omega$$

the term $i\omega\sigma$ in the first exponential being a direct result of the term $\sigma\partial/\partial t$ present in this operator. Noting that

$$\sqrt{(\omega^2/c^2)-i\omega\sigma} = i\sqrt{\left(\frac{i\omega}{c}+\frac{\sigma c}{2}\right)^2-\frac{\sigma^2c^2}{4}}$$

and letting $p=(i\omega/2)+(\sigma c/2)$, we obtain

$$\begin{aligned} G(R,\tau) &= \frac{c}{4\pi R}\exp(-\sigma c^2\tau/2)\frac{1}{2\pi i} \\ &\quad\times\int\limits_{\gamma-i\infty}^{\gamma+i\infty}\exp(-R\sqrt{p^2-\sigma^2c^2/4})\exp(pc\tau)dp \\ &= -\frac{c}{4\pi R}\exp(-\sigma c^2\tau/2)\frac{\partial}{\partial R}\left[I_0\left(\frac{\sigma c}{2}\sqrt{c^2\tau^2-R^2}\right)\right], \\ &\quad c\tau > R. \end{aligned}$$

5.8 In Cartesian coordinates

$$\mathbf{r} = \hat{\mathbf{x}}x+\hat{\mathbf{y}}y+\hat{\mathbf{z}}z,$$

$$\mathbf{r}_0 = \hat{\mathbf{x}}x_0+\hat{\mathbf{y}}y_0+\hat{\mathbf{z}}z_0.$$

Condition (i) implies that $\hat{\mathbf{k}} = \hat{\mathbf{z}}$, condition (ii) that $\hat{\mathbf{r}}_0 \simeq \hat{\mathbf{z}}$ and condition (iii) that $z=0$. Conditions (i) and (ii) give $\alpha \simeq \hat{\mathbf{n}}\cdot\hat{\mathbf{k}}+\hat{\mathbf{n}}\cdot\hat{\mathbf{r}}_0 \simeq 2$ (because in this geometry $\hat{\mathbf{n}} = \hat{\mathbf{z}}$) and with condition (iii), we obtain

$$u(x_0,y_0,z_0,k) = \frac{i}{\lambda}\frac{\exp(ikr_0)}{r_0}\int\limits_S\exp\left[-\frac{ik}{r_0}(xx_0+yy_0)\right]dxdy.$$

Assuming that the point of observation lies in a plane (the observation screen) located at a fixed distance z_0 from the aperture (which will be large compared with x_0 and y_0 since observations are made only at small angles),

$$\begin{aligned} r_0 &= \sqrt{x_0^2 + y_0^2 + z_0^2} = z_0\left(1 + \frac{x_0^2 + y_0^2}{z_0^2}\right)^{\frac{1}{2}} \\ &\simeq z_0 + \frac{x_0^2 + y_0^2}{2z_0}. \end{aligned}$$

Using this expression for r_0 in the exponent $\exp(ikr_0)$ but using $r_0 \simeq z_0$ elsewhere,

$$\begin{aligned} u(x_0, y_0) &= \frac{i}{\lambda}\frac{\exp(ikz_0)}{z_0} \\ &\times \exp\left(ik\frac{x_0^2 + y_0^2}{2z_0}\right)\int_S \exp\left[-\frac{ik}{z_0}(xx_0 + yy_0)\right] dxdy. \end{aligned}$$

If the aperture is described by $f(x, y)$ which by implication is of compact support, then

$$\begin{aligned} u(x_0, y_0, z_0, k) &= \frac{i}{\lambda}\frac{\exp(ikz_0)}{z_0}\exp\left(ik\frac{x_0^2 + y_0^2}{2z_0}\right) \\ &\times \int_{-\infty}^{\infty} f(x, y)\exp\left(-\frac{ik}{z_0}(xx_0 + yy_0)\right) dxdy. \end{aligned}$$

Finally, let

$$k_x = \frac{x_0}{z_0\lambda}$$

and

$$k_y = \frac{y_0}{z_0\lambda}$$

so that the intensity of the wavefield can be written as

$$I(x_0, y_0) = \frac{1}{\lambda^2 z_0^2} \mid \hat{F}[f(x, y)] \mid^2$$

where $\hat{F}$ is the Fourier transform operator:

$$\hat{F}[f(x, y)] = \int_{-\infty}^{\infty} f(x, y)\exp[-2\pi i(k_x x + k_y y)]dxdy.$$

5.9 Using the same approach as in the solution to Exercise 5.8,

$$u(x_0, y_0, z_0, k) = \frac{i}{\lambda}\frac{\exp(ikz_0)}{z_0}\exp\left(ik\frac{x_0^2+y_0^2}{2z_0}\right)$$

$$\times \int_{-\infty}^{\infty} f(x,y)\exp\left[-\frac{ik}{z_0}(xx_0+yy_0)\right]\exp\left[\frac{ik}{2z_0}(x^2+y^2)\right] dxdy.$$

Noting that

$$\frac{ik}{2z_0}(x_0^2+y_0^2)+\frac{ik}{z_0}(-xx_0-yy_0)+\frac{ik}{2z_0}(x^2+y^2)$$
$$=\frac{ik}{2z_0}[x_0^2-2xx_0+x^2+y_0^2-2yy_0+y^2]$$
$$=\frac{ik}{2z_0}[(x_0-x)^2+(y_0-y)^2]$$

this result can be written in the form

$$u(x_0, y_0, z_0, k) = \frac{i}{\lambda}\frac{\exp(ikz_0)}{z_0}$$

$$\times \int_{-\infty}^{\infty} f(x,y)\exp\left(\frac{i\pi}{\lambda z_0}[(x_0-x)^2+(y_0-y)^2]\right) dxdy.$$

5.10 (i) Since $c = c_0 + v$,

$$\frac{1}{c^2} = \frac{1}{(c_0+v)^2} = \frac{1}{c_0^2}\left(1+\frac{v}{c_0}\right)^{-2} = \frac{1}{c_0^2}\left(1-\frac{2v}{c_0}+\ldots\right)$$
$$\simeq \frac{1}{c_0^2}-\frac{2v}{c_0^3}, \quad \frac{v}{c_0} \ll 1.$$

The equation given then becomes

$$\left(\frac{\partial^2}{\partial x^2}+k^2-2k^2\frac{v}{c_0}\right)u(x,\omega)=0.$$

With $u = w + \exp(-ikx)$ we have

$$\left(\frac{\partial^2}{\partial x^2}+k^2\right)\exp(-ikx)$$
$$+\left(\frac{\partial^2}{\partial x^2}+k^2\right)w-2k^2\frac{v}{c_0}\exp(-ikx)-2k^2\frac{v}{c_0}w=0$$

which reduces to

$$\left(\frac{\partial^2}{\partial x^2}+k^2\right)w(x,k)=2k^2\frac{v}{c_0}\exp(-ikx)$$

because

$$\left(\frac{\partial^2}{\partial x^2} + k^2\right) \exp(-ikx) = 0$$

and since $v/c_0 \ll 1$ and $| w | \ll 1$, the term $2k^2 vw/c_0$ can be neglected.

(ii) The (outgoing) Green's function solution is

$$\begin{aligned} w(x_0, k) &= \frac{2k^2}{c_0}\frac{i}{2k}\int\limits_{-\infty}^{\infty} \exp[(ik \mid x - x_0 \mid)]v(x)\exp(-ikx)dx \\ &= \frac{ik}{c_0}\exp(ikx_0)\int\limits_{-\infty}^{\infty} v(x)\exp(-2ikx)dx, \qquad x_0 \to \infty \end{aligned}$$

(i.e. $x_0 > x \quad \forall x$). Now

$$\begin{aligned} ik\int\limits_{-\infty}^{\infty} v(x)\exp(-2ikx)dx &= -\frac{1}{2}\int\limits_{-\infty}^{\infty}\left(\frac{d}{dx}\exp(-2ikx)\right)v(x)dx \\ &= -\frac{1}{2}\int\limits_{-\infty}^{\infty}\frac{dv}{dx}\exp(-2ikx)dx \end{aligned}$$

assuming $v(x) \mid_{\pm\infty} = 0$. Thus

$$w(x_0, k) = -\exp(ikx_0)\frac{1}{2c_0}\int\limits_{-\infty}^{\infty}\frac{dv}{dx}\exp(-2ikx)dx.$$

Since $x = ct$ and $k = \omega/c$, we can write this result in the form

$$w(\tau_0, \omega) = -\exp(i\omega\tau_0/2)\int\limits_{-\infty}^{\infty}\frac{1}{2c_0}\frac{dv}{d\tau}\exp(-i\omega\tau)d\tau$$

where $\tau = 2t$. Taking the inverse Fourier transform and using the convolution theorem we obtain

$$w(\tau) = -\frac{1}{2c_0}\frac{dv}{d\tau} * \delta(\tau + t_0/2)$$

where $\tau_0 \to \infty$. The condition $v/c_0 \ll 1$ implies that v is a small perturbation of c_0. $| w | \ll 1$ implies weak or Born scattering. This expression for the impulse response function is obtained under the Born approximation – the conditions allowing a linearisation of the problem.

(iii) With $u(x,k)=\exp[iks(x)]$,

$$\frac{\partial u}{\partial x}=ik\exp(iks)\frac{ds}{dx}$$

and

$$\frac{\partial^2 u}{\partial x^2}=(ik)^2\exp(iks)\left(\frac{ds}{dx}\right)+ik\exp(iks)\frac{d^2s}{dx^2}$$

and the original equation transforms to

$$\frac{\omega^2}{c_0^2}\left(\frac{ds}{dx}\right)^2=\frac{\omega^2}{c^2}\quad\text{or}\quad\frac{ds}{dx}=\frac{c_0}{c}.$$

Hence, if $c=\alpha c_0/x$, then

$$\frac{ds}{dx}=\frac{x}{\alpha}\quad\text{or}\quad s=\beta+\frac{x^2}{2\alpha}$$

where β is a constant of integration. The condition $\omega\to\infty$ implies that the wavelength is much smaller than the characteristic variation of s.

5.11 With $u=ge^s$, the equation becomes

$$\nabla^2g+k^2g+2\nabla\cdot\nabla g+g\nabla\cdot\nabla s+g\nabla^2s=-k^2\gamma g-\delta^3.$$

Under the Rytov approximation (i.e. neglecting the term $g\nabla s\cdot\nabla s$), we have

$$g\nabla^2s+2\nabla s\cdot\nabla g=-k^2\gamma g$$

which, after the substitution of $s=w/g$, reduces to

$$\nabla^2w+k^2w=-k^2\gamma g-\frac{w}{g}\delta^3.$$

The Green's function solution to this equation at a point $\mathbf{r}_s$ say (assuming homogeneous boundary conditions) is

$$\begin{aligned}w(\mathbf{r}_s\mid\mathbf{r}_0,k)&=k^2\int\gamma(\mathbf{r})g(\mathbf{r}\mid\mathbf{r}_0,k)g(\mathbf{r}\mid\mathbf{r}_s,k)d^3\mathbf{r}\\&\quad+\int\frac{w(\mathbf{r},k)}{g(\mathbf{r}\mid\mathbf{r}_0,k)}g(\mathbf{r}\mid\mathbf{r}_s,k)\delta^3(\mathbf{r}-\mathbf{r}_0)d^3\mathbf{r}\\&=k^2\int\gamma(\mathbf{r})g(\mathbf{r}\mid\mathbf{r}_0,k)g(\mathbf{r}\mid\mathbf{r}_s,k)d^3\mathbf{r}\end{aligned}$$

since $1/g(\mathbf{r}_0\mid\mathbf{r}_0,k)=0$. Hence the solution is

$$\begin{aligned}u(\mathbf{r}_s\mid\mathbf{r}_0,k)&=g(\mathbf{r}_s\mid\mathbf{r}_0,k)\\&\quad\times\exp\left[\frac{k^2}{g(\mathbf{r}_s\mid\mathbf{r}_0,k)}\int\gamma(\mathbf{r})g(\mathbf{r}\mid\mathbf{r}_0,k)g(\mathbf{r}\mid\mathbf{r}_s,k)d^3\mathbf{r}\right]\\&=g(\mathbf{r}_s\mid\mathbf{r}_0,k)=k^2\int\gamma(\mathbf{r})g(\mathbf{r}\mid\mathbf{r}_0,k)g(\mathbf{r}\mid\mathbf{r}_s,k)d^3\mathbf{r}+\cdots.\end{aligned}$$

The back-scattered field is given by

$$u(\mathbf{r}_0, k) = \lim_{\mathbf{r}_s \to \mathbf{r}_0} [u(\mathbf{r}_s \mid \mathbf{r}_0, k) - g(\mathbf{r}_s \mid \mathbf{r}_0, k)] = k^2 \int \gamma(\mathbf{r}) g^2(\mathbf{r} \mid \mathbf{r}_0, k) d^3\mathbf{r}.$$

(i) Taking the Laplace transform, the equation transforms (using the convolution theorem) to

$$U(p) = \frac{1}{p^2} - \frac{1}{p^2} U(p).$$

Thus

$$U(p) = \frac{1}{p^2 + 1} \qquad \text{and} \qquad u(x) = \sinh x.$$

(ii) The Neumann series for this equation is

$$u_n(x) = x - \int_0^x (t - x) u_{n-1}(t) dt, \qquad n = 1, 2, 3 \ldots.$$

Thus

$$\begin{aligned}
u_0(x) &= x \\
u_1(x) &= x - \int_0^x (t - x) t dt \\
&= x - \left[\frac{t^3}{3} - x \frac{t^2}{2} \right]_0^x \\
&= x - \frac{x^3}{3} + \frac{x^3}{2} = x + \frac{x^3}{6} \\
u_2(x) &= x - \int_0^x (t - x) \left(t + \frac{t^3}{6} \right) dt \\
&= x - \left[\frac{t^3}{3} + \frac{t^5}{50} - x \frac{t^2}{2} - x \frac{t^4}{24} \right]_0^x \\
&= x + \frac{x^3}{6} + \frac{x^5}{120}
\end{aligned}$$

and by induction,

$$x + \frac{x^3}{6} + \frac{x^5}{120} + \cdots = \sinh x.$$

5.12 With $\mathbf{R} = \mathbf{r} - \mathbf{r}_0$ and $\tau = t - t_0$, solve

$$\left(\nabla^2 + \sigma \frac{\partial}{\partial \tau} \right) G(R, \tau) = -\delta^3(R) \delta(\tau), \qquad \tau > 0.$$

Take Laplace transforms

$$\nabla^2\bar{G}(R,p)+\sigma[p-G(R,0)]\bar{G}(R,p)=-\delta^3(R)$$

or

$$\nabla^2\bar{G}(R,p)+\sigma p\bar{G}(R,p)=-\delta^3(R),\qquad G(R,0)=0.$$

Solving this equation gives:

$$\bar{G}(R,p)=\frac{1}{4\pi R}e^{-\sqrt{\sigma p}R}.$$

Hence

$$\begin{aligned}
G(R,\tau) &= \frac{1}{4\pi R}\hat{L}^{-1}\left[e^{-\sqrt{\sigma p}R}\right]\\
&= \frac{1}{4\pi R}\frac{a}{2\sqrt{\pi}}\hat{L}^{-1}\left[\frac{2\sqrt{\pi}}{a}e^{-a\sqrt{p}}\right], \qquad a=R\sqrt{\sigma}\\
&= \frac{1}{4\pi R}\frac{\sqrt{\sigma}}{2\sqrt{\pi}}\tau^{-\frac{3}{2}}e^{-\frac{1}{4}R^2\sigma/\tau}\\
&= \frac{1}{\sigma}\left(\frac{\sigma}{4\pi\tau}\right)^{\frac{3}{2}}e^{-\frac{\sigma R^2}{4\tau}}H(\tau).
\end{aligned}$$

5.13 Consider

$$\left(\nabla^2-\sigma\frac{\partial}{\partial t}\right)G(\mathbf{r}|\mathbf{r}_0,t|t_0)=-\delta^3(\mathbf{r}-\mathbf{r}_0)\delta(t-t_0)$$

together with the time reversed equation

$$\left(\nabla^2+\sigma\frac{\partial}{\partial t}\right)G(\mathbf{r}|\mathbf{r}_1,-t|-t_1)=-\delta^3(\mathbf{r}-\mathbf{r}_1)\delta(t-t_1).$$

Multiply the first by $G(\mathbf{r}|\mathbf{r}_1,-t|-t_1)$ and the second by $G(\mathbf{r}|\mathbf{r}_0,t|t_0)$ and subtract, integrate over the region of interest V and over t from $-\infty$ to t_0. Using Green's Theorem gives:

$$\begin{aligned}
&\int_{-\infty}^{t_0}dt\oint_S[G(\mathbf{r}|\mathbf{r}_1,-t|-t_1)\nabla G(\mathbf{r}|\mathbf{r}_0,t|t_0)\\
&\quad G(\mathbf{r}|\mathbf{r}_0,t|t_0)\nabla G(\mathbf{r}|\mathbf{r}_1,-t|-t_1)]\cdot\hat{\mathbf{n}}d^2\mathbf{r}\\
&\quad -\sigma\int_V d^3\mathbf{r}\int_0^{t_0}\left[G(\mathbf{r}|\mathbf{r}_1,-t|-t_1)\frac{\partial}{\partial t}G(\mathbf{r}|\mathbf{r}_0,t|t_0)\right.\\
&\quad\left.+\,G(\mathbf{r}|\mathbf{r}_0,t|t_0)\frac{\partial}{\partial t}G(\mathbf{r}|\mathbf{r}_1,-t|-t_1)\right]dt\\
&= G(\mathbf{r}_1|\mathbf{r}_0,t_1|t_0)-G(\mathbf{r}_0|\mathbf{r}_1,-t_0|-t_1).
\end{aligned}$$

The first integral vanishes under the assumption that G statisfies the homogeneous boundary conditions. In the second integral we obtain

$$[G(\mathbf{r}|\mathbf{r}_1, -t| - t_1)G(\mathbf{r}|\mathbf{r}_0, t|t_0)]_{t=-\infty}^{t=t_0}$$

and since $G(\mathbf{r}|\mathbf{r}_0, t|t_0) = 0$ if $t < t_0$, $G(\mathbf{r}|\mathbf{r}_0, t|t_0)|_{t=-\infty} = 0$ and $G(\mathbf{r}|\mathbf{r}_1, -t| - t_1)|_{t=t_0} = 0$ for t in the range of integration. Hence

$$G(\mathbf{r}|\mathbf{r}_0, t|t_0) = G(\mathbf{r}|\mathbf{r}_1, -t_0| - t_1).$$

[illegible]

Bibliography

Ablowitz MJ and Clarkson PA (1991) Solitons, non-linear evolution equations and inverse scattering. Cambridge University Press.

Abramowitz M and Stegun IA (1964) Handbook of mathematical functions. Dover Publications.

Acheson DJ (1990) Elementary fluid dynamics. Clarendon Press, Oxford.

Apostol TM (1974) Mathematical analysis. McGraw Hill.

Atkin RH (1962) Theoretical electromagnetism. Heinemann.

Bleecker D and Csordas G (1995) Basic partial differential equations. Chapman and Hall.

Burkill JC (1962) Theory of ordinary differential equations. Oliver and Boyd Ltd.

Copson ET (1935) The theory of functions of a complex variable. Oxford University Press.

Evans GA (1996) Practical numerical analysis. John Wiley and Sons.

Evans GA, Blackledge JM and Yardley PD (1999) Numerical methods for partial differential equations. Springer-Verlag.

Garabedian PR (1964) Partial differential equations. John Wiley and Sons, New York.

Gustafson KE (1987) Introductions to partial differential equations and Hilbert space methods. John Wiley and Sons, New York.

Jeffrey A (1992) Complex analysis and applications. CRC Press, Boca Raton, Ann Arbor, London.

Lambert J (1990) Numerical solution of ordinary differential equations. John Wiley and Sons.

Logan JD (1994) An introduction to non-linear partial differential equations. John Wiley and Sons.

Lomen D and Mark J (1988) Differential equations. Prentice-Hall.

Marsden JE and Hoffman MJ (1987) Basic complex analysis. WH Freeman and Co, New York.

Miller RK (1987) Introduction to differential equations. Prentice-Hall.

Mizohata S (1973) The theory of partial differential equations. Cambridge University Press.

Morse PM and Feshbach H (1953) Methods of theoretical physics. McGraw-Hill, New York.

Nagle RK and Saff EB (1993) Fundamentals of differential equations. Addison-Wesley.

Needham T (1997) Visual complex analysis. Clarendon Press, Oxford.

Piaggio HTH (1950) Ordinary differential equations. Academic Press.

Redheffer R and Port D (1992) Introduction to differential equations. Jones and Bartlett Publications, Boston.

Renardy M and Rogers RC (1993) An introduction to partial differential equations. Springer-Verlag, New York.

Roach GF (1970) Green's functions. Van Nostrand Reinhold Co, London.

Sneddon IN (1957) Numerical solution of partial differential equations. McGraw Hill.

Stakgold I (1979) Green's functions and boundary value problems. John Wiley and Sons, New York.

Stewart I and Tall D (1983) Complex analysis. Cambridge University Press.

Strauss WA (1992) Partial differential equations, an introduction. John Wiley and Sons.

Titchmarsh EC (1932) The theory of functions. Oxford University Press.

Trim DW (1990) Applied partial differential equations. Prindle, Weber and Schmidt, Boston.

Vvedensky P (1993) Partial differential equations with Mathematica. Addison-Wesley.

Watson GN (1922) Theory of Bessel functions. Cambridge University Press.

Wloka J (1987) Partial differential equations. Cambridge University Press.

Wunsch AD (1994) Complex variable with applications. Addison-Wesley.

Zauderer E (1983) Partial differential equations of applied mathematics. John Wiley and Sons.

Index